Advanced Al[illegible]
Problem Strings

Pamela Weber Harris
with
Kara Louise Imm
B. Michelle Rinehart
Susan M. Simmons

Discovering
Mathematics

Discovering Algebra
Discovering Geometry
Discovering Advanced Algebra

Kendall Hunt
publishing company

The graphs were produced in Desmos.
Cover image © Shutterstock, Inc.

Kendall Hunt
publishing company

www.kendallhunt.com
Send all inquiries to:
4050 Westmark Drive
Dubuque, IA 52004-1840

ISBN: 978-1-5249-4395-0 (Coil bound)
978-1-5249-2128-6 (Perfect bound)

Printed in the United States of America

1 2 3 4 5 6 7 8 9 10 22 21 20 19 18 17

Table of Contents

*Lesson numbers correspond to lesson numbers in *Discovering Advanced Algebra*, 3rd edition.

Acknowledgments

We gratefully acknowledge the influence of the work in the following: *Young Mathematicians at Work*, *Contexts for Learning*, *Math In Context*, the Freudenthal Institute, NumberStrings.com, Dr. Rachel Lambert of Chapman University, Andrew Stadel, Clotheslinemath.com, Texas Instruments' Teachers Teaching with Technology, *Functions Modeling Change*, Frank Demana & Bert Waits and their *Precalculus Mathematics: A Graphing Approach*, and the *Math Vision Project*.

Special thanks to the following:

My husband, Daniel Harris, for supporting me in all of my crazy ventures. And for cooking dinners, cleaning house, and reminding me to sleep.

Cameron Harris for his thoughtful editing of the Introduction. Thanks for helping me put on paper what I mean.

Matthew Harris for the graphics production. His attention to detail and great advice is much appreciated.

Craig Harris who willingly acts the student and pushes back with "No one would do that."

Abigail Harris, my "I can't do it unless I understand why" girl, for pushing me to understand why like no one else.

Tim Pope at Kendall Hunt for so many things, including the idea and support for this project and suggestions for statistics and probability problem strings.

Kelly Fagan at Kendall Hunt for a fabulous job of editing in a tight timeline.

Debra Plowman for supporting me to produce the *Focus on Algebra* series of professional development workshops, my first foray into high school problem strings.

Kim Montague for answering the calls, "Kim, this isn't your content area, but is this a problem string? No? Then how can I make it one? How do you think about this concept/model/strategy? Really? That's a thing? I'm going to write a string so I can think that way too!"

Kathy Hale for providing me a venue to present to teachers and build my own content knowledge at the same time.

Scott Hendrickson for getting me started. He is the master teacher who has integrity in his teaching—it all fits together.

The myriad of teachers and students who have played with mathematics with me, opened their minds to real math, and given me great feedback.

Michelle Rinehart for running with my half-baked ideas and making them shine. You found the clever twist so many times that makes a string *fun*!

Sue Simmons for writing and editing and catching so many subtleties. And keeping me straight in so many areas of my life. You're a joy professionally and personally.
Kara Imm for joining me in this project that took on a life of its own and grew bigger than any of us thought. I continue to learn from you.

Jerry Murdock, Ellen Kamischke, Eric Kamischke as the original task sequencers for giving me the first and best glimpse of what it means to teach real math. When I described problem strings to them as a lesson structure, they replied, "Isn't that just good teaching?" Yes, yes it is.

Introduction

Teaching and Learning Mathematics

We believe that teaching mathematics is about mentoring mathematicians. This means helping students to mathematize experiences both in and out of school. To mathematize, according to Dutch mathematician Hans Freudenthal, means to structure, model, and interpret one's "lived world" mathematically. That is, to look for connections, make conjectures, seek to generalize and justify reasoning. As teachers of algebra, our job is not to distribute to students a set of prescribed rules and facts, but to help them to structure and schematize, thus creating mathematical relationships in their minds.

To help students construct mathematics in this way, we draw from a variety of lesson structures, including, but not limited to: inquiry, investigations followed by math discussions, and mini-lessons. The important mini-lesson structure that we highlight in this book is called *problem strings*. I, (Pamela Weber Harris) developed and introduced problem strings in my book *Building Powerful Numeracy for Middle and High School Students* (2011). But long before this, the idea of mental math routines focused on students' strategies originated in the work of Cathy Twomey Fosnot and Maarten Dolk, when they introduced the idea of *number strings*. Number strings were first described in their book *Young Mathematicians at Work: Constructing Number Sense, Addition and Subtraction* (2001), part of a series that would later extend through rational numbers. Later, number string resource books were written for teachers, starting with *Minilessons for Early Addition and Subtraction: A Yearlong Resource* (2007), and also extending through rational numbers. In this book, we deepen this well-known mathematical practice by bringing problem strings to the high school math class, supporting mathematicians to make sense of their algebra and advance algebra courses.

Of these lesson structures, investigations and math discussions typically involve messier, bigger questions that can be more involved and complicated, and therefore take longer to work with and solve. Students tackle these bigger, non-routine problems in context in small groups, working together to make sense of the math, and then deciding how to explain their findings to the class for review and comment. Problem strings, in comparison, tend to be shorter and much more targeted.

Discovering Advanced Algebra contains some of the best rich investigations to help students construct complicated concepts and skills. Yet often, when students begin solving these challenging problems in context, they stumble, struggle, get frustrated with themselves, or us, or give up. In these moments, it can be tempting to do one of three things:

1. Pre-teach the anticipated skills and ideas before introducing the investigation.
2. Use the tasks solely to engage students, but then simply show them how to solve them.
3. Or, abandon the use of investigations altogether.

Each of these pedagogical choices is an attempt to minimize the messiness of learning and the evidence of struggle. But without some struggle there is rarely deep learning. To learn is to continually re-organize one's ideas and encounter the unfamiliar until it becomes connected to our ideas, sensible, understood, and eventually familiar.

We believe that the regular use of the powerful routine called *problem strings* helps both students and teachers before and after investigations. Increasingly, problem strings are being used by teachers to:

- preview big ideas that will arise in an investigation,
- solidify the ideas and skills that came up in the investigation,
- create puzzlement, disequilibrium, and curiosity,
- invite students to prove or justify their ideas,

- describe and solidify strategies, and move towards efficient strategies,
- build students' efficacy at choosing strategies, and
- generalize an idea beyond the task at hand.

Problem strings allow students to struggle in a contained, guided, purposeful set of tasks. They compliment and support the work of investigations and math discussions, working together to foster conversations and form conclusions about relationships, structures, and repeated reasoning. This is the work worthy of teachers, to help students develop and grow into mathematicians.

What Is a Problem String

A problem string is a purposeful sequence of related problems, designed to help students mentally construct mathematical relationships. It is a powerful lesson structure during which teachers and students interact to construct important mathematical strategies, models, and concepts. The power of a problem string lies in the carefully crafted conversation as students solve problems, one at a time, and the teacher models student thinking and draws out important connections and relationships.

	Problem strings are...	**Problem strings are not...**
The problems	a purposefully chosen sequence of related problems.	random nor entirely predictable (15×1, 15×2, 15×3...).
Lesson format and timing	a mini-lesson that typically precedes or follows an investigation.	a substitute for other forms of problem solving, including inquiry, rich investigations, and math discussions.
Role of the teacher	guided by the teacher to bring out the students' mathematics; the teacher systematically nudges students toward more efficient and sophisticated strategies.	student-led—students do not hijack the conversation.
Type of instruction	facilitative—there is some explicit teaching only when related to social knowledge (mathematical terminology, notation, etc.).	teacher-centered with direct instruction; not an "I do, we do, you do" approach.
Strategies	developed by students. Over time and with lots of experiences, we nudge students to • develop a variety of strategies to draw upon and • choose the most efficient one for the problem at hand.	spaces for teachers to demonstrate their own strategies or thinking (e.g., "This is how I might solve this one..."); places to "practice" using the same strategy for every problem over and over again.
Concepts	based on the ideas noticed by students, though the teacher must be looking and listening for "glimmers" of these ideas during the routine so that they can be made public and then explored together as a class.	opportunities for teachers to explain the big ideas (and why they work) to students.

	Problem strings are...	Problem strings are not...
Pacing	short routines. Be purposeful, deliberate, respectful of the time it takes to think, but with some energy—don't put the kids to sleep! Teacher celebrates ideas, risk taking, and deep thinking—not speed.	a race or anything resembling a timed test.
Modeling	the teacher modeling students' thinking while the students articulate their ideas to the class; the teacher makes students' thinking visible so that it can be compared and discussed.	intended for students to model their own thinking for each problem—often students do not initially know how to model their thinking.
Engagement	mini-lessons where every student in the class is engaged in thinking about the mathematics.	sparse conversations due to the teacher calling only on students who have a right answer or an "effective" strategy.
Assessment	wonderful experiences to learn about how your students think, what they understand, and are still constructing.	"gotcha" moments where students are categorized as either "right" or "wrong."
Focus	focused on strategies, sense-making, ability to generalize, and convincing others of their ideas.	routines to reward students who get the "right" answers quickly or students who say they "just know it."
Student participation	all students solving every problem. The teacher chooses which strategies are to be shared, developing the students as mathematicians. Sometimes this can be sharing less sophisticated strategies or misconceptions so that students can discuss, compare, and ferret out common pitfalls.	every student sharing their thinking during the whole class discussion, especially if a strategy has already been shared or if it is not helpful to moving the mathematics forward.
Student experience	times for students to try new ideas, to be uncertain, and to question. Learning is happening and that does not always feel like you are on solid ground.	times for students to stick to one strategy just because it works.

Addressing 10 Common Misconceptions:

- Teachers should not present the entire problem string all at once. Teachers may be inclined to hand out the problems all together, list them on the board as students come in the room, or show them on a slide all at once. This approach is unlikely to support many of the goals of problem strings and limits the kind of deliberate and purposeful conversation about mathematical relationships.
- Problem strings are not a collection of random or unrelated practice problems. There may be legitimate space in your teaching practice for this activity, but that is not a problem string.
- Problem strings are not opportunities for a teacher to demonstrate a strategy for students to then "practice." Said differently, a problem string is not the place to introduce a traditional algorithm or a procedure. This is a time for teachers to listen and watch carefully, picking up on ideas and helping to bring them forward by noticing, questioning, and wondering. This noticing, questioning, and wondering is about what the students are saying, doing, and thinking about. If a teacher puts forth "the right" way, this often prevents students from trying out strategies and taking risks in the conversation. Students will pick up on this and wait a teacher out, knowing that "the teacher's way" will come. Mathematizing is not about "the" way; it's about using relationships and connections to solve problems.

- Problem strings are not spaces for direct teaching—with the exception of social knowledge (including mathematical terminology) that students may not have access to. For example, notation or technical language is a social construct that should be mentioned to students (e.g., "Yes, we call that cube root.") and recorded publicly so that students can see and hear the new ideas. Don't make students guess about convention! However, the logical mathematical knowledge that must be constructed cannot be passed down by simply telling, but must be experienced so that connections and relationships are constructed in the learner's mind.
- Teachers should not expect students to all use the same strategy. If you find that this occurs, you may be too prescriptive, leading, or rigid in your facilitation. Problem strings are designed with multiple entry and exit points, meaning that students with a variety of skills and understanding can access the problems. Likewise, the conversations in problem strings are designed to meet students where they are and nudge them along their mathematical journey. Hence, we expect multiple exit points, meaning students will not all construct the same relationships at the same time. The goal is to help grow judicious problem solvers who choose strategies based on the numbers or structure of the problem, or on what they infer and understand—based on relationships they own. When students take up this type of noticing-then-choosing-a-strategy approach, we find they are better able to tackle unfamiliar problems and tinker with new ideas. If students use the same strategy (regardless of the numbers or the structure of the problems), they are likely mimicking, not mathematizing. Our goal is thoughtful, flexible, creative problem solvers who possess a bank of strategies, not just one. Problem strings often have "sister strings" that allow students to revisit the big ideas on successive occasions. Students are enabled to construct the ideas when they are ready instead of on any one day.
- During the sharing time, students should not model their own thinking. Often the teacher models a student's thinking using a different representation than the student because the teacher understands that certain models can help bring relationships to light, help students make comparisons, and help models become a realized tool that students can begin to use. Many students do not know how to model their own thinking. Thus, this can be an opportunity for the teacher to help make the reasoning and relationships visible for the rest of the class.
- The teacher should not take time to figure out a student's unknown strategy. Be prepared so you will have thought deeply about possibilities, but if something takes you by surprise and you cannot make sense of it in a reasonable amount of time, don't leave the rest of the students hanging too long. Taking too long to decipher an unknown strategy at the moment (problem strings are short) can derail the discussion. Don't discount student thinking, but if it's taking you too long in the moment, take note, tell the student that you need to think about it (how cool is that—you have to think about what they've come up with!) and study it later. Bring it to the class later if it is generalizable or if it is an example of a common misconception for students to parse out.
- The sharing time is not an opportunity to share every student's strategy. We definitely want students engaged, solving, and sharing, but it is not a free for all. Allowing every strategy to be shared—regardless of its usefulness or novelty—incentivizes lazy repetition rather than constructive mathematizing. Conversely, we do not want to share only the 'right' strategy. By striking a balance between showing everything and showing one thing, you set the pattern you want your students to follow. You are demonstrating that students should be flexible enough in their problem solving to consider multiple approaches, but discriminating enough not to be satisfied with a strategy just because it comes to mind.
- It is not necessary to share every student's strategy to honor every student's thinking. By encouraging students to solve problems using their own thinking rather than requiring them to regurgitate yours, you are already honoring them far more than you would be otherwise. In addition, sharing every student's strategy regardless of its usefulness cheapens your regard. Not always, but often the question, "Did anyone do it another way?" (just trying to get lots of strategies on the board) indicates a teacher who has not thought deeply about the big ideas involved, the mathematical terrain ahead, or how to purposefully guide the conversation. This is different than a teacher who asks, "What does everyone think about the strategies we have on the board?"
- The student asked to share his or her strategy is not always or even often the student who "got it right" or had the clearest explanation. Sometimes a student who only got started can share an important beginning or can highlight the complexity of finding a starting point. Sometimes students who are a little muddy but on a fruitful track can bring the ideas to the class in such a way that the rest of the students get the opportunity to

weigh the ideas and make sense of them as the class works out the ideas together. Sometimes a student can be the "canary in the mine shaft" and bring a strategy for the class to weigh, compare, and bring to light misconceptions. Such instances must be handled with the utmost respect and gratitude. While the student may have gotten something wrong, his or her blunder has created the opportunity for the entire class to better understand the topic. Likewise, questions from students can provide similar opportunities. This can be done with strategic turn and talks and by creating a community where students are encouraged to ask questions, challenge ideas, and try new things.

Modeling and Models

Within mathematics, models and modeling have always played a prominent role. Yet, the terrain of models and modeling can be complex and hard to understand. The terms can be used within (and beyond) mathematics in several different ways, making it hard to distinguish what is meant by each in its context. The idea of "model" itself can be used both as a noun and a verb, sometimes describing the actions of the teacher and other times describing students' mathematical activity.

Yes, a model is a person walking down a runway in fashionable clothing. And yes, to model good behavior, we copy or mimic what others are doing. But these are not the meanings we intend here. When leading problem strings, we use the words model and modeling very specifically in *two different, but related ways*:

Model students' thinking—an action, performed by the teacher

Here we are describing what a teacher is doing during a problem string. To begin, we are listening and trying to make sense of students' strategies. Then, using what we understand, we make that thinking visible for the community. Sometimes we say, "I'm going to make a picture of your idea so that we can all see it and hear it at the same time." This is a deliberate and essential part of what it means to lead problem strings—and part of what makes them powerful. As students speak, we are capturing their ideas in a visual image that allows more students to make sense of the ideas, and may even help the "authoring" student to better understand his or her own idea. To see one's idea made public can be a powerful boost of status to any member of a community.

Use a model—a noun or object, introduced by the teacher and eventually taken up by students

When teachers are modeling students' thinking (e.g., making it visible) we are not just drawing whatever we choose, nor are we writing down a symbolic transcript of what the student says. Instead, we are typically using well-known mathematical models. These are generalizable mathematical representations that help us think about and solve problems in more than one way and one context, and communicate our ideas to other mathematicians. Examples of the models we draw from include, but are not limited to:

- open number lines
- open double number lines
- open arrays
- ratio tables
- rectangular diagrams
- graphs
- tables
- expressions
- equations
- functions

By drawing upon a menu of mathematical models we are helping students to see their thinking in new ways.

How do models move from teachers to the minds of students?

We share the belief with our colleagues Fosnot & Dolk (see *Young Mathematicians at Work* series, 2001) and the research traditions of the Freudenthal Institute that mathematical models typically begin as *models of thinking*. When a teacher "makes a picture of my thinking" he or she is gently suggesting that what I am saying and this particular model are related. When students see their teachers using models to represent their thinking over and over again—and when they are allowed to investigate rich contexts where these models arise naturally—they begin to transition from *model of thinking* to *model for thinking*. That is, students begin to embody or envision the model on their own, without our prompting, and use that mental model to solve new problems. When students begin to say things like, "I thought about it on a number line," or "I imagined the graph of $y = x$ and shifted it up two places," we know that students are moving towards *models for thinking*. This, of course, is our eventual goal, that these mathematical models are constructed in the minds of students and used strategically by them whenever they encounter new and unfamiliar problems.

Facilitating Problem Strings

The first time you try a problem string in class might feel like sailing into the uncharted ocean of student thinking and reasoning, and to some extent it is. But it is important to realize, that whether we know it or not, students have always been trying things, thinking of alternatives, or wishing they could see the big picture so it could all make sense.

Facilitating a problem string requires careful attention to the mathematics as well the ability to really listen to students and model their thinking for all to see. Thinking is often in development—not fully polished or formed and sometimes idiosyncratic or just tricky to understand. Because we believe in a mathematical community that includes all learners, our role is to bring before the class what is helpful for development—whether it is clear, clean, and polished, or messy, incomplete, and developing, or even incorrect. The goal is always to give students the chance to articulate their ideas and to see each others' thinking—to give the class the chance to respond to, challenge, and make sense of someone else's strategy or idea. This mean that as teachers we are withholding our authoritative stamp of approval and giving the mathematics back to kids to reflect on and sort out. This requires both restraint and the artful use of questions.

As teachers of algebra, there is often institutionalized pressure to teach kids "the steps"—to give students access to algebra by offering generic, one-size-fits-all strategies that "work," particularly in testing situations. We may find ourselves working in isolation from our peers, focused on "getting our kids to…." [say the right answer, do the right strategy] and not willing to explore a variety of strategies because we believe we do not have the time. Teaching under these conditions is not easy, but we invite you to resist many of these tendencies, which are informed by a culture of testing more so than a culture of learning. Too often we encounter students who mimic the teacher's mathematics, sometimes using procedures when they are not relevant, and often having no opportunity to explore why they do (or do not) make sense. We want to offer a different mathematical experience and we believe that the regular use of problem strings provides:

- a chance for kids to make sense of algebra,
- opportunities to build algebraic relationships that will extend past our particular course, and
- a way to build a classroom community where knowledge is explored, validated, and constructed together.

Our belief in problem strings stems partially from the idea that telling or showing students mathematics does not produce learning—and in fact, it never has. Disrupting this pattern of "delivering" or "showing" students mathematics is a bold undertaking that will require planning, restraint, and real trust in students' ability to think for themselves. We must believe that our students are full of interesting mathematical ideas, insights, and questions—and that what they offer is enough to begin the work of formal algebra together. Allowing students to solve problems any way they want, asking students to share their thinking, and pushing students to justify can be foreign, new and unsettling. And, downright fun. In the next section, we offer some pedagogical "moves" that will support you on your journey.

Sample Dialogs

The sample dialogs were written to help you get a "vivid image" of what the string might look like and sound like in a classroom. We tried to highlight the important parts of the mathematics, the questions to have ready to ask, and the ways you might model student thinking. These transcripts are based on our own experiences leading problem strings with students, as well as observation of our colleagues leading strings with their students.

We varied the sample dialogs because we know different teachers may need different types of support. There are three formats of sample dialogs:

	What It Is	Why We Included It	What You Might Find
Full dialog	A sample exchange (resembling a transcript) between the teacher and students, with accompanying modeling of strategies.	We wanted to provide a clear window into the classroom interactions. Sometimes there were new or atypical strategies present. Or we felt that it would be helpful to study a possible way of steering the conversation. Or, we wanted to highlight how the timing of the certain questions and modeling might be important to the success of the string. All of the full dialogs offer an important window into the possibility of a vibrant community of learners.	You will find purposeful examples of students expressing ideas in vague, non-mathematical, student vocabulary where the teacher responds by restating, agreeing, or questioning the ideas using precise mathematical language and helping students name important concepts.
Partial dialog	Many of the questions a teacher might ask alongside the modeling of student strategies.	We provide these examples when the modeling was more important than every interaction between the teacher and students. Perhaps the models were new, or we are suggesting juxtaposing two models or strategies.	Sometimes partial dialogs accompany the second string in a series of strings so that the interaction has already been established in the first string, but the problem-by-problem modeling remains still important.
Important questions	A list of central, critical questions that can help guide the facilitation of the problem string.	We wrote these for problem strings where teachers tended to need less guidance, especially on the problem-by-problem modeling.	These often accompany strings in a series, where related strings have already been discussed through full or partial dialogs, and many of the pedagogical moves have already been explored.

Preparing

Leading your first few problem strings requires equal parts courage, curiosity, spontaneity, and preparation. Despite your careful planning, anticipate that you will hear ideas and strategies that you do not expect, may not fully understand, or may struggle to model. In these moments, you want to just listen and then you can pivot to the class—"Who understands what Henry is saying and can put his strategy in their own words?" Work as a class, share the task of understanding kids' thinking with your class, and do not feel that just because you understand it that they will.

As you gain experience, you will develop greater confidence. You will begin to know which ideas from students will lead you away from the mathematical goals of the problem string, and which ideas are central to building new mathematical ideas. Even if you do not choose to pursue an idea you can always respectfully honor students' contribution to the conversation. It's acceptable to tell a student, "That's interesting. Let me think about that and get back to you." And then choose to model the strategies you were planning on.

It's ideal to prepare a problem string with a colleague or other adult who can "play" the student as you try to model their thinking, or who can hear your thinking as you think aloud with them. If you have never led the string before, it is essential that you take time to model students' thinking in advance—yes, take time to draw what different strategies will look like. This will allow you to feel more comfortable and confident as you are leading the problem string with students. As you prepare, ask yourself the following questions:

- How is each problem related to the previous problem(s)? Describe the progression of the string. What's going on here?
- Why these numbers? These problems?
- What big ideas could emerge during this string? How might I encourage kids to articulate these big ideas as they are solving problems?
- What strategies do I expect students to use to solve the problems? How will I model, or represent, this thinking?
- What kinds of questions or prompts might I use to encourage students to consider and explore the big ideas?
- Are there problems I want to insert or add? Why?
- Would a context support my students to reason about the mathematics? If so, what context? Why?
- What is the "rhythm" of the string? Where is the energy? Where will I slow down, speed up, or really try to engender good conversation?

Questioning

You will notice that often the dialogs have examples of questions that seem open but are actually purposefully focused. A mistake that teachers sometimes make is to ask questions that are too open. This can end up with students confused about what you are asking. It can also result in students just trying to guess what you're after. This is not about playing Twenty Questions. The questions should be specific enough where the answers can be based on reasoning, not guessing. Following are some things to think about when you plan the questions you will ask.

If You Want to...	You Might Prompt With or Consider...
Nudge towards more efficient strategies	After putting the strategies on the board, you might ask, "What about the problem or the numbers make this an efficient or not very efficient strategy?" "Which one was more helpful here and why?" This is very different than asking a vague, "Explain your thinking." Or "How did you do it?" Or even, "Which is the best strategy?" Note: Best is a subjective term. We want to get out of personal preference and into the space of justifying a strategy choice based on some insight about the problem. Using words like clever, efficient, and elegant can help convey a sense of searching for sophistication in thought, not a mimicking of procedure nor an absolute "best" strategy.
Move quickly through some problems to get to some more interesting mathematics	Do not belabor the conversation in the beginning of a problem string. Acknowledge that the questions are not so tricky, and that more challenging problems are on their way. You might say something like, "Okay, it seems like you solved this one quickly, so let's keep going." or "Tell us what you're thinking when you see this one. Yep. Onto the next one." If you wait too long or ask for reasoning or multiple strategies on the simpler problems, you may lose students' interest in the string. Make sure you have planned where the "energy" of the string should be—meaning which problems to spend more time discussing because there is something more interesting to discuss.

If You Want to...	You Might Prompt With or Consider...
Juxtapose two strategies	"Did anyone think of the problem this way?" If no students volunteer, you might ask, "*Could* you think about the problem this way?" "These are two really different ways to think about this problem. What is the same about them and what is different?" Alternatively, sometimes you can follow with, "Last year, I had a student who used this relationship. What do you think about that?" You can also insert a problem that might prompt that strategy or plan to write a "sister" string for the next day that would bring that strategy to the forefront.
Build a community of learners	Remove yourself as the mathematical authority in the room by changing your pronouns. Go from "tell me, show me" to "show <u>us</u>, tell <u>us</u>." Along with your body language, this deliberate use of a collective pronoun can dramatically change the dynamics of the conversation. One measure of the strength of a classroom community is the extent to which kids are talking to each other, unfiltered by the teacher, about each other's ideas. Whenever you are puzzled by a student's ideas, you can begin by asking, "What do <u>we</u> think of this idea?"
Move from "algebraic rules" to sense-making	Algebra has been commonly associated with an unknown mathematical authority (usually the teacher, sometimes a more abstract figure) who provides a set of often confusing laws or rules to follow. This has prevented students from making sense because they are more worried about whether they can or cannot "do" something or whether "it is allowed." We want to move away from the paradigm of "Is this allowed?" and into the space of "What makes sense to us?" So we invite students to paraphrase—"Who understood that idea/strategy and could say it again for us?"—as a way to get students to really listen to each other as sense-makers. Additionally, we include questions such as: • Would this make sense? Say why it makes sense to you... • What would this mean? • Are we okay with this? • Is this helping us to (solve, graph, envision, simplify)? • Do you think this would help us? • How do we know that this is an equivalent form? • Will this always work? • What allows us to do this? • Are these still equivalent? How do we know?
Bring the string to a close	At the end of the problem string, we suggest that you question students to help them reflect and pull together what they learned during the string. We provide suggested answers that can help you focus the problem string as you facilitate it.

Sample Display

Each problem string includes a sample display section. This is not meant to be prescriptive, but as a possible end display when planning the string. It might not seem important to plan how your display will end up, but the very nature of a problem string is enhanced when things are displayed in such a way as to help suggest connections, patterns, and relationships. You can encourage sense making as you model strategies using deliberate color, organization, labeling, and placement. You will notice that some of the strings suggest a rather large area. Try to arrange for sufficient space for each string. If you must write on paper under a document camera, write in pencil so you can erase if needed. Plan ahead so that you write in such a way that students can see as much as possible. Keeping the record of the work in front of students encourages students to use what they've learned in prior problems, find patterns between problems or answers or strategies, and connect multiple representations.

Problems in the String

Answers

Color coded strategeies

Graphs, usually of the last problem or of several problems together.

$x^2+3x+2=0$

$x=-2,-1$

$(x+1)(x+2)=0$

$x+1=0 \quad x+2=0$

$x=-1,-2$

$x(x+3)=-2$

$x=-1,-2$

$3x^2+5x+20=0$

$x=\dfrac{-5\pm i\sqrt{215}}{6}$

$$x=\frac{-5\pm\sqrt{(5)^2-4(3)(20)}}{2(3)}=\frac{-5\pm\sqrt{25-240}}{6}=\frac{-5\pm\sqrt{-215}}{6}=\frac{-5\pm i\sqrt{215}}{6}$$

$x^2-4x+1=0$

$x=2\pm\sqrt{3}$

$$x=\frac{-(-4)\pm\sqrt{(-4)^2-4(1)(1)}}{2(1)}=\frac{4\pm\sqrt{16-4}}{2}=\frac{4\pm\sqrt{12}}{2}=\frac{4\pm2\sqrt{3}}{2}=\frac{2(2\pm\sqrt{3})}{2}=2\pm\sqrt{3}$$

$x^2-4x=-1$

$x^2-4x+4=-1+4$

$(x-2)^2=3$

$x-2=\pm\sqrt{3}$

$x=2\pm\sqrt{3}$

$-4.9x^2+76.2x+9.67=0$

$x\approx-0.1, 15.7$

$x^2+8x=9$

$x=-9,1$

$x(x+8)=9$

$x=-9,1$

$(x+9)(x-1)=0$

$x+9=0 \quad x-1=0$

$x=-9,1$

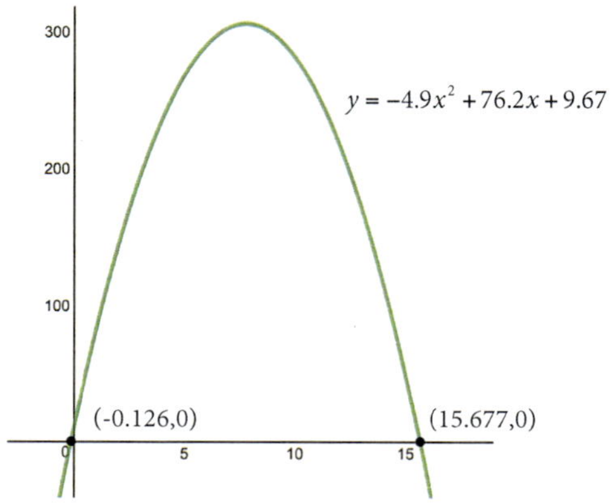

Sample final display from Lesson 5.5 Solving Quadratic Equations.

Sample Facilitation Notes

The facilitation notes at the end of each string are meant to be an abbreviated version of notes for you to use at the time of doing the string. When you prepare to lead, you might study the entire problem string, first learning the math and noting the relationships, and then go back over the string a second time attempting to capture the modeling, the flow, and the possible questions. Finally, you might review the entire plan, noting the important changes and questions from problem to problem, and how you will model student strategies using the sample display. Compare your notes with our facilitation notes and add any that you need.

Problems in the String

Things to do

Notes about timing

$x^2 + 4x + 5 = 0$	*Elicit and model factoring, complete the square, quad formula. Quickly graph with display calc.*
$x^2 - 4x + 5 = 0$	*Elicit and model complete the square, quad formula. Add graph to previous. Connect to previous problem.*
$-x^2 - 4x - 5 = 0$	*Repeat. Quad form preferred because of $-x^2$? How is this the same, different from previous two? Predict graph, then add.*
$-x^2 + 4x - 5 = 0$	*Repeat. How is this the same, different from previous two? Predict graph, then add.*

Things to ask

Sample Facilitation Notes from Lesson 5.4 Complex Numbers.

Sample Anchor Charts

Some of the problem strings have sample anchor charts. An anchor chart is a semi-permanent fixture in the classroom designed to codify some learning that the class has developed together. While they are typically "teacher-made" they are full of students' ideas and often give citations and credit to the student or students who first articulated the ideas (e.g., "George's conjecture", "Amaia's strategy"). Anchor charts are useful when students have a number of strategies and are beginning to determine which strategy to use when. In this way, having these strategies named and clearly modeled with sample problems allows students to refer to them during any act of problem solving.

An anchor chart might also be about big ideas or conjectures. When a student has an important mathematical idea that the group is wrestling with, or is convinced of, a nice way to honor this thinking and to reinforce this practice is to make a poster of this big idea or conjecture for all students to see.

Unlike images on an electronic board that tend to disappear quickly, anchor charts are designed to be displayed in such a way that every student can see and refer to them throughout the unit or year. Teachers can refer to them, "We are working on solving quadratics right now, and we all know where there is a wall of strategies that we can draw upon." These anchor charts evolve over time as the class brings new strategies on board.

Choosing a Smart Strategy	
If it's easy use factor pairs or factor.	$x^2 + 3x + 2 = 0$ $x(x+8) = 9$
Ugly coefficients? Graph, especially if you're going to get approximations anyway.	$-4.9x^2 + 76.2x + 9.67 = 0$
If a = 1 and b is nice, complete the square.	$x^2 - 4x + 1 = 0$
You can always use the quadratic formula, but it might not be worth it. Try other things first.	$3x^2 + 5x + 20 = 0$

Sample Anchor Chart from Lesson 5.5 Solving Quadratic Equations.

Problem String Structures

Our colleague Rachel Lambert (Chapman University) noted that strings have various structures. If you have used problem strings you know that there isn't a single "formula" for designing strings. However, there are some general structures that we use. In this section we highlight a few of those structures—giving some insight into how and why certain sequences of problems were chosen. The examples below are provided to give you some ideas, if or when you begin to write your own problem strings.

Name of Structure	What Is It	An Example
Helper-Challenge OR **Helper-Challenge-Clunker**	Providing access to all students at the beginning of the routine is not only a smart pedagogical move towards building a classroom community, but it also speaks to issues of equity in mathematics. Typically this problem is followed by one whose structure is similar, but with a small new twist, the *challenge problem*. This *helper-challenge* structure can be repeated throughout the problem string. We have found that when students first begin to notice this pairing, they later make use of it—drawing upon something in the helper problem to make sense of the challenge problem. Sometimes this structure will conclude with what we playfully call *the clunker*—chosen to be challenging and not to immediately resemble previous problems. This sudden unfamiliarity forces students out of the habit of looking back to a previous problem, and nudges them to think about a helpful strategy to use.	**String 5.1—At a Glance** $y = x^2$ — Helper problem $y = 6x^*$ — Optional helper $y = x^2 + 6x$ — Challenge problem $y = x^2 - 6x$ — Related challenge problem $y = x^2 + 6x + 7$ $y = x^2 - 6x + 7$ — These two are new, but related problems—just a vertical translation of the previous. $y = x^2 + 4x + 5$ — Clunker—The new numbers but similar structure of the function may make this problem feel challenging for students. It can nudge them to look up and consider the previous relationships in the problem string. This is often a great opportunity for formative assessing how students are taking up the relationships at hand.
Equivalence	Problems in this structure may result in the same solution. Equivalent problem strings—such as doubling and halving in multiplication, constant difference in subtraction, keeping the ratio constant in division, or yielding the same solution to equations—are designed to give students opportunities to construct big ideas. "Why are we continuing to get the same answer here?," a teacher might ask, nudging students into the space of wondering and puzzlement. When students encounter problems that have the same answer, they tend to pay attention and wonder. This shifts the mathematical goal away from answer-seeking towards investigating the relationships that would explain this phenomenon. Later, they may use strategies related to equivalence—such as solving the equivalent reciprocal equation—to solve new messy problems.	**String 6.9—At a Glance** $\frac{17}{4} = x$ $\frac{4}{17} = \frac{1}{x}$ — The first two problems are equivalent by design. This supports students to ponder why each results in a solution of $x = 4.25$ and how the format of the equations relate to each other. $\frac{x}{2} = x + 1$ $\frac{2}{x} = \frac{1}{x+1}$ — The third and fourth problems are also equivalent and are placed in the sequence to solidify the big idea that reciprocal equations have equivalent solutions (considering domain restrictions). $\frac{4}{2x+3} = \frac{1}{x}$ — The last problem is designed to encouraged students to take up this big idea or its related strategy of solving an equivalent reciprocal equation.

Name of Structure	What Is It	An Example
Building a Model	For some problem strings, the purpose is to help students build a model as a mental construct. These are important models that begin as models of student thinking that can then transition as tools with which to reason and use as a mechanism for thinking. In other words, these strings build a model to represent thinking so it can later become a model for thinking. These problem strings consist of what may seem to be random problems, but the problems are based on specific relationships. As students solve the problems, a model begins to emerge. The order of the problems matter—they are usually not in order (too predictable, not intriguing), but are purposefully sequenced so that as answers are plotted, graphed, or even listed, a visual representation emerges. Students begin to realize that as their brains make and use certain relationships, those connections can be represented outside of their brain. These representations built and synthesized from mental relationships begin to solidify into tools that help with other problems. These *building a model* problem strings are intended as introductions. We should not expect mastery of all aspects or even facility with the model after the first string. Rather, continue to build the model with more rich experiences.	**String 4.5—At a Glance** $f(4) = \_\_$ $f^{-1}(2) = \_\_$ $f(\frac{1}{2}) = \_\_$ $f(\_\_) = 0$ $f(\_\_) = 1$ $f(\frac{1}{4}) = \_\_$ $f^{-1}(3) = \_\_$ $f^{-1}(x) = \_\_$ In this problem string, students start with a graph of the function in red. Students have begun to learn about inverse functions, so to help cement inverse functions and notation, the problems ask for values on the *f* and on the inverse of *f*. As the teacher plots students' answers on the display, a familiar function emerges, the blue exponential function. This is actually the last problem of the string, for students to identify the inverse of *f*. Students are now primed to learn about the inverse of exponential functions, they've just dealt with one, logarithms. 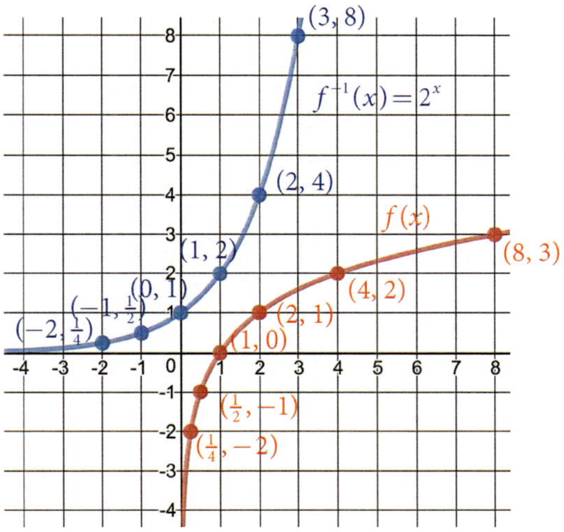

Challenges

As we've mentioned, problem strings take some preparation and practice to engender the kinds of vibrant mathematical conversations we envision. If you study the problem strings in advance, you will typically have a good sense of the mathematics that are likely to emerge. The conversation should never feel like a complete surprise. However, students' ideas can often be idiosyncratic, unformed, and even confusing. Knowing what to do when the problem string you've planned begins to feel different than the one that is unfolding takes experience and insight. In the table below we frame some of the most common challenges you may experience and offer a few suggestions about what you might say or do in these moments.

The Challenge	What You Might Consider	What You Might Say or Do	Try to Avoid
Students are not talking much, if at all.	• Have you given students something interesting to think about? • Is the math accessible to the students? • Have students done this math before and are not being asked to think about it in a new way? • Are students with partners that are "just right" for them? • Are you asking them about a problem that is too easy? • Have you created a class climate of respect and a low cost of failure, while rewarding risk taking?	• Ask students to turn and talk to their partner and listen to their pair talk. Then invite partners to share together, or ask a pair of students if you can tell the class what they discussed. • Change the conversation from "What's the answer?" to "Let's not talk about the answer yet. Who can get us started?" This may make some students feel safer and more equipped to share their ideas. • Get students paying attention to the logic of the string: "What problem do you think I'm going to put up next and why?" • Circulate and see what math students have recorded, if any. Ask individuals about insights.	• Saying for students what you hoped they would say. • Telling or showing students how you would solve the problem. • Randomly calling on students who may not be ready to participate in the routine.
You do not fully understand what a student is saying when describing a strategy.	• Do other students seem to be making sense of it? • Would hearing the strategy again help you to make sense of it? • Is what you are understanding going to help the whole class?	• "Who can put Dominic's strategy in their own words so that we can all try to make sense of it?" • "Who understands this idea? Will you say it in your own way?" • "Let's slow down for a moment. What parts of Dominic's ideas do you understand?" Then later, "What questions do you have for Dominic? What doesn't make sense to us yet?" • "That seems like a really different strategy. Will you hold onto it and see if it helps you with the next few problems?"	• Devaluing the idea just because it is not straight-forward. • Thinking that you are the only one who needs to understand the strategy—remember you are working as a community.

The Challenge	What You Might Consider	What You Might Say or Do	Try to Avoid
You understand a student's idea but are not sure how to model it.	• What about the strategy makes it challenging to model? • If you had a moment to think, would you be able to model it? • Is it worth taking the time now to model it—will you lose the interest and attention of the class as a result?	• Acknowledge the complexity of making a picture of the thinking: "Okay, I'm going to try to make a model of Monica's idea. It's a little new for me so let's see if I can capture it. Be thinking about whether this model shows what Monica just said to us." • "Turn and tell your partner what parts of Monica's strategy make sense to you. While you are doing that I'll make a model of it." • Decide that even when modeled, the strategy or idea won't help most of the students. Encourage the student to hold the idea or strategy to test out with other problems, or huddle with this student right after the string.	• Passing the pen to a student to model his or her own thinking. • Asking the class how to represent it (they are likely still making sense of the model for themselves). • Dismissing the idea entirely.
The energy of the conversation is low.	• Are you clear about what students understand and are still constructing? • Are you and one student engaged in a conversation that excludes everyone else? • Are students just solving but perhaps not being challenged to think about the relationships in new ways? • Are students waiting for you to solve the problem for them? Or for you to choose a student with the "right" strategy?	• Use partner talk strategically. • Try to intrigue students or get them wondering about whether something is true in other cases, or why something is happening. Your energy can be contagious. • Play a skeptic yourself, "You all seem pretty convinced that this strategy will always work, but what about with fractional or negative coefficients? Will it work then? I'm just not sure. Who thinks it might and wants to try to convince us?" • Take a very quick "stand and stretch" break and perhaps ask students (with their partners) in the back of the room to take a seat at the front of the room and vice versa.	• Taking over the conversation yourself. • Calling on the same students all the time. • Staying at the board the whole time instead of circulating. • Emphasizing students' note taking over thinking and reasoning.
The string is taking too long.	• What part of the string took longer than you expected and why? • Are you letting every student have the chance to share their strategy (not the goal)? Or are you purposely choosing a few strategies to model, based on what you are seeing and hearing? • Are you trying to have all students master everything during one problem string? • Are you clear what the goal of the string is?	If it's appropriate you can: • jump ahead in the problem string to a more challenging problem. • decide to end the string wherever you are and return to it another day, or not at all. • move into generalizing or summarizing with students instead of getting to the end of the string. • pass out index cards and ask students to record a question they have about the string and a strategy that is making more sense to them. • post the next problem in the string and ask students to think about it but not solve it (come back to it tomorrow).	• Feeling the need to record every single idea and strategy that is in the room. • Taking too long on easy problems that are only meant to set the stage, not be belabored.

The Challenge	What You Might Consider	What You Might Say or Do	Try to Avoid
Some students are done working the problem quickly while other students need more think time.	• Have all students gotten a good start? • Would inserting a helper problem help or be too pointed? • Are quicker finishers focused only on getting an answer? • Would it benefit slow starters to clarify the question?	• Ask early finishers to consider efficiency, connections to other problems, or to begin to generalize. • "Now that you've solved it, look back to see if you can identify any relationships you could use to solve it more efficiently." • "How does this problem connect with the previous problems?" • "Will your strategy work all of the time? How do you know? When is it a great strategy and when might it not be very efficient?"	• Moving on before the majority of students have had time to get a good start on the problem or get far enough to make the conversation fruitful. • Telling slower students how to work the problem.
Students do not make the connections or offer up the strategies you are intending.	• Are you modeling their work in a way that connects to the big idea? • Are your students still working with other ideas and not ready to construct your target idea? • Are the students moving forward even if it does not match your anticipated goal?	• "Did anyone use the previous problem to help them? Could you? Explain how." • Use a follow up "sister" string later to come back to it.	• Telling students what you wanted to hear from them. • Skipping the big idea of the string because no one offered it up.

PROBLEM STRINGS
LESSON

1.0 Division as Ratio

At a Glance

$4 \times \_\_\_ = 20$

$4 \times \_\_\_ = 21$

$4 \times \_\_\_ = 23$

$\times\_\_$
5, 20, ____

$\times\_\_$
5, 22.5, ____

Term #	Term
1	12
2	6

$\times\_\_$

Term #	Term
1	12
2	9

$\times\_\_$

Objectives

The goal of this string is to help students connect multiplication and division in such a way that they will be better able to recognize that they can use division to find a common multiplier in a geometric sequence.

Placement

This string could come before the work on arithmetic and geometric sequences, getting students ready to consider sequences defined by repeated multiplication (geometric).

You can use this string to preface the work on geometric sequences in textbook chapter 1 and on exponential functions in chapter 4.

Guiding the Problem String

The first three problems are meant to ground students in the connection between multiplication and division, but also to model student thinking about division as fractions. As you work with students, keep the discussion about equivalence. Don't let it be too pointed toward "what to do." Write $5.25 = 5\ \frac{1}{4} = \frac{21}{4}$ and acknowledge them as equivalent. The division notation is social knowledge so treat it appropriately: "I can model your thinking this way." "We can write division like this, with fraction notation." The next two problems are written as sequences, and the last two are in table format where the term number is paired with the term. These can serve as precursors to finding the common ratio of messy data in a table.

About the Mathematics

Sequences and division can be notated (by convention) in different ways. The notation is social knowledge and can be successfully told and shown to students. How to think about the relationships between multiplication, division, and ratios is logico-mathematical knowledge and needs to be worked out by students as they make connections by solving problems, looking for patterns, and using relationships.

Sample Interactions

Use the following as you plan how to elicit and model student strategies. This is not meant as a script, but as a view into the relationships involved and the intent of the problem string.

Teacher: *Today's problem string starts with this problem, an easy one: 4 times what is 20?*	$4 \times \_\_\_ = 20$
Teacher: *Right, I know that's not hard—what is that? Yep, 4 times 5 is 20. So, if 4 people have \$20, then they each had \$5? Four times \$5 is \$20.*	$4 \times \underline{\ 5\ } = 20$

(continued)

Teacher: *What if I told you that 4 times something is 21? Those 4 people had a total of $21. How much did each have? Did anyone get something more than 5? What did you find?* **Student:** *Five and a quarter.* **Teacher:** *How?* **Student:** *You just need a dollar more, and so they each get a quarter.* **Teacher:** *I might model your thinking like this. We can write division like this, with fraction notation. You were thinking about 21 divided by 4 as 20 divided by 4 and then that leftover 1 divided by 4, so 5 and a quarter. And we can write that as* $^{21}/_{4}$ *or* $5\frac{1}{4}$ *or 5.25.*	$4 \times \underline{\quad} = 21$ $4 \times \underline{5\frac{1}{4}} = 21$ $\frac{21}{4} = \frac{20}{4} + \frac{1}{4} = 5 + \frac{1}{4} = 5\frac{1}{4} = 5.25$
Teacher: *Next problem: 4 times what is 23?* **Student:** *Five and three quarters. It's like you have three of those one-quarters from before.* **Student:** *Or I thought about sharing 3 dollars by 4, that 3 over 4.* **Teacher:** *So, sharing $3 with 4 people is like 3 divided by 4, or three-fourths? Nice. Did anyone use 24 divided by 4? Could you?* **Student:** *Ahhh, yeah, you could have 24 divided by 4, that's 6, but it's one too many, so 6 minus a quarter is* $5\frac{3}{4}$ *or 5.75.*	$4 \times \underline{\quad} = 23$ $4 \times 5\frac{3}{4} = 23$ $\quad \frac{23}{4} = \frac{20}{4} + \frac{3}{4} = 5\frac{3}{4}$ $\frac{23}{4} = \frac{24}{4} - \frac{1}{4} = 6 - \frac{1}{4} = 5\frac{3}{4} = 5.75$
Teacher: *You've had problems before where you were given a list of numbers and you had to find the next number in the pattern, right? So, the next problem is one of those. Start with 5 and multiply to get to 20. What would that factor be? What times 5 is 20?* **Teacher:** *Again, that's an easy one. That's what? Yep, 4. So 5 times 4 is 20. Did anyone think of that as 20 divided by 5 is 4? Could you? I'll write that too.*	$\times 4$ (from 5 to 20) 5, 20, ____ $5 \times 4 = 20$ $\quad \frac{20}{5} = 4$
Teacher: *The next problem is in that same list format. The first term is 5 and you multiply that by something to get the next term, 22.5. So 5 times what is 22.5?* The teacher pauses, then elicits and models student responses.	$5 \times 4.5 = 22.5$ $\quad \frac{22.5}{5} = \frac{20}{5} + \frac{2.5}{5} = 4\frac{1}{2}$
Teacher: *The next problem shows up in a table like this. If I told you that you got to the second term by multiplication, how could you get from the 12 to the 6?* **Students:** *"Double", "Times 2", "One-half", "Minus 6".*	Term # | Term 1 | 12 2 | 6 $\Big\}\times$__

Teacher: *How can we make sense of this? If you know this is multiplicative, how can you get from the 12 to the 6 with multiplication?*

Student: *All I can think of is divide by 2.*

Student: *Yeah, but that's like times one-half.*

Teacher: $12 \times \frac{1}{2} = 6$*? Does it make it easier to think about if we use the commutative property,* $\frac{1}{2} \times 12$*, is that 6? Yes? What if I record the division? If it's 12 times something is 6, then 6 divided by 12 is that something. What is 6 divided by 12? I'll record it this way.*

Teacher: *Ah, and* $\frac{6}{12}$ *simplifies to* $\frac{1}{2}$ *which we can write as a fraction and decimal.*

Term #	Term
1	12
2	6

$\times \underline{0.5}, \ \times \frac{1}{2}$

$$12 \times \underline{\quad} = 6$$

$$\frac{6}{12} = \frac{1}{2} = 0.5$$

Teacher: *So this is interesting. All of these other problems before resulted in a bigger number with multiplication. What do you notice about those multipliers?*

Student: *They are greater than 1.*

Teacher: *And now this is a fraction between 0 and 1? For this problem, 12 times a number between 0 and 1, 0.5, resulted in a smaller number than 12.*

Teacher: *Okay, the last problem of the string shows up in another table like this:*

Teacher: *It's decreasing again! How can you get from 12 to 9 with multiplication? Did anyone use the previous problem to help?*

Term #	Term
1	12
2	9

$\times \underline{\quad}$

Student: *I did. Half of 12 is 6 and a quarter of 12 is 3. Since it's 9, it's just* $\frac{3}{4}$*.*

Teacher: *I might model your thinking like this: To get from 12 to 9, you could undo that by division, 9 divided by 12. Nine divided by 12 you found by using the* $\frac{6}{12}$ *from before,* $\frac{1}{2}$*, and then thought about the left over 3, and 3 divided by 12 is* $\frac{1}{4}$*. So* $\frac{1}{2}$ *and* $\frac{1}{4}$ *is* $\frac{3}{4}$*.*

Teacher: *Again, with division, we found the multiplier by dividing the term by the previous term. Great work. I wonder if these relationships might come in handy in this new chapter.*

Term #	Term
1	12
2	9

$\times \underline{0.75}, \times \frac{3}{4}$

$$6+3=9, \frac{9}{12} = \frac{6}{12} + \frac{3}{12} = \frac{1}{2} + \frac{1}{4} = \frac{3}{4} = 0.75$$

Teacher: *How would you summarize some of the things that came up in this string today?*

Elicit the following:

- *Multiplication and division are related.*
- *To find the multiplier, you can divide a term by the previous term.*
- *If the term is greater than the previous, the multiplier is greater than one.*
- *If the term is less than the previous, the multiplier is between zero and one.*

(continued)

Sample Final Display

Your display could look like this at the end of the problem string:

$4 \times \underline{\;5\;} = 20$

$4 \times \underline{5\frac{1}{4}} = 21 \qquad \frac{21}{4} = \frac{20}{4} + \frac{1}{4} = 5 + \frac{1}{4} = 5\frac{1}{4} = 5.25$

$4 \times \underline{5\frac{3}{4}} = 23 \qquad \frac{23}{4} = \frac{20}{4} + \frac{3}{4} = 5\frac{3}{4}$

$\times \underline{4}$

$5, 20, \_\_\_\_ \qquad 5 \times 4 = 20 \qquad \frac{20}{5} = 4$

$\times \underline{4.5}$

$5, 22.5, \_\_\_\_ \qquad 5 \times 4.5 = 22.5 \qquad \frac{22.5}{5} = \frac{20}{5} + \frac{2.5}{5} = 4\frac{1}{2}$

Term #	Term
1	12
2	6

$) \times \underline{0}.5, \; \times \frac{1}{2} \qquad 12 \times \_\_\_ = 6 \qquad \frac{6}{12} = \frac{1}{2} = 0.5$

Term #	Term
1	12
2	9

$) \times \underline{0}.75, \times \frac{3}{4} \qquad 6 + 3 = 9, \frac{9}{12} = \frac{6}{12} + \frac{3}{12} = \frac{1}{2} + \frac{1}{4} = \frac{3}{4} = 0.75$

Facilitation Notes

This version of the problem string lists short notes for important teacher moves during the string. After you've done the string yourself and studied the relationships involved, you might make similar notes for the things you want a reminder of or deem important.

4 × ___ = 20	*Quickly. Mention money context.*
4 × ___ = 21	*How do you know? Model 21/4*
4 × ___ = 23	*Three of those 1/4's? Back one 1/4?*
×__ 5, 20, ___	*What if it's written this way? Quickly.*
×__ 5, 22.5, ___	*22.5 = 20 + 2.5* *Write 1/2 x 5 = 2.5, x 1/2 and x 0.5*
Term # | Term 1 | 12 2 | 6) ×__	*What if it's written this way? A bit longer.* *Wait, how can we get a smaller number? 12 to 6 by multiplication?* *Write "x 1/2" and "x 0.5"*
Term # | Term 1 | 12 2 | 9) ×__	*Linger.* *Decreasing again. Hmmm... 12 to 9 by multiplication?* *Ellicit using previous and simplifying.* *Multiplication can result in a smaller number, huh? Interesting.*

PROBLEM STRINGS LESSON

1.1 Difference versus Removal

At a Glance

42 – 37

61 – 4

505 – 398

9 – 6 = 3

6 – 9 = –3

6 – (–4)

–4 – 6

–4.03 – (–1.82)

Objectives

The goal of this problem string is for students to build facility using both the difference and removal meaning of subtraction so that when they work with arithmetic sequences and linear functions, differences can be represented with subtraction using the meaning of distance and not just take away.

Placement

This string could come any time near the beginning of your course to establish the different meanings of subtraction so that students will recognize and find meaning in formulas and relationships that are represented by subtraction.

You could deliver this string any time before textbook lesson 1.2 so that students can think about common differences in arithmetic sequences or lesson 1.6 as students work with the slope formula, the ratio of the difference in y-values to the x-values.

Guiding the Problem String

Since students are about to deal with differences in both arithmetic sequences and linear functions, this problem string can help with that underpinning. To that end, do not get lost in the numbers in this string. Help students focus on developing (or reviewing) one of the two big meanings of subtraction: difference. The first problem, since the numbers are close together, suggests finding the difference, especially with the *sports-score* context. The second problem, because the numbers are so far apart, suggests removal. Contrast the thinking. Because the numbers in the third problem are neither particularly close together nor far apart, students can continue to contrast the two approaches, clarifying as they go. The next two partner problems set up the idea of considering direction with subtraction. The next two bring in subtraction of negative numbers. The last problem, the clunker, students encounter in the problem set of textbook lesson 1.1, question 5d.

About the Mathematics

Subtraction can be thought of as *removal*, but it can also be conceptualized as the *difference* between the numbers.

9 – 6 = 3

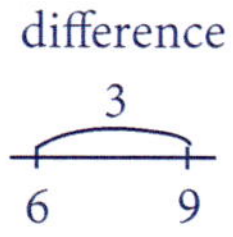

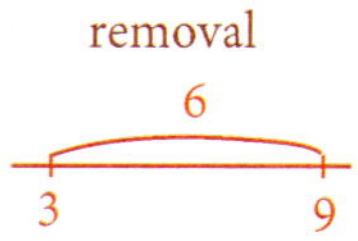

Removal ceases to be a helpful strategy with subtracting negative numbers. Difference remains useful but you must consider both the *distance* between the numbers and *direction*. When from a number you subtract a smaller number, the result is positive. When from a number you subtract a larger number, the result is negative. Often when dealing with subtracting integers, teachers help students add the opposite, which is a fine strategy, but that can leave students without the conceptual underpinning of difference. The conceptualization of subtraction as *difference* is essential for understanding things like the distance formula, the formula for the slope of a line, and many other higher math ideas, including formulas in calculus.

Sample Interactions

Use the following as you plan how to elicit and model student strategies. This is not meant as a script, but as a view into the relationships involved and the intent of the problem string.

Teacher: *Let's warm up our brains today with a problem string. No paper and pencil needed. I encourage you to really think about the problems, visualizing the relationships. In this first problem we are playing a game and the score is 42 to 37. How much to we need to score to catch up?*	42 −37
Student: *Five.* **Teacher:** *How do you know?* **Student:** *You need 3 points and 2 points.* **Teacher:** *Why would you need 3 points?* **Student:** *To get to 40.* **Teacher:** *Ahhh, to get to that friendly number 40. Nice. And then 2 more?* **Student:** *Yes, to get to 42. So 3 and 2 is 5.*	42 −37 = 5 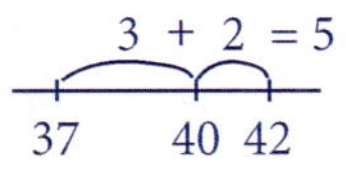
Teacher: *The next problem is a different game, where the score is 62 to 4. Eek! How much do we need to score to catch up?*	62 − 4
Student: *58 points.* **Teacher:** *You think we need to score 58 points to catch up? Why?* **Student:** *You just minus 4 from 62 and that's 58.* **Teacher:** *Did anyone else think about subtracting 4 from 62, taking 4 away from 62? I see lots of nods. How?* **Student:** *Just subtract.* **Teacher:** *Did anyone start with 62 and subtract 2?* **Student:** *Yeah, because you can just minus 2 to get 60 and then minus 2 more is 58.* **Teacher:** *Great. I'll model your thinking like this.*	62 − 4 = 58 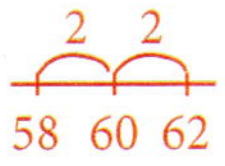

Teacher: *Did you all use the same thinking for these two first problems? They seem a little different to me.*

Student: *They are both subtraction.*

Student: *But I didn't think about the first one like subtraction. I added up to find the points we needed, like you have on the board.*

Teacher: *Does everyone agree that for this first problem many of you were finding how far you need to go from 37 to 42? What about the second problem? Were you finding how far to go from 4 to 62?*

Student: *Well, kind of. We found how many points from 4 to 62 but we found it by subtracting 4 from 62.*

Student: *You wouldn't have to, but I think most of us did.*

Teacher: *What do you mean you wouldn't have to? You wouldn't have to remove 4 from 62?*

Student: *Yeah, you could go from 4 to 62, but that would be a lot to find. It's easier to just subtract.*

(continued)

Teacher: *So, this idea of finding out how far apart numbers are, finding the difference between them—I wonder if you could think about this next problem like that? What is 505 subtract 398? What is the difference, the distance, between 398 and 505?*

Students work and the teacher circulates, asking students nudging questions when needed and looking for students using a difference and a removal strategy.

505 – 398

Teacher: *I saw you start on 505. Tell us about your thinking?*

Student: *I didn't want to subtract 398, so I subtracted 400.*

Teacher: *And what is 505 minus 400?*

Student: *That's just 105. But I subtracted too much, so I had to fix it by 2, so 107.*

Teacher: *Why did you go to the right 2, not left? Isn't this subtraction? Shouldn't you remove the 2?*

Student: *No, because the jump of 400 is too big. I really only had to go back 398, so that's two to the right.*

Teacher: *Where is the 398? Someone else, can you find the 398?*

Student: *It's in between the 107 and the 505.*

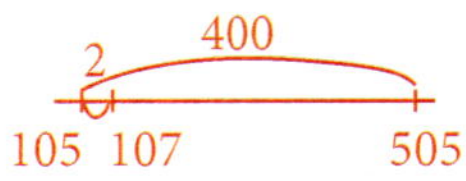

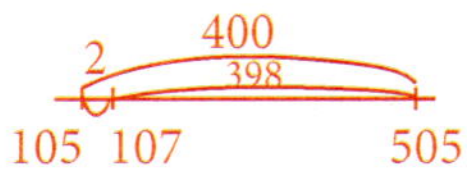

Teacher: *Is that the same as finding the distance between 398 and 505?*

Student: *No, that's subtracting. I found the distance between them by putting them both on the number line. Then it's just 2 to 400, 100 to 500, and 5 more to 505. That's 107 total.*

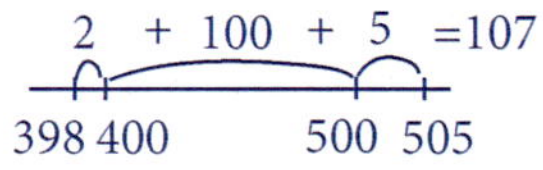

Teacher: *What questions do you have? How many of you removed, like this red strategy? How many of you found the distance between the numbers in the problem, like this blue strategy? I'm going to write those words up here to help us all remember what the thinking was all about.*

42 – 37 = 5

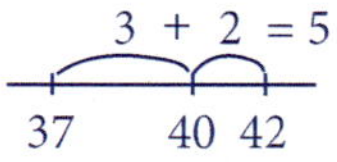

difference

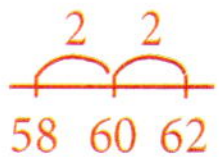

removal

62 – 4 = 58

505 – 398 = 107

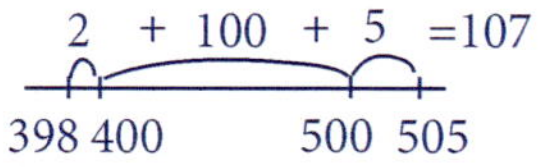

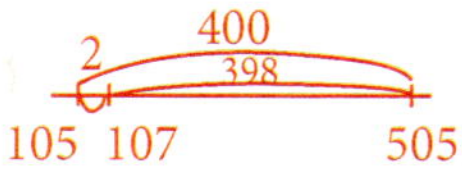

Teacher: *This next problem is just to set us up for more work, so I know it's easy. You've known the answer to this since you were in elementary school. What is 9 minus 6?*

Students: *Three.*

Teacher: *What would that look like, both as difference and removal? While you're thinking, I'll model that quickly. Which one is difference? Removal? How do you know?*

9 – 6

9 – 6 = 3

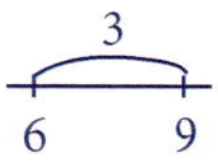

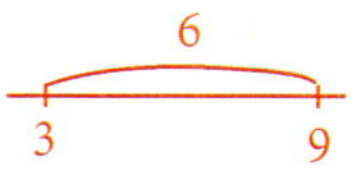

Teacher: *And just so we're on the same page. When you were subtracting like that as young kids, from 9 are you subtracting a smaller number or a bigger number?*

Students: *Smaller.*

Teacher: *And that was true all the way until you got to integers, negative numbers, right? Now, I'm not sure how you dealt with negative numbers in your past, but I wonder how what we're doing today might impact negative numbers?*

Teacher: *Here's the next problem, 6 subtract 9. How do you think about that? First of all, what is it?*

Students: *Negative three.*

$6 - 9$

Teacher: *How do you know? Did anyone think about it by starting on 6 and getting rid of, or removing the 9?*

Student: *I think I did. If you start on 6 and go back 9, you'll land on negative 3.*

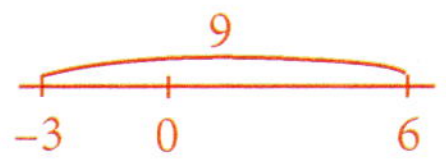

Teacher: *Did anyone find the difference? Can you?*

Student: *So, if you put them both on the number line, 6 and 9, that's positive three, not negative three.*

Student: *But it's 6 minus 9, not 9 minus 6. Since 9 is bigger than 6, it has to be negative.*

Teacher: *Why?*

Student: *It's like debt. If you only have 6 dollars, but you spend 9, you're in debt 3.*

Student: *You have to think about both distance and take away.*

Teacher: *You have to consider both distance and removal? Find the distance but then think about the sign? From the first number, if you're subtracting something larger it will be negative. That sounds important. Let's write that up here so we can continue to think about it.*

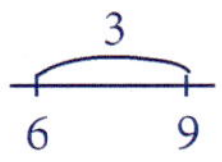

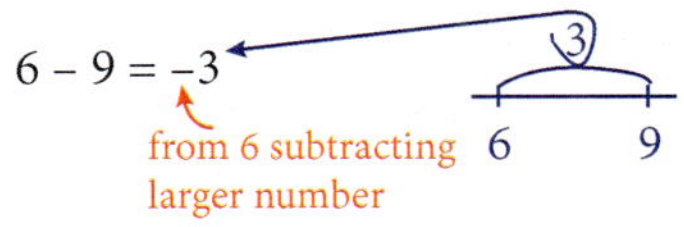

Teacher: *So, how do you think about subtraction in this problem, 6 subtract negative 4?*

Brief think time.

Teacher: *What is 6 subtract negative 4?*

Student: *Ten.*

$6 - (-4)$

$6 - (-4) = 10$

Teacher: *How do you know?*

Student: *I learned that when you're subtracting, you add the opposite.*

Teacher: *Have any of you learned that too? I'll write that up here.*

$6 + (+4) = 6 + 4 = 10$

add the opposite

(continued)

<table>
<tr><td>Teacher: Can we make sense of this with finding the difference, looking at the distance and considering the direction to find the sign?
Student: Put −4 and 6 on the number line. So, that's 4 to 0 and then 6 to 6, so that's 10 total.
Teacher: So, either way, we get that the difference, or the solution, is 10.</td><td>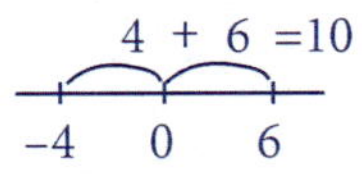
</td></tr>
<tr><td>Teacher: Here's the next one, negative 4 subtract 6. What's that?
Brief think time.
Student: I think it's negative 10 because if you add the opposite, you get negative 4 plus negative 6, that's negative 10.</td><td>$-4 - 6$
$-4 + -6 = -10$</td></tr>
<tr><td>Teacher: What about difference?
Student: If they are both on the number line, that's like what we have before, so the distance is 10. But now we're subtracting something bigger than the first number, so it has to be negative.
Teacher: So, the distance can tell us the number and the direction tells us the sign? Nice work. It looks like we can find the difference if we consider both distance and direction.</td><td>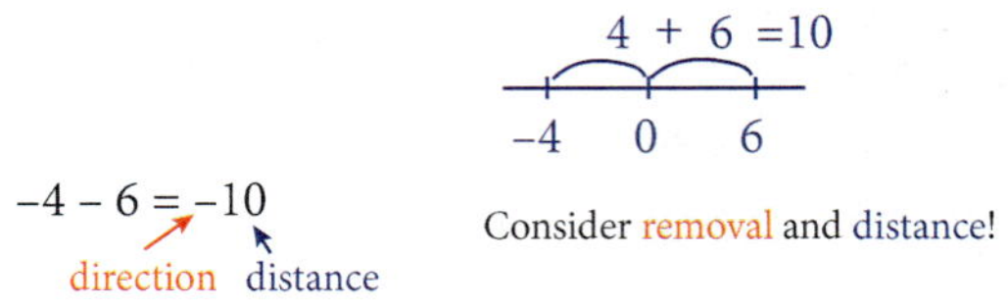
</td></tr>
<tr><td colspan="2">Teacher: What about this last lovely problem, −4.03 subtract negative 1.82?
After think time, the teacher calls on both a difference strategy and an add the opposite strategy.
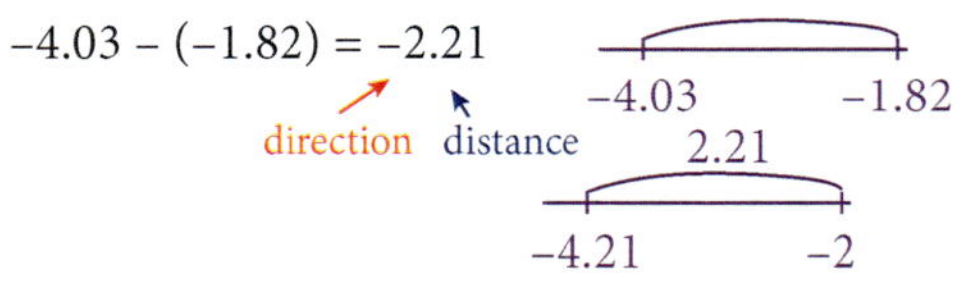

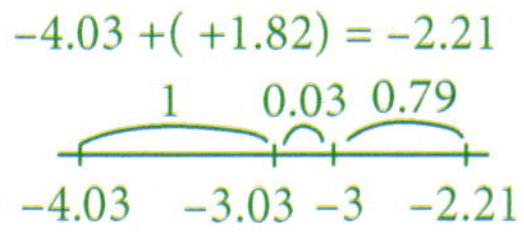
</td></tr>
</table>

Teacher: *How would you summarize some of the things that came up in this string today?*

Elicit the following:

- *Subtraction can mean removal, but it can also mean the difference between the numbers.*
- *If you are finding the difference with integers, you need to consider both distance and direction.*

Sample Final Display

Your display could look like this at the end of the problem string:

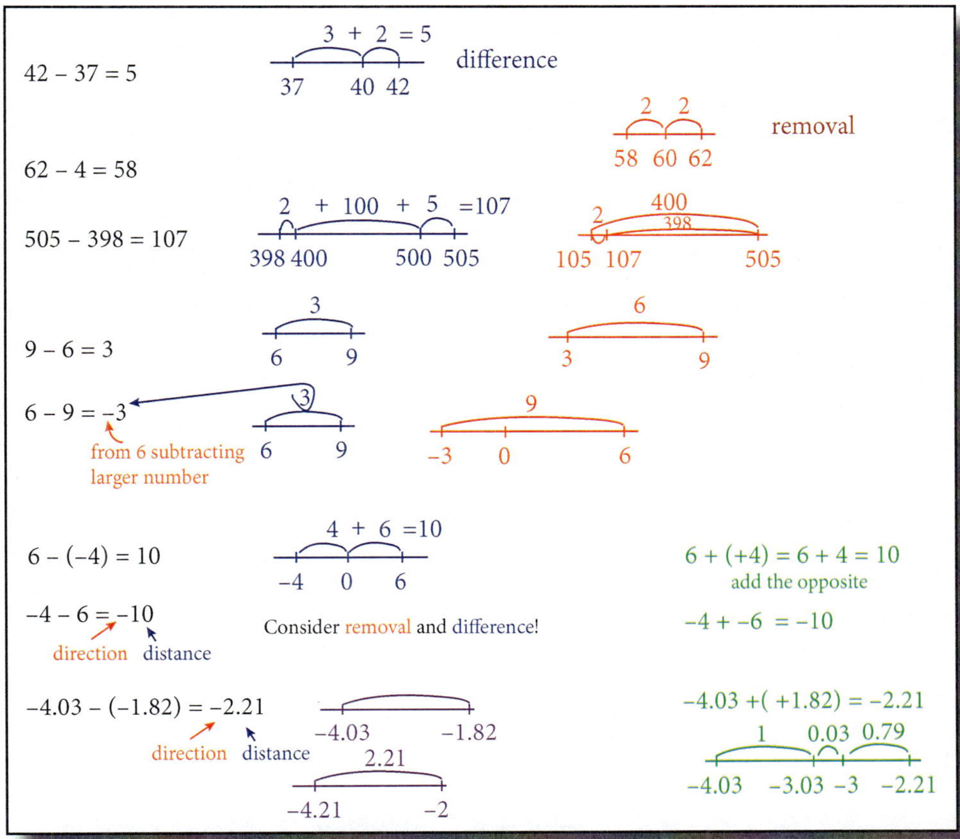

Facilitation Notes

This version of the problem string lists short notes for important teacher moves during the string. After you've done the string yourself and studied the relationships involved, you might make similar notes for the things you want a reminder of or deem important.

42 – 37	*What if the score was 42 to 37, how much to catch up?* *Model difference on one side of display in one color.*
61 – 4	*What if the score was 61 to 4 at half time, how much to catch up?* *Model removal on other side of display in different color.* *What's going on? Why different?*
505 – 398	*Difference or removal? How?* *Write "difference" and "removal".*
9 – 6	*So if subtraction can mean difference, what is difference? Removal?* *What does that look like?*
6 – 9	*What is the difference now? What about removal?* *What about adding the opposite?* *There is a difference between the distance between and the answer.*
6 – (–4)	*So if subtraction can mean difference, what is difference? Consider direction.* *What about adding the opposite?*
–4 – 6	*What does this expression mean? Removal, difference, add opposite.* *With integers, consider difference (distance) and removal (direction).*
–4.03 – (–1.82)	*Clunker! Difference? Add the opposite?*

PROBLEM STRINGS LESSON

1.2 Rebound

At a Glance

$u_0 = 1.0$ and $u_n = 0.64 \cdot u_{n-1}$

$u_0 = 3.0$ and $u_n = 0.64 \cdot u_{n-1}$

$u_0 = 1.0$ and $u_n = 0.8 \cdot u_{n-1}$

$u_0 = 1.0$ and $u_n = 0.4 \cdot u_{n-1}$

$2, 1.5, 1.125, 0.844$

Objectives

The goal of this problem string is to solidify the learning about geometric sequences that took place in the Looking for the Rebound Investigation in which students used data of the maximum heights of the bounces of a ball to find the rebound ratio and write a recursive rule.

Placement

This string could come right after or the day after students have used real data to write a recursive formula for the geometric sequence of the bounce heights versus bounce number of a bouncing ball.

You could deliver this string any time after the Looking for the Rebound Investigation in textbook Lesson 1.2 Modeling Growth and Decay.

Guiding the Problem String

The first problem is based on the data from the answers in textbook lesson 1.2 teacher edition. You could change the numbers to match the data your class collected. The purpose of the first problem is to provide an anchor to the investigation that students can refer to throughout the string. Use the display grapher to graph the sequences.

Take a little time on the second problem to predict and discuss similarities and differences. Push on answers for justification. The next two problems should go quickly.

The last problem requires students to find rebound ratios. Encourage students to think about the relationship between 2 and 1.5 before they just start dividing, because 1.5 is $\frac{3}{4}$ of 2. Then $\frac{3}{4}$ of 1.5 is $\frac{9}{8}$ or 1.125.

About the Mathematics

The sequence of the maximum heights of each bounce of a ball create a geometric sequence. When graphed as (bounce number, max height of bounce), the points create an exponential function with a restricted domain, in this case with a domain of whole numbers.

Caution, the function of maximum heights of a bounce over time do not create an exponential function. In order for the sequence to be geometric, the dependent variable is bounce number, not time.

Rebound ratios of $0 < r < 1$ create decreasing sequences.

Sample Interactions

Use the following as you plan how to elicit and model student strategies. This is not meant as a script, but as a view into the relationships involved and the intent of the problem string.

Teacher: *Let's all recall the bouncing ball investigation that we did. Remember we found (something like) this recursive formula?*
Remind us all of what the numbers mean in terms of the bouncing ball.

Student: *The 1.0 is the height of the first bounce, or bounce 0. That's why it's* u_0*. And the 0.64 is how much it rebounded each time.*

Student: *Yeah, so if the first bounce was 1 meter, then the height of the next bounce is 0.64 meters and then next would be 64% of that.*

Teacher: *How did you know that the height of the next bounce is 0.64 meters?*

Student: *Because it's easy to find 0.64 times 1.*

Teacher: *Let's put a graph of this sequence of bounce heights in the grapher.*

$u_0 = 1.0$ and $u_n = 0.64 \cdot u_{n-1}$

1, 0.6, 0.4, 0.3

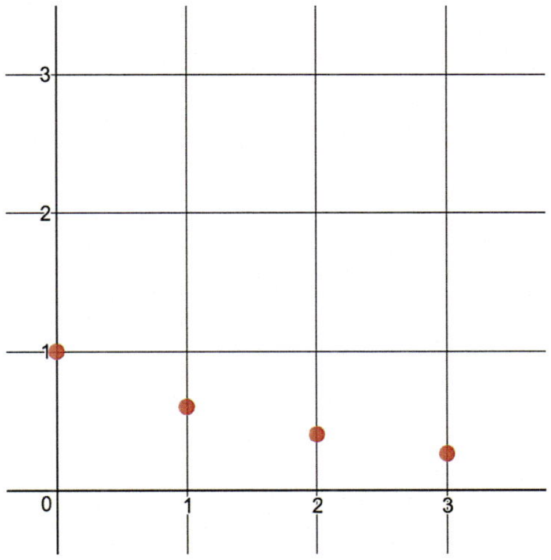

Teacher: *Great, okay the next problem in our string today is this one. Who can describe the bouncing ball that this recursive formula represents?*

Teacher: *What is the same as the first scenario? What is different? Let's write out a few terms.*

Teacher: *Predict how the graph of this sequence will compare with the first graph.*

Teacher: *Let's put both graphs up and see how our predictions fair.*

$u_0 = 3.0$ and $u_n = 0.64 \cdot u_{n-1}$

3.0, 1.9, 1.2, 0.8

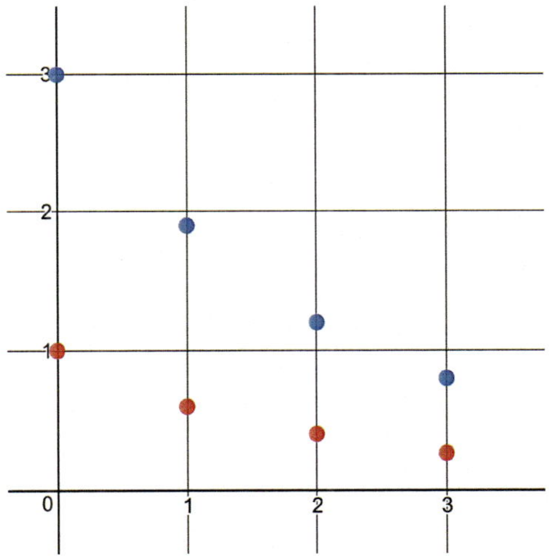

(continued)

Teacher: *In our next problem, we go back to a start value of 1 but this number changes to 0.8. How does that affect everything?* **Teacher:** *Predict the graph, find some terms, and turn and talk about the similarities and differences.* **Teacher:** *Who is willing to compare all three of these?*	$u_0 = 1.0$ and $u_n = 0.8 \cdot u_{n-1}$ 1, 0.8, 0.6, 0.5 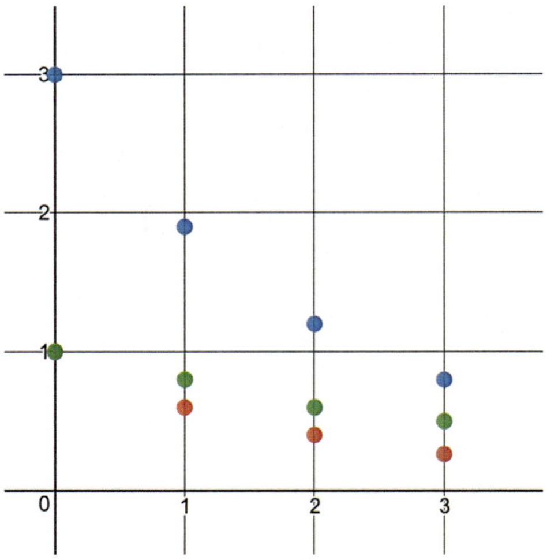
Teacher: *In this problem, the first bounce height of 1 stays the same again, but this time the rebound ratio changes to 0.4. What affect will that have? Predict, find some terms, and let's talk.*	$u_0 = 1.0$ and $u_n = 0.4 \cdot u_{n-1}$ 1, 0.4, 0.2, 0.1 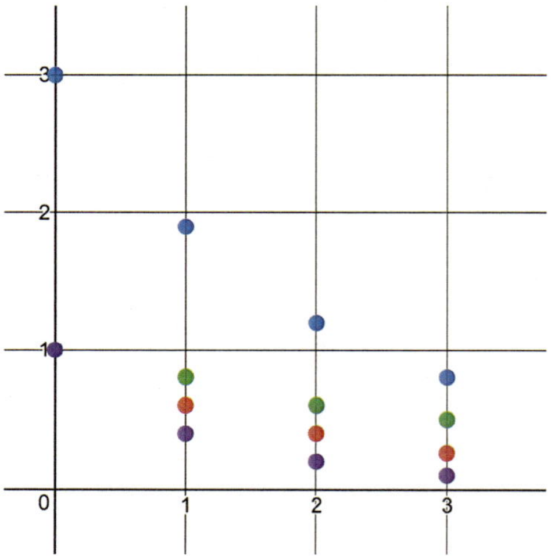

Teacher: *And for the last problem, what if we just have the max heights of a ball? Here are the heights of four bounces. Write the recursive rule for this bouncing ball and predict how it relates to our others.* **Teacher:** *What rebound ratio did you find? How? Did anyone think of the fact that 1.5 is ¾ of 2?* **Teacher:** *Did anyone notice that the start height is halfway in between our previous heights of 1 and 3, but check out the second height, should it be halfway in between too? Why or why not?*	$2, 1.5, 1.125, 0.844$ $\frac{1.5}{2} = 0.75, \frac{1.125}{1.5} = 0.75$ $u_0 = 2$ and $u_n = 0.75 \cdot u_{n-1}$ 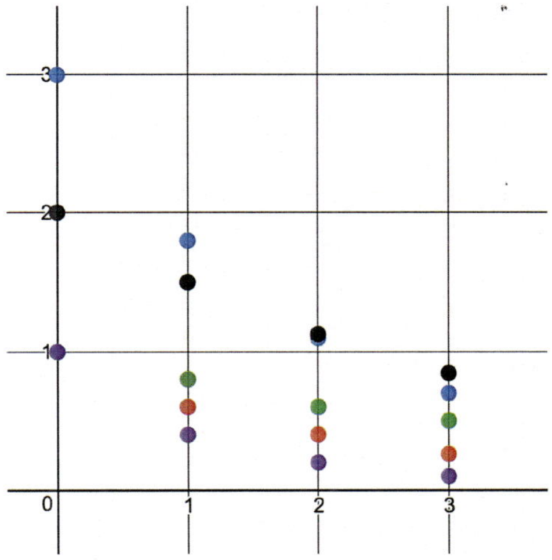
Teacher: *Hopefully everyone is getting a little more confident in the recursive rules for geometric sequences.*	

Teacher: *How would you summarize some of the things that came up in this string today?*

Elicit the following:

- *The u_0 term is the first term in the sequence.*
- *The number in the rule that is being multiplied every time is the ratio of the terms. It is the constant multiplier.*
- *Ratios (multipliers) in a geometric sequence that are in between 0 and 1 create decreasing sequences.*
- *Higher ratios (multipliers) make a decreasing sequence decrease slower than lower ratios.*

(continued)

Sample Final Display

Your display could look like this at the end of the problem string:

height of 1st bounce → $u_0 = 1.0$ and $u_n = 0.64 \cdot u_{n-1}$ ← *rebound ratio*

$u_0 = 1.0$ and $u_n = 0.64 \cdot u_{n-1}$ — $1, 0.6, 0.4, 0.3$

$u_0 = 3.0$ and $u_n = 0.64 \cdot u_{n-1}$ — $3.0, 1.9, 1.2, 0.8$

$u_0 = 1.0$ and $u_n = 0.8 \cdot u_{n-1}$ — $1, 0.8, 0.6, 0.5$

$u_0 = 1.0$ and $u_n = 0.4 \cdot u_{n-1}$ — $1, 0.4, 0.2, 0.1$

$2, 1.5, 1.125, 0.844$ — $\frac{1.5}{2} = 0.75, \frac{1.125}{1.5} = 0.75$

$u_0 = 2$ and $u_n = 0.75 \cdot u_{n-1}$

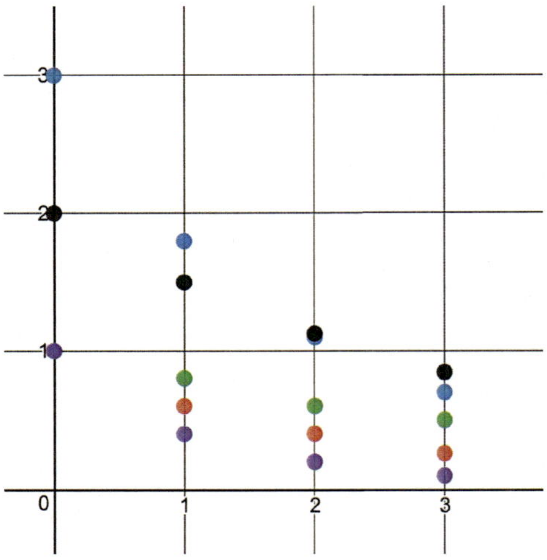

Facilitation Notes

This version of the problem string lists short notes for important teacher moves during the string. After you've done the string yourself and studied the relationships involved, you might make similar notes for the things you want a reminder of or deem important.

$u_0 = 1.0$ and $u_n = 0.64 \cdot u_{n-1}$	*Remember the ball bouncing investigation?* *What does this rule mean? What do the numbers mean?*
$u_0 = 3.0$ and $u_n = 0.64 \cdot u_{n-1}$	*What if we change the rule? Tell us about this bouncing ball?* *Predict the graph. Compare.*
$u_0 = 1.0$ and $u_n = 0.8 \cdot u_{n-1}$	*How about this rule? Describe this ball.* *Predict the graph. Compare.*
$u_0 = 1.0$ and $u_n = 0.4 \cdot u_{n-1}$	*And this one? Describe this ball.* *Predict the graph. Compare.*
$2, 1.5, 1.125, 0.844$	*What if we just have the max heights of each bounce?* *What is the rebound ratio? Compare.* *Write the rule, predict and sketch the graph.*

PROBLEM STRINGS
LESSON

1.3 Difference or Ratio

At a Glance	
4, 2, ___ 1, 5, ___ 3.2, 6.4, ___ 0.1, 0.01, ___	**Objectives** The goal of this string is to help students solidify the difference between a constant difference in arithmetic sequences and a constant multiplier in geometric sequences, for both increasing and decreasing sequences.
	Placement This string can happen during the work on arithmetic and geometric sequences. This string could also happen before textbook Lesson 1.3 Applications and Other Sequences to help students solidify arithmetic and geometric sequences.
	Guiding the Problem String This problem string can help students consider both an additive next term and a multiplicative next term. Help students realize the connection between addition and subtraction for arithmetic sequences by representing the difference as subtraction, even when the problems are easy enough to just "see" the difference. Similarly, represent the ratio of the two terms to find the multiplier so that students may use that suggestion to find multipliers with messier data.
	About the Mathematics Arithmetic sequences increase or decrease by a constant difference. This difference can be represented by subtraction of consecutive terms. The graphs of arithmetic sequences form a line, where the domain is restricted to the whole or natural numbers. Geometric sequences increase or decrease by a constant multiplier. This multiplier can be represented by the ratio of consecutive terms. The graphs of geometric sequences form an exponential curve, where the domain is restricted to the whole or natural numbers.

Sample Interactions

Use the following as you plan how to elicit and model student strategies. This is not meant as a script, but as a view into the relationships involved and the intent of the problem string.

Teacher: *Today's problem string starts with a short sequence. The first term is 4 and the next term is 2. What kind of sequence is this, arithmetic or geometric?*	4, 2, ___
Student: *Arithmetic—no, geometric—it could be either.*	
Teacher: *Some of you said arithmetic. If so, what's the next term?*	
Student: *It would be 0.*	
Teacher: *What's the difference?*	
Student: *Two, you're subtracting two. So the next term is zero.*	−2 −2 4, 2, 0

(continued)

Teacher: *Some of you said geometric. Still think so?* **Student:** *Yes, though I see how they found an arithmetic one. But it could also be half, so the next term would be 1.* **Teacher:** *Can they both be right?* **Student:** *Looks like it.*	$\times\frac{1}{2}$ $\times\frac{1}{2}$ 4, 2, <u>1</u>
Teacher: *Is this sequence increasing or decreasing? With decreasing sequences, what is the difference like and what is the multiplier like?* **Student:** *The difference is negative. The multiplier is a fraction. The fraction is between 0 and 1.* **Teacher:** *I'm going to sketch both of these quickly. While I do, predict what each sequence as ordered pairs of (term number, term) looks like.*	4, 2, ___ -2 -2 4, 2, <u>0</u> $\times\frac{1}{2}$ $\times\frac{1}{2}$ 4, 2, <u>1</u> 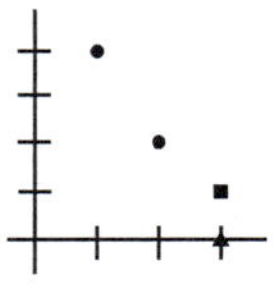
Teacher: *Next problem: first term is 1, next term is 5. This time, let's assume both kinds of sequences and write the next term if this sequence is arithmetic and then write the next term if the sequence is geometric.* The teacher facilitates the same kind of conversations, but this problem should go quickly, especially since the sequence is increasing. **Teacher:** *How would you describe the look of the arithmetic sequence? The geometric sequence?* The teacher draws out comments like: The arithmetic are in a line—linear. The geometric is not linear; it is going up (or down) faster.	1, 5, ___ +4 +4 1, 5, <u>9</u> ×5 ×5 1, 5, <u>25</u> 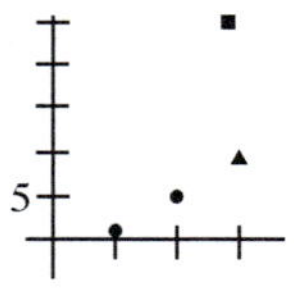
Teacher: *What do you think might be the next terms in an arithmetic and geometric sequence if they start this way? Again, this could be either sequence. Write the next term for both.* Encourage students to think and reason about the numbers; suggest thinking about money. *Could you think about 3 dollars and 20 cents, and 6 dollars and 40 cents?* **Teacher:** *What's the next term if the sequence is arithmetic? And what's the difference?* **Student:** *The next term is 9.6 because the difference is 3.2.* **Teacher:** *I'm going to write that the difference between 6.4 and 3.2 is 3.2 by writing it as a subtraction sentence. And is it also true that it is equivalent to 9.6 – 6.4?*	3.2, 6.4, ___ +3.2 +3.2 3.2, 6.4, <u>9.6</u> 6.4 – 3.2 = 3.2 = 9.6 – 6.4

Teacher: *What about a geometric sequence?*

Student: *It's just double, so times 2 and the next term is 12.8.*

Teacher: *I'm going to model that as the ratio of the two terms.*

Student: *When you write it that way, you could scale up to have the equivalent 64 divided by 32.*

Student: *Oh yeah, 64 divided by 32 is also 2.*

$\times 2$ $\times 2$

3.2, 6.4, <u>12.8</u>

$\frac{6.4}{3.2} = \frac{64}{32} = 2$ ($\times 10$ above and below)

Teacher: *And again we see the arithmetic sequence sort of marching along, plotting the next term at the same difference, every time. But the geometric sequence is curvy and changing, in this case swooping up. Do you think that will always be true?*

3.2, 6.4, ____

+3.2 +3.2

3.2, 6.4, <u>9.6</u>

$\times 2$ $\times 2$

3.2, 6.4, <u>12.8</u>

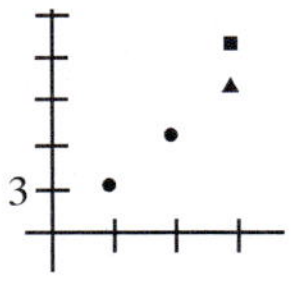

$6.4 - 3.2 = 3.2 = 9.6 - 6.4$

$\frac{6.4}{3.2} = \frac{64}{32} = 2$ ($\times 10$ above and below)

Teacher: *For the last problem, again find the next term as if it was arithmetic and then if it was geometric. The first term is one-tenth or point one or you could even think of it as a dime. The second term is one-hundredth or point zero one. What would this second term be in reference to money?*

Student: *A penny.*

Students work to find the difference.

0.1, 0.01, ______

Teacher: *Many of you look a bit stymied. I wonder if it would help to keep thinking about these as money?*

Student: *I thought about going from a dime to a penny, you'd have to subtract 9 pennies. So the difference is −0.09 and so the next term is −0.08.*

Teacher: *What if you thought about it as subtraction, like we did above, with the second term minus the first term, 0.01 − 0.1? Could you find the difference between a penny and a dime? Sure enough, that's also 9 cents. And since it's a penny minus a dime, that's 0.09, so you land on −0.08.*

0.1, 0.01, ____

−0.09 −0.09

0.1, 0.01, <u>-0.08</u>

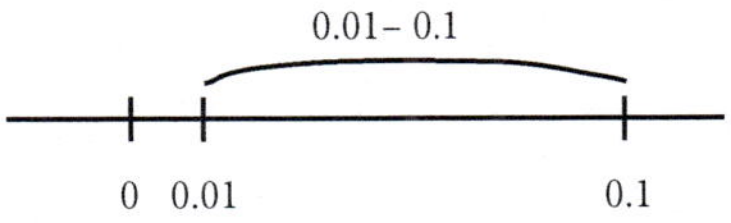

distance is 0.09
from 0.01, you are subtracting a bigger number, so 0.01 − 0.1 = −0.09

(continued)

Teacher: *Okay, so for the geometric sequence, what are you thinking?*

Student: *I'm not sure how to multiply to get from* 0.1 *to* 0.01.

Teacher: *Is the sequence increasing or decreasing?*

Student: *Decreasing.*

Student: *I thought about it like a place value shift, from one-tenth to one-hundredth is divided by* 10 *which is the same as times one-tenth.*

Teacher: *Did anyone think about it using division like we wrote in the previous problem?*

Student: *I took the second term divided by the first term. Then scaled up to create an equivalent problem, one-tenth.*

Teacher: *So, what is* 0.01 *times* 0.1*? Is that another place value shift? And what would the graphs look like?*

0.1, 0.01, ____

−0.09 −0.09

0.1, 0.01, −0.08

× 0.1 × 0.1

0.1, 0.01, 0.001

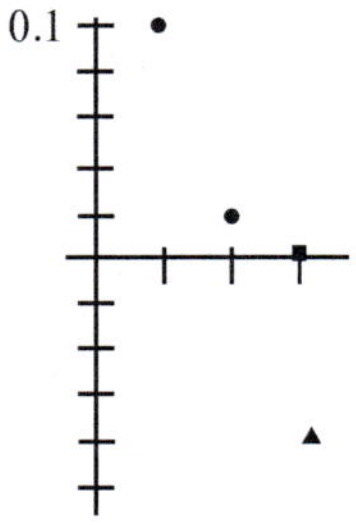

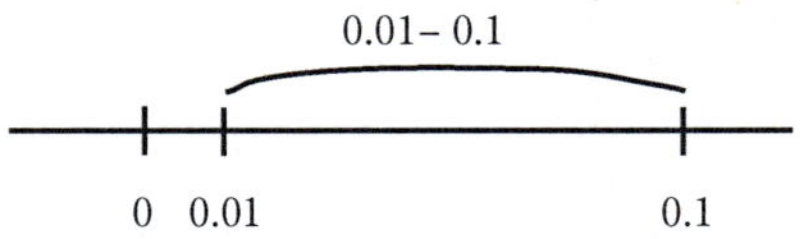

distance is 0.09

from 0.01, you are subtracting a bigger number, so 0.01 − 0.1 = −0.09

$$\frac{0.01}{0.1} = \frac{1}{10} = 0.1 \quad (\times 10 \text{ top and bottom})$$

Teacher: *How would you summarize some of the things that came up in this string today?*

Elicit the following:

- *With only two terms, you could have either kind of sequence.*
- *You can find the difference or the ratio and use that to find the next term.*
- *Arithmetic sequences are linear and geometric are not linear.*

Sample Final Display

Your display could look like this at the end of the problem string:

4, 2, ____

−2 −2

4, 2, <u>0</u>

×½ ×½

4, 2, <u>1</u>

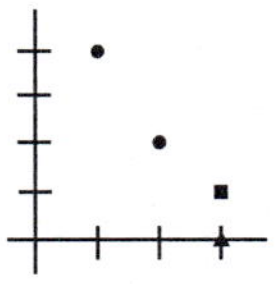

1, 5, ____

+4 +4

1, 5, <u>9</u>

×5 ×5

1, 5, <u>25</u>

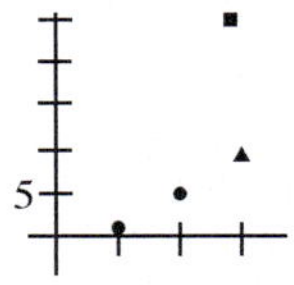

3.2, 6.4, ____

+4.2 +4.2

3.2, 6.4, <u>9.6</u>

×2 ×2

3.2, 6.4, <u>12.8</u>

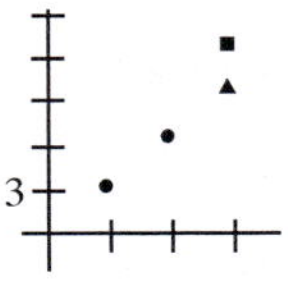

6.4 − 3.2 = 3.2 = 9.6 − 6.4

×10

$\frac{6.4}{3.2} = \frac{64}{32} = 2$

×10

0.1, 0.01, ____

−0.09 −0.09

0.1, 0.01, <u>−0.08</u>

× 0.1 × 0.1

0.1, 0.01, <u>0.001</u>

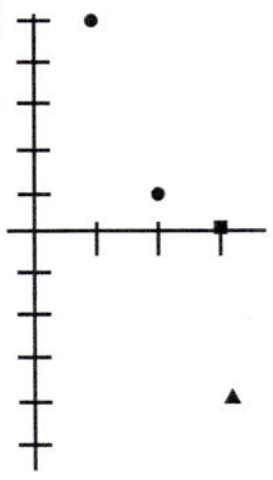

distance is 0.09

from 0.01, you are subtracting a bigger number, so 0.01 − 0.1 = −0.09

×10

$\frac{0.01}{0.1} = \frac{1}{10} = 0.1$

×10

Facilitation Notes

This version of the problem string lists short notes for important teacher moves in the string. After you've done the string yourself and studied the relationships involved, you might make similar notes for the things you find important or want a reminder of.

4, 2, ____	*Arithmetic? Geometric? Next term? When decreasing, describe multiplier. Graph quickly.*
1, 5, ____	*Quickly, describe graphs. Linear versus non-linear.*
3.2, 6.4, ____	*Decimals ☺ Keep reasoning! Represent difference and ratio.*
0.1, 0.01, ____	*Linger. Money? Ellicit difference and ratio.*

PROBLEM STRINGS LESSON

1.4 Missing Terms

At a Glance	
1, 64, ... 1, ___, 64, ... 1, ___, ___, 64, ... 5, ___, ___, 40, ... −2, ___, ___, −250, ... 1, ___, ___, ___, 81, ...	**Objectives** The goal of this string is to help students continue to build understanding of the multiplicative, exponential nature of geometric sequences by finding the multiplier when consecutive terms are missing. **Placement** This string can happen during the work on geometric sequences, helping students to reason about the repeated multiplication involved in exponentiation. Use this string as early as textbook Lesson 1.2 Modeling Growth and Decay when you want to help students focus on geometric sequences. **Guiding the Problem String:** Help students realize the multiplier must be the same throughout the sequence. Also, help students connect the social notation of $4 \cdot 4 = 4^2$. Consider skipping the first one or two problems for more advanced students. **About the Mathematics** The common multiplier or common ratio in a geometric sequence can be found be dividing a term by the previous term. When a term is missing in a geometric sequence, you can find the multiplier by dividing the term by a previous given term and then consider how many terms are missing in between them. Since geometric sequences are built with repeated multiplication, the multiplier when there are n missing terms will be the $(n + 1)$th root of the ratio of the given terms. For example, if there are 2 missing terms, such as 1, ___, ___, 64, the ratio is 64:1. Since the multiplier happens three times, find the cube root of the ratio. $1, \overset{\times 4}{\frown}\ \underline{4}, \overset{\times 4}{\frown}\ \underline{16}, \overset{\times 4}{\frown}\ 64, \ldots$ $\quad 4 \times 4 \times 4 = 4^3 \quad \sqrt[3]{64} = 4 \quad a, \overset{\cdot m}{\frown}\ a \cdot m, \overset{\cdot m}{\frown}\ a \cdot m^2, \overset{\cdot m}{\frown}\ a \cdot m^3, \ldots$

Sample Interactions

Use the following as you plan how to elicit and model student strategies. This is not meant as a script, but as a view into the relationships involved and the intent of the problem string.

Teacher: *We've been working with geometric sequences. Today's problem string starts with a really short geometric sequence:* 1, 64. *What is the multiplier if this sequence is geometric?*	$1, \overset{\times ?}{\frown}\ 64, \ldots$
Teacher: *That's easy right, just* 64.	$1, \overset{\times 64}{\frown}\ 64, \ldots$

Teacher: *The next sequence is slightly longer and it's missing a term: 1, don't know, 64. What is the missing term? How do you know?* **Students:** 32, 4, 8 **Teacher:** *So, which one? Which works? Why does 8 work? What does 8 have to do with 64? 8 times 8 can be written* 8^2. **Teacher:** *Is positive 8 the only possibility?*	$\times ?\ \times ?$ 1, ___, 64, ... $\times 8\ \times 8$ 1, 8, 64, ... $8 \times 8 = 8^2$ $\times 8\ \times 8$ 1, 8, 64, $8 \times 8 = 8^2$ also $\times(-8)\ \times(-8)$ 1, −8, 64, ...
Teacher: *Next problem: 1, blank, blank, 64. What goes in these blanks? What relationships are you thinking about? I wonder what else you know about 64?* **Student:** *I think it's 4 because 1 times 4 is 4 and 4 × 4 is 16 and 16 × 4 is 64. Yep, it's 4.* **Teacher:** *The factor is 4. And we write 4 times 4 times 4 as 4 cubed. Is positive 4 the only option? How do you know?*	$\times ?\ \times ?\ \times ?$ 1, ___, ___, 64, ... $\times 4\ \times 4\ \times 4$ 1, 4, 16, 64, ... $4 \times 4 \times 4 = 4^3$
Teacher: *Okay, here's another geometric sequence that's missing some terms. The first term is 5, then we don't know the next term, nor the next, but the term after that is 40. What are the missing terms? How do you know?* The teacher circulates and works with students, asking questions. *What kind of sequence is it? Since it's geometric, what do you know?* *You think the multiplier is 8? Does that work? What would the third term be?* *Ah, so there's something about 8, how can you break up 8 multiplicatively? Exponentially?*	$\times ?\ \times ?\ \times ?$ 5, ___, ___, 40, ... $\times 2\ \times 2\ \times 2$ 5, 10, 20, 40, ... $40 \div 5 = 8 = 2 \times 2 \times 2 = 2^3$
Repeat for the last two problems. $\times 5\ \times 5\ \times 5$ −2, −10, −50, −250, ... $-250 \div (-2) = 125 = 5 \times 5 \times 5 = 5^3$ $\times 3\ \times 3\ \times 3\ \times 3$ 1, 3, 9, 27, 81, ... $81 = 3 \times 3 \times 3 \times 3 = 3^4$ also $\times(-3)\ \times(-3)\ \times(-3)\ \times(-3)$ 1, −3, 9, −27, 81, ...	

Teacher: *How would you summarize some of the things that came up in this string today?*

Elicit the following:

- *Geometric sequences really are all about repeated multiplication of the same factor.*
- *Repeated multiplication is represented by exponentiation.*
- *The multiplier can be negative.*

(continued)

Sample Final Display

Your display could look like this at the end of the problem string:

×64
1, 64, ...

×8 ×8
1, <u>8</u>, 64, ... $8 \times 8 = 8^2$ also ×(−8) ×(−8) 1, <u>−8</u>, 64, ...

×4 ×4 ×4
1, <u>4</u>, <u>16</u>, 64, ... $4 \times 4 \times 4 = 4^3$

×2 ×2 ×2
5, <u>10</u>, <u>20</u>, 40, ... $40 \div 5 = 8 = 2 \times 2 \times 2 = 2^3$

×5 ×5 ×5
−2, <u>−10</u>, <u>−50</u>, −250, ... $-250 \div (-2) = 125 = 5 \times 5 \times 5 = 5^3$

×3 ×3 ×3 ×3
1, <u>3</u>, <u>9</u>, <u>27</u>, 81, ... $81 = 3 \times 3 \times 3 \times 3 = 3^4$ also ×(−3) ×(−3) ×(−3) ×(−3) 1, <u>−3</u>, <u>9</u>, −27, <u>81</u>, ...

Facilitation Notes

This version of the problem string lists short notes for important teacher moves during the string. After you've done the string yourself and studied the relationships involved, you might make similar notes for the things you want a reminder of or deem important.

×? 1, 64, ...	*What's the multiplier? Ratio? Quick*
×? ×? 1, ____, 64, ...	*Has to be the same multiplier... We write 8 x 8 as* 8^2 *Is positive 8 the only possibility?*
×? ×? ×? 1, ____, ____, 64, ...	*Still has to be the same multiplier each time...* $4^3 = 64$. *Is positive 4 the only option?*
×? ×? ×? 5, ____, ____, 40, ...	*There's something about* $40 \div 5 = 8$*?* $8 = 2^3$
×? ×? ×? −2, ____, ____, −250, ...	$-250 \div -2 = 125$, $125 = 5^3$
×? ×? ×? ×? 1, ____, ____, ____, 81, ...	$81 = 3^4$ *Is positive 3 the only possibility?*

PROBLEM STRINGS LESSON

1.5 Finding Slope

<table>
<tr><td rowspan="4">

At a Glance

(0, 5) (2, 9)
(0, 5) (2, 1)
(−14, 5) (−10, −5)
(−14, 5) (−10, 15)
(−37, −2) (−40, −5)
(−37, −2) (?, ?)

</td><td>

Objectives
The goal of this string is to help students understand and reason through the process of finding slope. With several strategies available to students, the goal is that they will develop an intuition of which strategy is most efficient given specific points.

</td></tr>
<tr><td>

Placement
This string can happen during the work of finding slope given two points. It could also come after students have been introduced to the slope formula.

You could use this string to strengthen students' notions about slope during or after textbook Lesson 1.5 Linear Equations and Arithmetic Sequences.

</td></tr>
<tr><td>

Guiding the Problem String
The strategies that should come out during this string are: choosing an efficient point to represent (x_1, y_1) in the slope formula, thinking about the ratio of the distances between the corresponding coordinates of the points, and considering direction—the positive or negative sign of the slope.

To encourage students to think about the vertical and horizontal distances, you might sketch the points on an open coordinate plane (without grid lines). This encourages students to visualize the horizontal and vertical distances instead of counting (a relatively inefficient strategy).

To support students in using precise mathematical language, use the terms "finding differences," "ratio of the differences," and "horizontal or vertical differences." If students refer to the vertical changes as up or down, repeat the question, "So what was the vertical change?" We want to emphasize finding the difference between the values, not up and down since that is dependent on your starting point.

</td></tr>
<tr><td>

About the Mathematics
The use of a "pivot point" in the pairs of problems encourages students to recognize the difference between positive and negative slope.

The use of points in quadrants two and three make the subtraction more challenging and, therefore, can nudge students toward thinking of subtraction as distance rather than removal.

</td></tr>
</table>

Sample Interactions

Use the following as you plan how to elicit and model student strategies. This is not meant as a script, but as a view into the relationships involved and the intent of the problem string.

Teacher: *Today's problem string starts with two points:* (0, 5) *and* (2, 9)*. Can you picture the line that goes through these points? What is the slope of a line passing through these points? What do you know about the line?*	(0, 5) (2, 9)

(continued)

<table>
<tr><td colspan="2">

Student: *I can picture the line going up.*

Student: *Yeah, so the slope would be positive.*

Teacher: *Anything else? How steep?*

Student: *Greater than one.*

Teacher: *The slope would be greater than one? How do you know?*

Student: *If you start at (0, 5) and go over two and up two, you'd be at 7 but you're up at 9.*

Student: *I used the slope formula and got 2.*

Teacher: *Which point did you use to start the formula? And does it matter which one you use first?*

</td></tr>
<tr><td>

Student: *It doesn't matter which one goes first, as long as you are consistent. I used (2, 9) first. So 9 minus 5 over 2 minus zero. That's 4 over 2, so 2.*

Teacher: *Okay, I am going to write that up here on the board. 9 minus 5 divided by 2 minus zero. That is 4 halves, or 2.*

</td><td>

$$\frac{9-5}{2-0}=\frac{4}{2}=2$$

</td></tr>
<tr><td>

Teacher: *Did anyone start with the (0, 5)?*

Student: *I did. 5 minus 9 divided by 0 minus 2.*

Teacher: *And that is −4 divided by −2. That's still 2, but which one would you rather think about: 4 divided by 2, or −4 divided by −2?*

Student: *I don't care, they are both easy.*

Student: *I'd rather have the positives.*

</td><td>

$$\frac{5-9}{0-2}=\frac{-4}{-2}=2$$

</td></tr>
<tr><td>

Teacher: *Let's sketch those points on a grid. Some of you were thinking about the graph. Tell us.*

Student: *I did. It was up 4 and over 2, so a slope of 2.*

Teacher: *So vertical change of 4 for a horizontal change of 2? For every two you go over, you go up 4?*

Student: *It's the same as for every one you go over, you go up 2.*

Teacher: *The ratio of the differences is 2. No negatives to deal with here. Just looking at how far apart they are.*

</td><td>

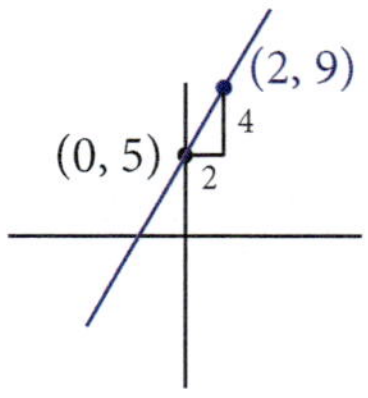

</td></tr>
</table>

Teacher: *Our next points are (0, 5) and (2, 1). Can you picture that line? What's the slope of the line passing through these?*

Think time.

Teacher: *Someone who used the formula share it with us. Tell us which point you started with.*

Student: *I used (0, 5) because I didn't want to get negative numbers like last time. Five minus 1 is 4, and 0 minus 2 is −2. So I got a negative anyway.*

Student: *I used (2, 1) first. One minus 5 is −4, and 2 minus 0 is 2. So I got a negative number too, −2.*

Teacher: *Interesting. It didn't matter which point was first, we still had to divide with a negative number. But the slope was −2, not 2.*

$$\frac{5-1}{0-2}=\frac{4}{-2}=-2$$

$$\frac{1-5}{2-0}=\frac{-4}{2}=-2$$

Teacher: *Any thoughts on how that will look on this graph?*

Student: *I saw that the first point was the same as in the first problem. The second point was down 4 instead of up 4.*

The teacher sketches the points.

Teacher: *So they were still 4 apart vertically, but down instead of up. How does that affect the slope?*

Student: *It means the slope is negative.*

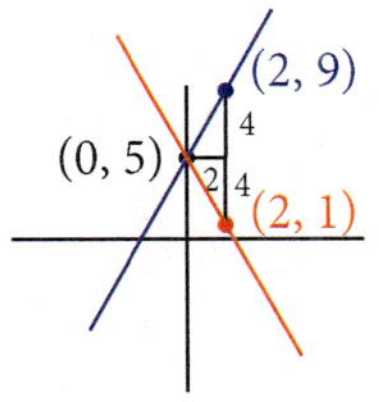

Teacher: *Who can describe what we're doing here?*

Student: *Instead of using the formula, we're looking at the distances and then deciding if it's going up or down.*

Teacher: *So you found the ratio of the differences, and then considered the sign, whether the line is increasing or decreasing.*

Teacher: *Next up are the points (−14, 5) and (−10, −5). Which point do you want to use first in the formula? Or I wonder if it would be easier to use a graph?*

Students work.

(−14, 5) (−10, −5)

Teacher: *Okay, what did you find for the slope? Someone who used the formula go first. What point did you use first?*

Student: *I used (−14, 5) first. 5 minus −5 is 10. Then −14 minus −10 is −4. And 10 divided by −4 is* $-\frac{10}{4}$.

Teacher: *That's a lot of negatives. Did you consider using the other point first?*

Student: *Yeah, but I didn't want to do −5 minus 5, or −10 minus −14.*

$$\frac{5-(-5)}{-14-(-10)}=\frac{10}{-4}=-\frac{5}{2}$$

$$\frac{-5-5}{-10-(-14)}=\frac{-10}{4}=-\frac{5}{2}$$

(continued)

Teacher: *So earlier someone said it didn't matter which point goes first. Do you still agree?*

Student: *Yeah, either point can go first and we still get the same slope.*

Student: *Yes, but sometimes it is easier to subtract if a certain point goes first.*

Teacher: *Does that make sense to everyone? Some interesting things to think about.*

Teacher: *Who didn't use the formula and sketched it instead? How did the points change vertically? Horizontally?*

Student: *I saw that the vertical change was 10 and the horizontal change was 4, so it's a slope of* $^{10}/_{4}$*. But it went down so it's negative.*

Teacher: *Okay, that was pretty quick. The points are in different quadrants. How do you know the vertical change was 10?*

Student: *I just know it was 5 to the x-axis and then 5 more.*

Teacher: *Ahhh, using zero. Nice. Did you count by 1s?*

Student: *No, I just used the numbers.*

Student: *So, you just used the graph, finding distances and knowing the line was going down?*

Student: *Yeah, it can be easy and you don't have to deal with all of those negatives, not really. Just know where they are.*

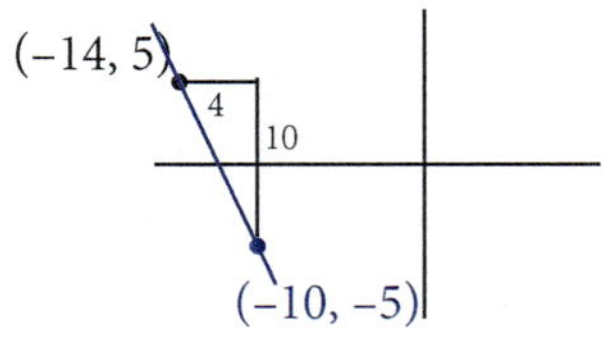

Teacher: *Up next,* (−14, 5) *and* (−10, 15). *Go.*

Student: *That's too much subtracting, I just want to look at the graph. The vertical change was 10 and the horizontal change was 4 again. But it is a positive slope. So, it's the same but positive,* $^{10}/_{4}$ *or 2.5.*

Teacher: *Does anyone want to talk through the formula?*

Student: *No, that's easy to see. It's just the same slope but decreasing.*

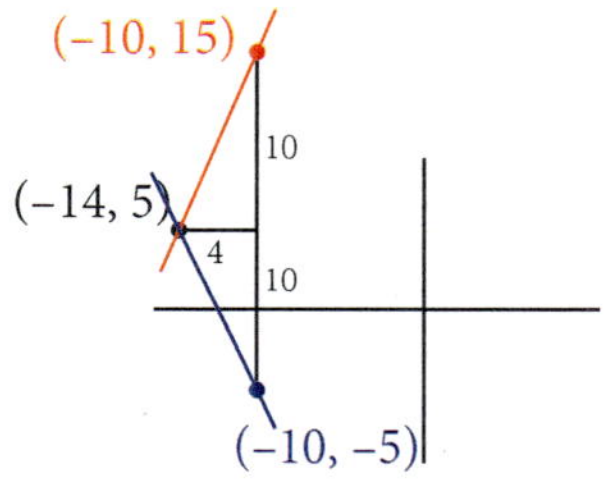

Teacher: *We are almost done,* (−37, −2) *and* (−40, −5). *You are welcome to do whatever you want to find the slope of the line that contains these points. If you use the formula, consider which point is easier to start with and tell us why.*

Student: *Well, we could use* (−40, −5) *first because* −5 *minus* −2 *seemed easier that* −2 *minus* −5.

Student: *I don't like either of those. I'd rather think about the graph.*

The teacher sketches the points.

Teacher: *Okay, here it is. What do you see? What is the vertical change? The horizontal change?*

Student: *It's* 3 *and* 3. *That's a slope of* 1.

Teacher: *You found the ratio of the differences. Is it positive or negative? Why?*

Student: *It moves up as it goes to the right, so it is positive.*

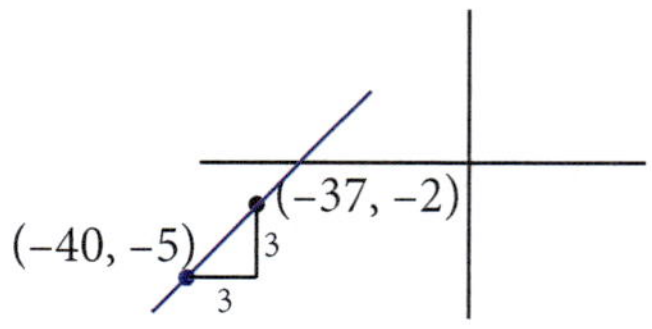

Teacher: *Last question. If we used the same point (−37, −2), what would be a point that has the same slope, but negative? And why? Turn and talk about where you would put that point.*

Students turn and talk while the teacher listens in.

Teacher: *What did you decide?*

Student: *It is* (−40, 1), *because it should go down 3 instead of up.*

The teacher sketches the point (−40, 1).

Student: *Now it still has a slope of* 1, *up* 3 *for every back* 3, *but the slope is negative because the line is going down.*

Teacher: *That's one choice. Nice. I wonder if there are any others? Great work today. Let's summarize what was important from today's discussion.*

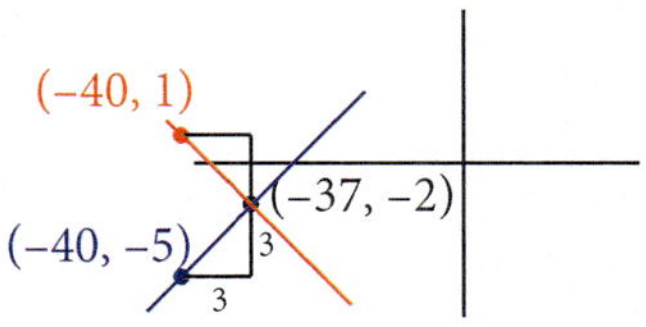

Teacher: *How would you summarize some of the things that came up in this string today?*

Elicit the following:

- *It doesn't matter which point you use first in the slope formula, but sometimes one is easier than the other for subtracting.*
- *You can sketch the points and see the vertical and horizontal differences instead of subtracting.*
- *You can find the ratio of the differences and then think about if the slope is positive or negative.*

Sample Final Display

Your display could look like this at the end of the problem string:

(0, 5) (2, 9) $m = 2$ $\quad \frac{9-5}{2-0} = \frac{4}{2} = 2 \quad \frac{5-9}{0-2} = \frac{-4}{-2} = 2$

(0, 5) (2, 1) $m = -2$ $\quad \frac{5-1}{0-2} = \frac{4}{-2} = -2 \quad \frac{1-5}{2-0} = \frac{-4}{2} = -2$

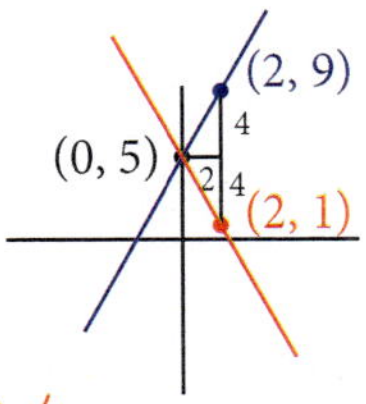

(−14, 5) (−10, −5) $m = -2.5$ $\quad \frac{5-(-5)}{-14-(-10)} = \frac{10}{-4} = -\frac{5}{2}$

(−14, 5) (−10, 15) $m = 2.5$

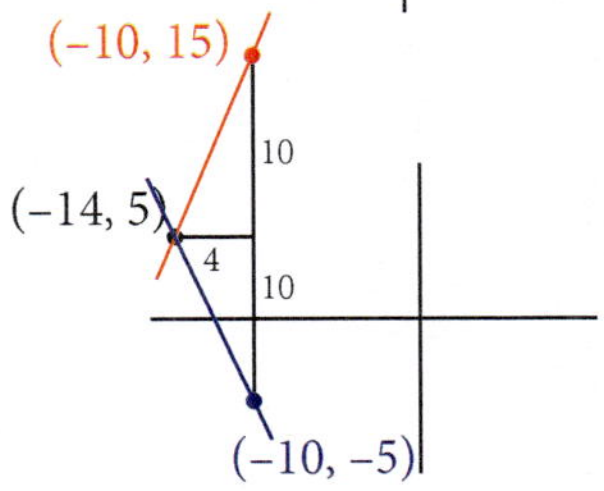

(−37, −2) (−40, −5) $m = 1$

(−37, −2) (?, ?) (−40, 1) $m = -1$

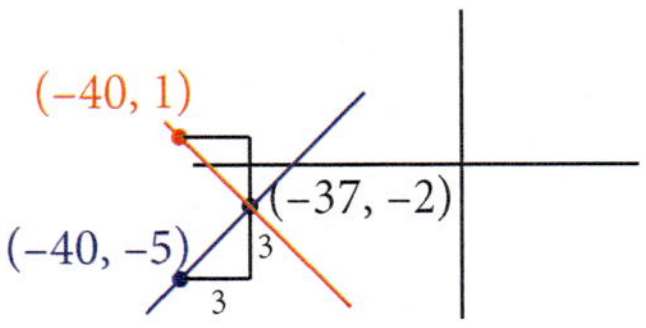

(continued)

Facilitation Notes

This version of the problem string lists short notes for important teacher moves during the string. After you've done the string yourself and studied the relationships involved, you might make similar notes for the things you want a reminder of or deem important.

(0, 5) (2, 9)	*Quick. Formula twice with each point first.* *Sketch, visualize vertical, horizontal differences.* *Ratio of differences.*
(0, 5) (2, 1)	*Formula twice again.* *Which point is better first?* *Sketch. Discuss positive versus negative slope.*
(−14, 5) (−10, −5)	*Formula? Which point is better first?* *Sketch. How are you finding the differences?*
(−14, 5) (−10, 15)	*Visualize. What quadrant(s)?* *Formula, or ratio of difference and direction?*
(−37, −2) (−40, −5)	*Repeat.*
(−37, −2) (?, ?)	*What point has opposite slope, same magnitude but opposite sign?*

PROBLEM STRINGS LESSON

1.6 Equations of Lines 1

At a Glance

rate of 2, start at 0 distance

rate of 2, (3, 7)

rate of 3, (2, 4)

Objectives

The goal of this problem string is to help students develop strategies for finding the slope y-intercept form of the equation of a line.

Placement

This string could happen as students are learning about the slope y-intercept form of the equation of the line. We have written the dialog as an example of work as students are making the transition from recursive formulas for arithmetic sequences to explicit equations. You could deliver this string before or after textbook Lesson 1.6 Modeling with Intercepts and Slope.

Guiding the Problem String

The first problem should be quick. By having students briefly consider the proportional relation of $y = 2x$, there is the potential that they might use it to find the explicit formula for the second problem. Setting the problems in the context of walking in front of a motion detector is purposeful. It potentially suggests reasoning using rates to find the "starting point" which is the y-intercept.

Allow students ample time to work on the second problem. Circulate and listen carefully to students. Look for students using these three strategies.

- Seek students who are walking back in time one second at a time. Model that unit rate strategy by drawing one jump at a time on the graph or table.
- Look for students who are reasoning about a bigger jump, 3 seconds. Model that non-unit rate strategy by drawing a big jump of 3 seconds on the graph or table.
- Listen for students who are focusing on the rate of 2 meters per second and how that should be 6 meters but it's not, the location is 7 meters at 3 seconds. Sketch the line $y = 3x$ and plot the point (3, 7) and wonder aloud how they might move the line to include (3, 7), while keeping the rate the same.

If you do not find all three strategies, wonder about the missing one(s) and ask students to help make sense of them.

As students work on the last problem, encourage students to make sense of the problem first, but then to look back and wonder if they can make sense of a different strategy. Wonder which strategy might be more efficient in which cases. At the end of the string, ask students to describe the three strategies.

About the Mathematics

Arithmetic sequences are linear functions with a restricted domain. The notation we are using for arithmetic sequences is $u_0 = a, u_n = u_{n-1} + b$. This corresponds with the explicit formula for the linear function $y = a + bx$ for the domain of whole numbers. The recursive rule for the arithmetic sequence defines a discrete sequence whereas the explicit rule defines a continuous function. The explicit function is a better model for the continuous measurement context of walking in front of a motion detector.

Graphs and tables are models. On these models, we can represent strategies—how students use relationships to find the y-intercept.

(continued)

Sample Interactions

Use the following as you plan how to elicit and model student strategies. This is not meant as a script, but as a view into the relationships involved and the intent of the problem string.

Teacher: *We have been working with arithmetic sequences. Today I want you to consider a sequence that describes a walk in front of a motion detector. For this walk, we know the walker started right at the motion detector and is walking 2 meters per second. Where would the walker be at each second? I know this is easy. Let's just get some points up here quickly.*

Student: *At time 0, the walker is at 0 meters.*

Student: *Then at 1 second, they are at 2 meters and 2 more meters at 2 seconds.*

Teacher: *What's a recursive rule for the position at each second?*

Student: *Start at 0 and the nth term is the term before plus 2.*

Teacher: *What if we wanted to model more than just the position at each second, but the position in between the seconds? What explicit rule could we write to model the position at any time?*

Student: *I think $y = 2x$?*

Teacher: *I am going to graph that along with the points. Looks like a match!*

$a_0 = 0, a_n = a_{n-1} + 2$

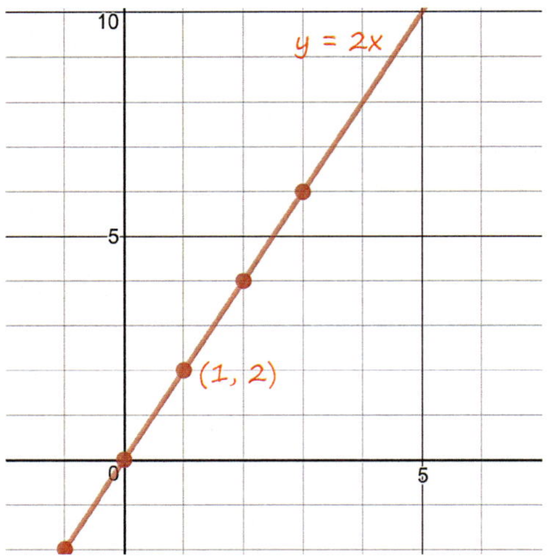

Teacher: *The next problem I want you to think about is still a person walking in front of a motion detector. This person is also walking away at 2 meters per second. You don't know where the person started, but at 3 seconds, the person was 7 feet away. What is an explicit formula to model this motion?*

Students work and the teacher circulates, looking for a unit rate strategy, a non-unit rate strategy, and a transformation strategy. To help some stumped students the teacher asks:

Teacher: *What do you know? How can you use that? What would be helpful to figure out? Can you use where the person is to work backwards? Can you use the first problem to help you?*

Teacher: *Some of you are using the rate of 2 meters per second, starting from the point (3, 7). Tell us about that.*

Student: *Well, since we know the walker was at 7 meters at 3 seconds, I went back in time to 2 seconds, 5 meters and then 1 second 3 meters, and so they must have started at 1 meter.*

Teacher: *I am going to model that on the graph like this. Does that represent what you did?*

Student: *Yes.*

Teacher: *It could also look like this in a table. Right?*

Student: *Yes, that's how I thought about it—not in the graph, but in a table.*

Teacher: *We can model that thinking, using those relationships either on a graph or in a table. So, what is the explicit formula?*

Student: *Now that we know the starting point of 1, I think of starting at 1 and adding 2 meters times the number of seconds,* $y = 1 + 2x$.

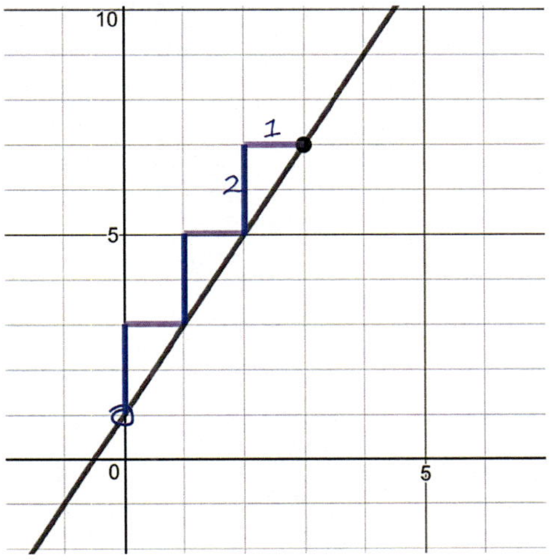

x	y
0	1
1	3
2	5
3	7

(x increases by 1 each row; y increases by 2 each row)

$y = 1 + 2x$

(continued)

Teacher: *Did anyone do something similar, starting with the (3, 7) but use the rate in a different way? By thinking about how far the walker went in all 3 seconds?*

Student: *Yeah, I started there too, but I knew that since they were walking at 2 meters per second and they were walking for 3 seconds, they covered 6 meters.*

Teacher: *That sounds important, but different than our other strategy. Who can repeat that for us?*

Student: *Since they walked for 3 seconds at 2 meters per second, they went a total of 3 times 2 is 6 meters in 3 seconds.*

Teacher: *I will model what you said on the graph like this. Go back all 3 seconds at once, and back all 6 meters at once. And check it out, we end up with the same starting point. So, you get the same explicit formula.*

Teacher: *Interesting, so some of you went back one unit at a time, using the unit rate. And some of you went back 3 seconds all together, so all 6 meters at once, using the non-unit rate of 6 meters per 3 seconds. Nice.*

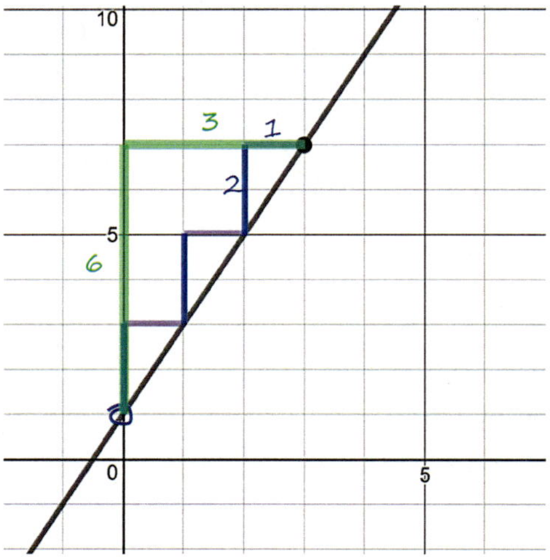

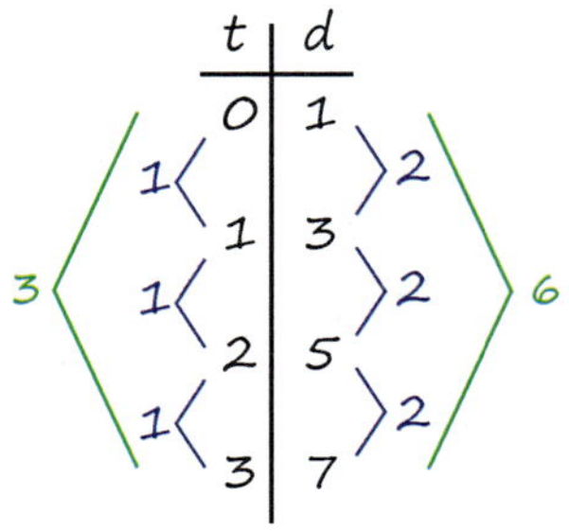

y = 1 + 2x

Teacher: *But I saw some of you using the first problem. Who can tell us about that?*

Student: *I was thinking about walking at 2 meters per second. If you did that, it would look like the first problem. But then at 3 seconds, you would be at 6 meters, not 7.*

Teacher: *I will model that by sketching the* $y = 2x$ *and the point (3, 7). Hmmm... that's interesting. Can anyone else see where you might go from here?*

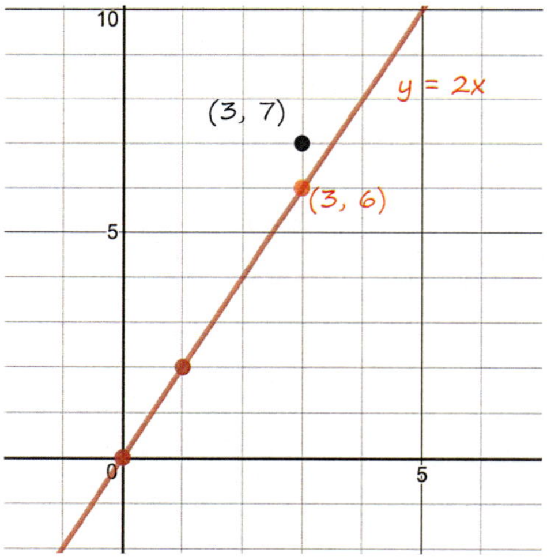

Student: *Well, since the line is just 1 unit too low, I thought about shifting it all up 1, so that would be $y = 2x + 1$.*

Teacher: *Are there any questions about this strategy? No? Then how might you describe it?*

Student: *Think about the rate and where you would be if you started at 0. Then shift that line to where the walker actually was.*

Teacher: *I noticed that if you thought about the starting point, the y-intercept first ,you might have thought of the line as $y = 1 + 2x$, but if you thought about the rate first and then shifting, you may have thought of the line as $y = 2x + 1$. Interesting.*

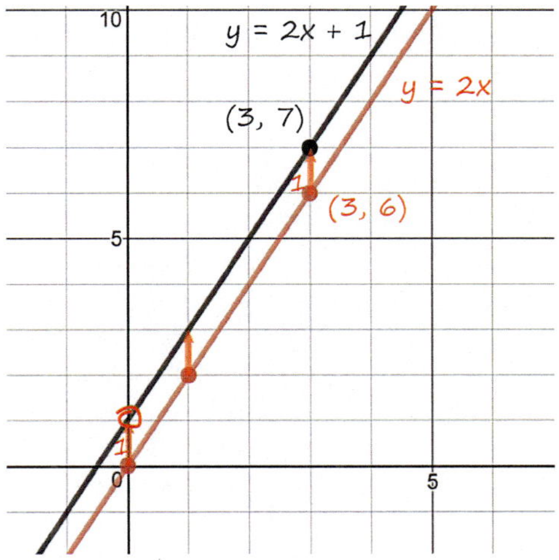

Teacher: *The next problem in our string is has the walker walking at 3 meters per second. You know that the walker was at 4 meters at 2 seconds.*

Students work, the teacher circulates and then asks students to share a unit rate, a non-unit rate, and a transformation strategy. Help students figure out that the student was behind the motion detector when the time started.

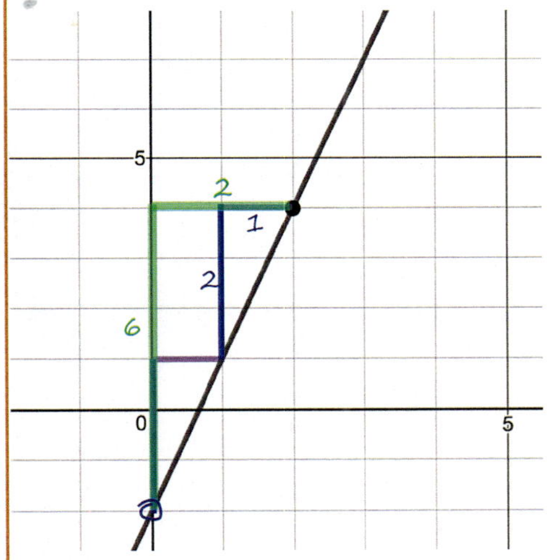

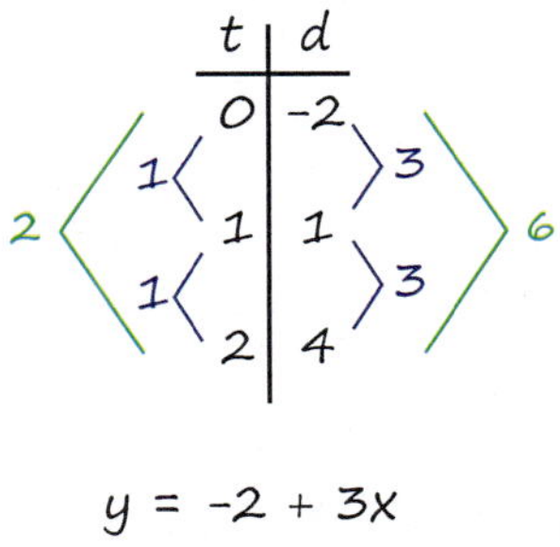

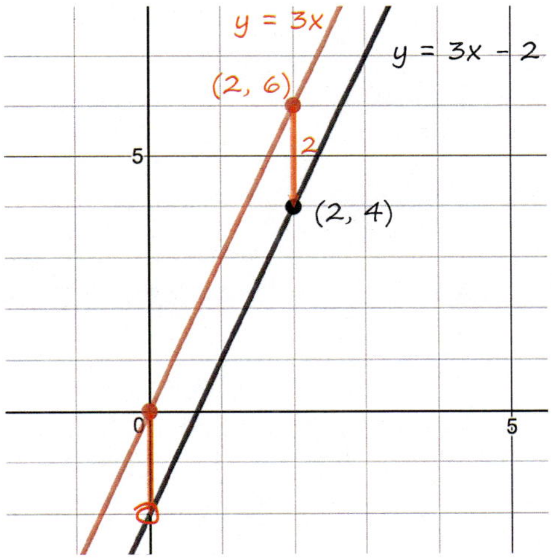

Teacher: *How would you summarize some of the things that came up in this string today?*

Elicit the following:

- *Once you know the rate and the y-intercept, you can write an explicit formula for a line.*
- *You can use the unit rate to walk back in time to find the y-intercept.*
- *You can use a non-unit rate, how far the walker went in a range of time, to walk back in time to find the y-intercept.*
- *You can use the rate b to write $y = bx$, and then the amount you need to shift the function up or down is the y-intercept.*

(continued)

Sample Final Display

Your display could look like this at the end of the problem string:

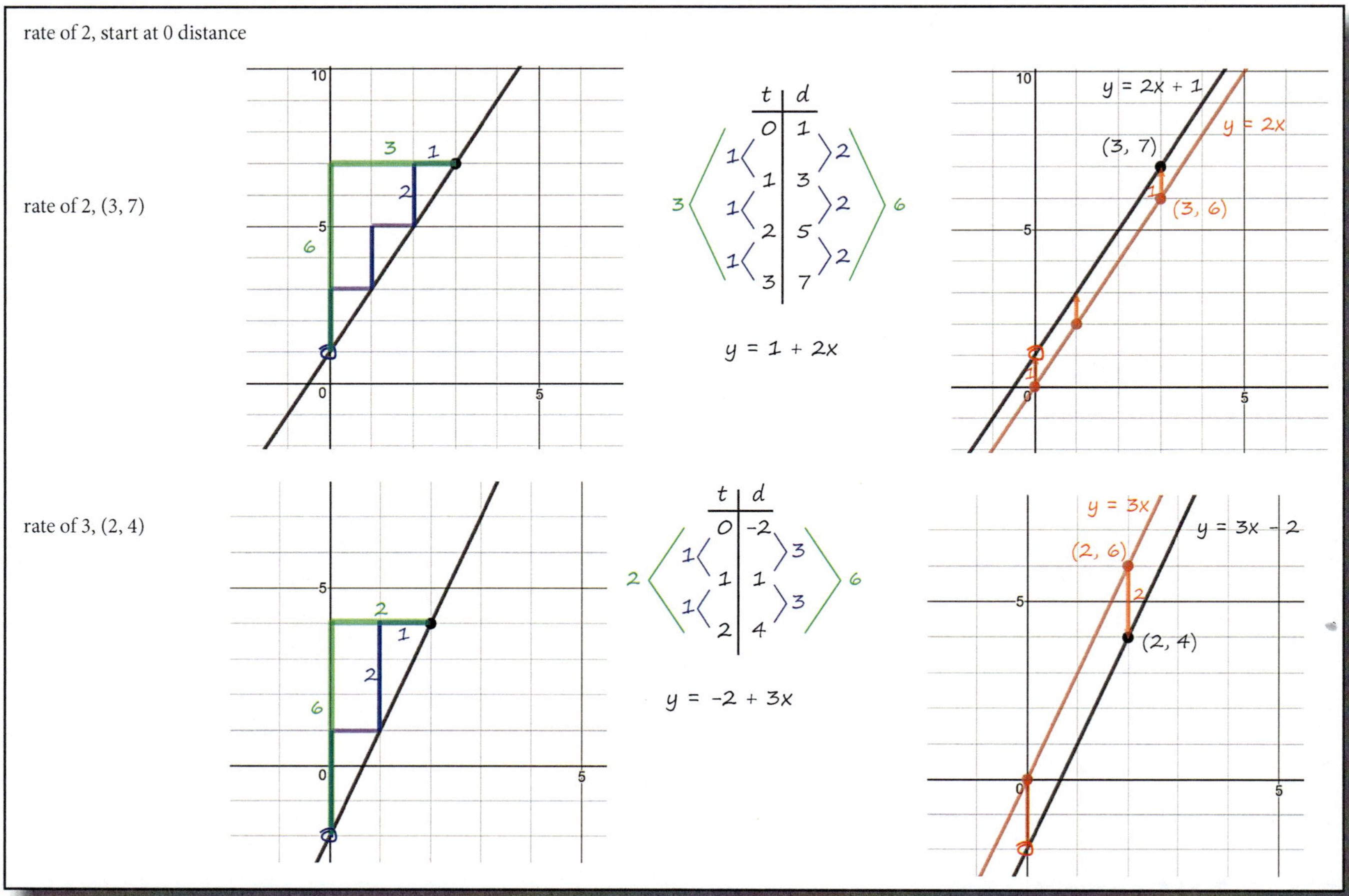

Facilitation Notes

This version of the problem string lists short notes for important teacher moves during the string. After you've done the string yourself and studied the relationships involved, you might make similar notes for the things you want a reminder of or deem important.

rate of 2, start at 0 distance	*What if this represents a walk away from a motion detector?* *Quick. Plot points. Write y = 2x.*
rate of 2, (3, 7)	*Still walking. Listen for unit rate, non-unit rate, transformation strategies.* *What do you know? What are you looking for? What's the equation of a line?* *Can you work backwards in time? Can you use the first problem to help?* *Linger. Repeat back. Explain. Use color.*
rate of 3, (2, 4)	*Repeat.* *Can you step out of the problem, look at relationships, and challenge yourself to try a different strategy?* *Connect y-intercept first to rate first approach,* $y=-2+3x=3x-2$

PROBLEM STRINGS
LESSON

1.7 Equations of Lines 2

At a Glance

rate of 3, (20, 64)

rate of −2, (1, 6)

rate of −5, (−10, 35)

Objectives

The goal of this problem string is to help students further develop strategies for finding the slope y-intercept form of the equation of a line.

Placement

This string is the second in a series of Equations of Lines. Use it to help students continue to work with rates and a point to find the equation of a line in slope y-intercept form.

You could deliver this string before textbook Lesson 1.7 Models and Predictions.

Guiding the Problem String

The three problems should take about the same amount of time. As always, let students solve each problem any way they choose, but promote efficiency and sophistication by celebrating it. Help students see connections between the strategies, how each is using the unit rate. The first problem involves the point (20, 64) which should cause students to think again if they try to walk back to the y-intercept using the unit rate. Bring out the algebraic way of modeling the transformation strategy so that when students try the second problem with its negative rate, they have other options. Highlight that algebraic strategy and also the efficient unit rate strategy. For the last problem, with its negative rate and point in the second quadrant, let students try everything and compare.

About the Mathematics

The first problem, rate of 3, (20, 64) is purposefully related to the last problem in the previous string, rate of 3, (2, 4). The similarity in the numbers may prompt students to use similar relationships. Because the point is so far out from the y-axis, it is no longer efficient to use the unit rate strategy many students were using in the previous string and so this may prompt students to try something more efficient.

In this problem string, we connect the transformation strategy (see textbook lesson 1.6) to an algebraic solution using the equation $y = mx + b$. You might be inclined to make this procedural (take the equation, substitute stuff in, solve for b), but if it's connected to students' experience with the transformation strategy and the non-unit rate strategy, students will not only own more connections and relationships, but they will also forget "how to do it" less and be more inclined to think and reason instead of reaching for rote memorized information and procedures.

Graph, tables, and equations are models. On these models, we can represent strategies—how students use relationships to find the y-intercept. We represent the transformation strategy in this string on two models—graphs and equations.

(continued)

Sample Interactions

Use the following as you plan how to elicit and model student strategies. This is not meant as a script, but as a view into the relationships involved and the intent of the problem string.

Teacher: *In our last problem string, we found the equations of lines to model motion walking away from a motion detector. Today's string is going to build on that. The first scenario is a walker going away at 3 meters per second. You know that at 20 seconds, she is 64 meters away. Write the equation of the line that models her walk.*	rate of 3 meters per second away, (20, 64)

Students work and the teacher circulates, noting students using a unit rate strategy, a non-unit rate strategy, and a transformation strategy. To help some stumped students the teacher asks:

Teacher: *What do you know? How can you use that? Do you remember any ideas we used in the last string? How fast is she walking? For how long?*

Teacher: *Let's get started by noting what equations you found.*

Student: *I got* $y = 4 + 3x$.

Student: *The same, just written differently,* $y = 3x + 4$.

Teacher: *I saw some of you using the unit rate to count back one second at a time. That is a lot of work. Great perseverance. Some of you were using the rate in bigger jumps, by thinking about how far the walker went in all 20 seconds. Please tell us about your thinking.*

Student: *I knew that since she was walking at 3 meters per second for 20 seconds, she would have walked 60 meters.*

Student: *She went a total of 3 times 20, so 60 meters in 3 seconds.*

Teacher: *I will model what you said on the graph and table like this. Go back all 20 seconds at once and back all 60 meters at once, and there we are at the same starting point, the y-intercept.*

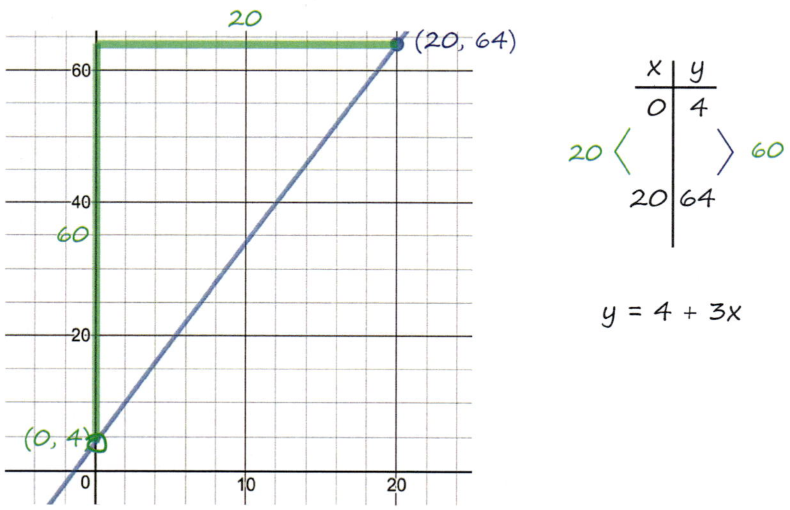

Student: *Wow. That could've saved me some time. I was going back one at a time.*

Teacher: *Yesterday's problems were just about as efficient either using the unit rate or the non-unit rate, but for this problem, you are saying that the non-unit rate is more efficient. Nice insights.*

Teacher: *But I saw some of you using that other strategy we worked with in the last string, where the line shifted. Who can tell us about that?*

Student: *I was thinking about walking at 3 meters per second. If you did that, that's $y = 3x$. So then at 20 seconds, you would be at 60 meters. But at 20 seconds, you should be at 64 meters, so shift the whole line up 4.*

Teacher: *So the $y = 3x$ shifted up 4, that sounds like the other version of the equation, $y = 3x + 4$. Does anyone have any questions about that?*

Teacher: *Turn and talk about these two strategies.*

Students talk and then the teacher calls them back together.

Student: *I am noticing that they both thought about the 20 and the 20 times 3, 60.*

Student: *I noticed that if you thought about the starting point first, you said the line is $y = 4 + 3x$, but if you thought about the rate first and then shifting, you said the line is $y = 3x + 4$.*

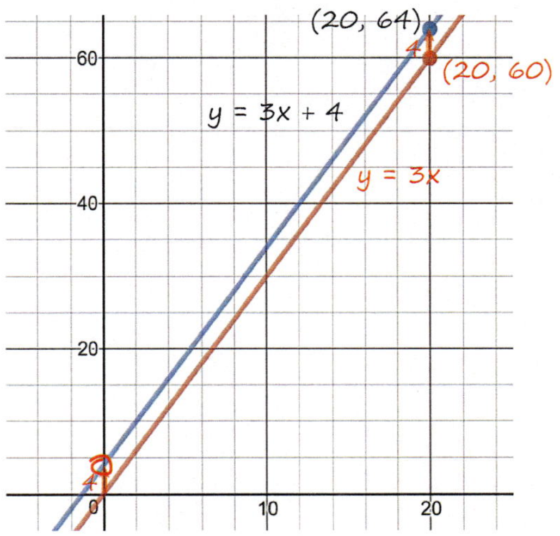

Teacher: *I'd like to model this transformation strategy a little differently. Each time we've talked about that strategy, you start out thinking about the rate and the line that represents that rate if the walker started at 0, right? So, I've been modeling that on a graph. Let's look at it with equations. So, starting with that rate of 3 and starting at 0, you have the line $y = 3x$, right? But you know that at 20 seconds, it was supposed to be at 64 meters. So, you know the line will have a y-intercept. So, where is $y = 3x$ at $x = 20$? At 60. So, you'll have to shift the 60 up 4 to get to the 64. So the shift is 4. What do you think? Can this algebraic stuff model your thinking too?*

$y = 3x$	the line with rate 3
$(20, 64)$	at 20 seconds, should be 64 meters
$y = 3x + b$	know the line will have y-intercept
$64 = 3(20) + b$	at 20, y=3x is y=3(20)=60
$64 = 60 + b$	ah, must have to shift it up 4
$b = 4$	

Teacher: *The next problem in our string is a bit different. The walker is walking toward the motion detector, so a rate of −2, and at 1 second he is 6 meters from the motion detector.*

rate of −2, (1, 6)

(continued)

Students work, the teacher circulates, and then asks students to share a transformation strategy.

Teacher: *This one was a little weird, yes? A rate of −2? But some of you made that work. What was the equation?'*

Student: *I got* $y = 8 - 2x$.

Teacher: *Did anyone get anything different? No? Okay, tell us how you made that work.*

Student: *I thought about walking toward the motion detector at a rate of 2 and that would look like* $y = -2x$. *So, at 1 second, that would be down at −2, but he was supposed to be at 6 meters, so shift the whole thing up 8,* $y = -2x + 8$.

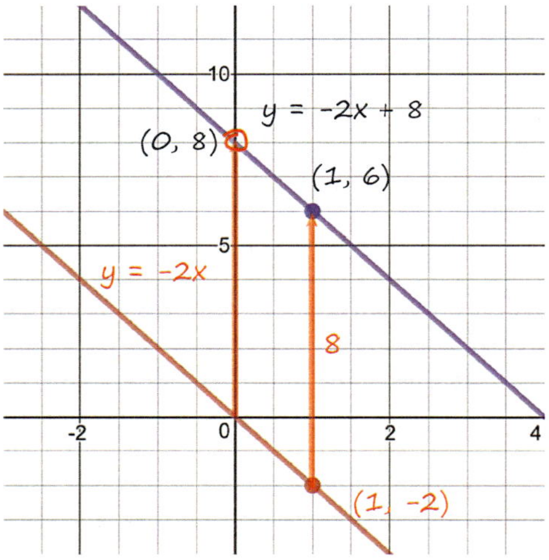

$y = -2x$

$(1, 6)$

$y = -2x + b$

$6 = -2(1) + b$

$6 = -2 + b$

$b = 8$

Teacher: *I'm a little curious. Did anyone try walking back by the unit rate?*

Student: *I did and for this problem, it was quite easy. You know* (1, 6) *and a rate of* −2, *so just back up* 1 *to* 0 *and back up* 6 *to* 8!

Teacher: *So, that's noteworthy. Mathematicians seek patterns and connections, but also efficiency. Let's try to keep all of our strategies in mind, all of the relationships we've been building to decide how to tackle this next problem.*

t	d
0	8
1	6

1 (change in t), −2 (change in d)

$y = 8 - 2x$

Teacher: *I wonder how you will think about the next problem in our string: a rate of* −5, *and the point* (−10, 35)*. Does it make sense for me to talk about a walker with the point* (−10, 35)*? Does it need to make sense? I wonder how you might find an equation of the line that has a rate of* −5, *and contains the point* (−10, 35)*?*

Students work, the teacher circulates and then asks a student to explain the algebraic transformation strategy and the non-unit rate strategy.

Student: *I thought kind of like before. I need the equation of a line,* $y = mx + b$*. And we know the rate is* −5 *so we know* $y = -5x + b$*. But we need to shift the line to the point* (−10, 35)*. Rather than graphing anything, I just put 35 has to be 50 plus the b, so that's just* −15*.*

Student: *I used the non-unit rate of* −50 *to 10. I thought about getting from* −10 *to 0 to find the y-intercept. Since the rate is* −5*, that's 10 times* −5 *equals* −50*. So back 50 from 35 is* −15*.*

rate of −5, (−10, 35)

$y = -5x$

$(-10, 35)$

$y = -5x + b$

$35 = -5(-10) + b$

$35 = 50 + b$

$b = -15$

$y = -5x - 15$

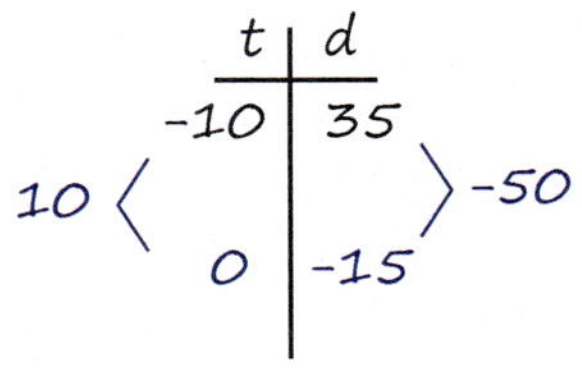

y = -15 - 5x

Teacher: *Why didn't anyone use a unit rate strategy?*

Student: *I thought about it, but you'd have to do it 10 times. Instead you can just do 10 times* −5 *and go back 10 and* −50 *all at once.*

Teacher: *Did anyone think about a transformation strategy on a graph but then solve it algebraically?*

Student: *Well, I thought about it that way to help me figure out the algebraic part.*

Student: *I didn't because thinking about negative time was weird.*

Student: *You can think about where a walker was before the data collection started, but since I had other ways to think about it, I chose not to this time.*

Teacher: *How would you summarize some of the things that came up in this string today?*

Elicit the following:

- *If you know the rate and the y-intercept, you can write the equation for a line.*
- *You can use the unit rate to walk back in time to find the y-intercept. This can be efficient if it's easy to get from the point to the y-intercept with the unit rate.*
- *You can use a non-unit rate, how far the walker went in a range of time, to jump forward or back in time to find the y-intercept.*
- *You can use the rate m to write* $y = mx$ *and then the amount you need to shift the function up or down is the y-intercept on a graph.*
- *You can use* $y = mx + b$ *and the given rate and point to algebraically find b. This can be efficient when the numbers are big or unwieldy.*

(continued)

Sample Final Display

Your display could look like this at the end of the problem string:

rate of 3, (20, 64)

$y = 4 + 3x$

rate of –2, (1, 6)

$y = 8 - 2x$

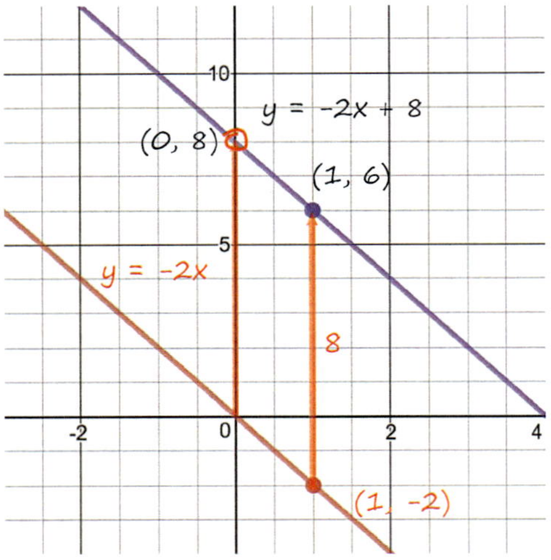

$y = -2x$

$(1, -2)$

$y = -2x + b$

$6 = -2(1) + b$

$6 = -2 + b$

$b = 8$

t	d
0	8
1	6

1 (t), –2 (d)

$y = 8 - 2x$

rate of –5, (–10, 35)

$y = -15 - 5x$

$y = -5x$

$(-10, 35)$

$y = -5x + b$

$35 = -5(-10) + b$

$35 = 50 + b$

$b = -15$

$y = -5x - 15$

t	d
-10	35
0	-15

10 (t), –50 (d)

$y = -15 - 5x$

Facilitation Notes

This version of the problem string lists short notes for important teacher moves during the string. After you've done the string yourself and studied the relationships involved, you might make similar notes for the things you want a reminder of or deem important.

rate of 3, (20, 64)	*Remember last string? Walking away at rate of 3 m/s. At 20 secs, 64 m away. Share non-unit, transformation. Why not use unit rate? Model transformation with y = mx + b, algebraically substituting in (20, 64).*
rate of –2, (1, 6)	*Walking toward. Listen for unit rate, transformation strategies: graph and algebraic models. Which is more efficient? Why?*
rate of –5, (–10, 35)	*Can you use what you know when walking doesn't make as much sense? Find non-unit and algebraic transformation. Why not the other strategies?*

PROBLEM STRINGS LESSON

1.8 Equations of Lines 3

At a Glance

x	y
0	27
1	21
2	15

x	y
−3	−2
0	4
3	10

x	y
3	25
4	20
5	15

x	y
2	5
6	17
10	29

Objectives

The goal of this problem string is to help students determine when it is efficient to use the point-slope form of the equation of a line and when it might be more efficient to use the given data and the slope y-intercept form.

Placement

This string is the third in a series of finding the Equations of Lines. It follows two strings that help students find the equation of a line given a rate and a point. In this string, students are given points, so they must find the rate in order to write an equation of the line that contains the points.

You could deliver this string any time after textbook Lesson 1.7 Models and Predictions.

Guiding the Problem String

The problem string is designed to encourage students to look at the data and choose an efficient strategy to write an equation that fits. For each problem, help students pay attention to what is given before they decide what to do. Help students realize each time how they found the rate and y-intercept. Encourage students to think about the given points and decide how to proceed, rather than automatically using the point-slope form every time.

About the Mathematics

In the first problem, the x-values are consecutive integers so the rate is obvious and the y-intercept is given. In the second problem, students must find the rate but the y-intercept is given. In the third problem, the rate is again obvious, but the y-intercept is not given. In the last problem, students must find both the rate and either the y-intercept or use the point-slope form of the equation of a line.

Important Questions

Use the following as you plan how to elicit and model student strategies.

- *Write an equation of the line that contains these points. Before you do, take a look at the points and decide which points you think you will use.*
- *Why did you choose those points?*
- *What information is rather obvious if you take the time to look?*
- *How does this problem compare to the previous? Does that impact your strategy choice?*
- *Look back at the problem string in order. Why those strategies for those problems?*
- *When might you choose to use the slope y-intercept form of the equation of a line? Why?*
- *When might you choose to use the point-slope form? Why?*
- *What do you look for when choosing a good/efficient strategy?*

(continued)

How would you summarize some of the things that came up in this string today?

- *If the y-intercept is one of the points, use it!*
- *If the slope is obvious because the x-values are increasing by one, use it!*
- *If you can easily find the slope and the y-intercept, you can use the slope y-intercept form.*
- *If the slope and y-intercept are not obvious, you might want to use the point-slope form.*

Sample Final Display

Your display could look like this at the end of the problem string:

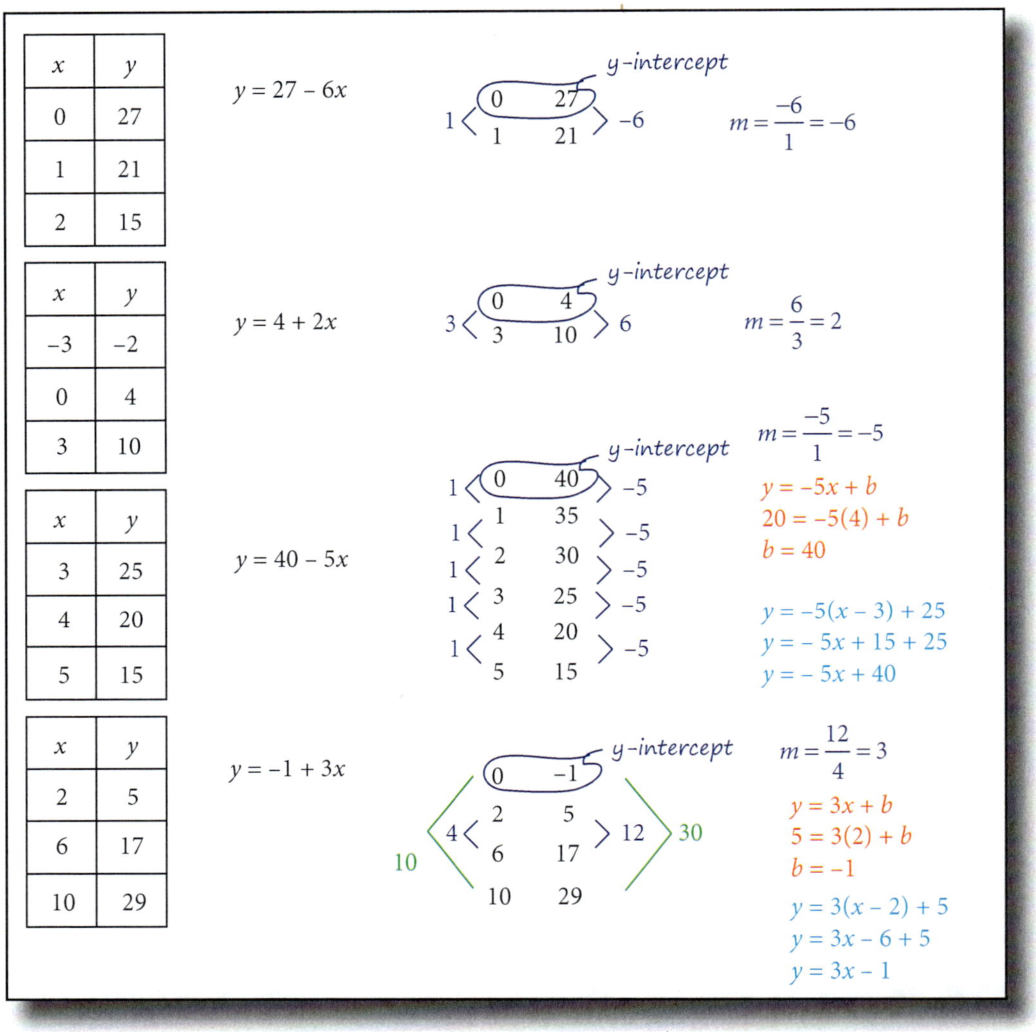

Facilitation Notes

This version of the problem string lists short notes for important teacher moves during the string. After you've done the string yourself and studied the relationships involved, you might make similar notes for the things you want a reminder of or deem important.

x	y
0	27
1	21
2	15

Quick.
Easy, right? Humor me.
What's the rate? The y-intercept?
How do you know?

x	y
–3	–2
0	4
3	10

Quick.
Which points did you use? Why?
What's the rate? How?
The y-intercept? How?

x	y
3	25
4	20
5	15

How does this compare to previous?
Which points did you use? Why?
What's the rate? How?
The y-intercept? How?
Did anyone use point slope? Why, why not?

x	y
2	5
6	17
10	29

How does this compare to previous?
Which points did you use? Why?
What's the rate? How?
The y-intercept? How?
Did anyone use point slope?
Which: slope intercept or point slope? Why?

PROBLEM STRINGS LESSON

1.9 Equations of Lines 4

At a Glance

x	y
−1	1
0	0
1	−1
2	−2

x	y
−2	3
−1	2
0	1
1	0

x	y
−2	1
−1	0
0	−1
1	−2

x	y
0	−2
1	−1
2	0
3	1

Objectives

The goal of this problem string is to help students develop a strategy to find the equation of a line in standard form when given points that readily suggest a relationship between x, y, and a constant.

Placement

This string could come near the middle or end of your work with linear functions. Before this string, students need experience representing changing amounts with variables. It may be helpful if students have experience with x- and y-intercepts or you could use this string to help build some connections between x- and y-intercepts and the standard form of the equation of a line.

You could deliver this string any time during textbook chapter 1, but it will have the biggest impact after students have found the equations of lines using the slope y-intercept form from exact and non-exact data.

Guiding the Problem String

Each of the problems have data with relationships that suggest the equation of the line in standard form. For the first problem ask students to use what they know to find the equation of the line that contains the given points. Have students share a few strategies and then ask if anyone sees any patterns between the x- and y-values. If not, ask students to put the equation they found into standard form and ask again if they see any patterns. If not, repeat with the second problem. When students begin to see a pattern across the table, such as all of the x's plus the y's in the first problem add to 0, write their insights as equations, such as $x + y = 0$. Thereafter, encourage students to look at the data before deciding on a strategy. When sharing emphasize the equivalence between the forms.

The first two points in each table use values that don't offer up the relationship as readily as the third and fourth points. This is purposeful, so that the patterns don't jump out too soon or easily. Don't point it out, but listen for students who are noticing it and ask them about it.

About the Mathematics

The standard form of the equation of a line is $ax + by = c$. In this form, it can be easy to solve for either the x- or y-intercept by substituting zero for either x or y.

This strategy of looking for patterns within the data works well when the rate is either 1 or −1. When the x- and y-values are always related, other slopes muddy the relationship so it is harder to see. As with all strategies, we consider the numbers and the relationships before deciding on an efficient strategy. This particular strategy might have limited use, but exploring it serves to strengthen students' understanding of the relationship between the inputs and outputs of a given function.

Important Questions

Use the following as you plan how to elicit and model student strategies.

- *Here is a table of data. What is the equation of the line that contains all of these points?*
- *Did anyone see any patterns in the data that suggested an equation?*
- *What form is your equation in? What does that look like in standard form? Are they equivalent?*
- *Do you notice any connection between the equation in standard form and the data in the table?*
- *Do you notice any patterns between the x's and the y's in the table?*
- *Do all the equations represent the same line?*
- *How are you finding it helpful to look for patterns in the table?*
- *What about these data made using patterns possible? Something about the rates of change? Something about* 1 *and* −1*?*

How would you summarize some of the things that came up in this string today?

- *Patterns in the data can be helpful to find the equation of lines in standard form.*
- *You can use whatever jumps out at you to find the equation of a line, but maybe look for patterns first.*
- *Be open to looking for patterns in data.*
- *Forms of the equations of lines found using different strategies are equivalent.*

(continued)

Sample Final Display

Your display could look like this at the end of the problem string:

x	y	
−1	+ 1	= 0
0	0	y-intercept
1	−1	
2 +	−2	= 0

Change in x: 1; change in y: −1

$y = -1x + 0$

$y = -x$ All of the y's are the opposite of the x's

$x + y = 0$ Every x + every y is 0.

x	y	
−2	3	
−1	2	
0	1	y-intercept
1 +	0	= 1

Change in x: 1; change in y: −1

$y = -1x + 1$

$x + y = 1$ Every x + every y is 1.

x	y	
−2	1	
−1	0	
0	−1	y-intercept
1 +	−2	= −1

Change in x: 1; change in y: −1

$y = -1x - 1$

$x + y = -1$ Every x + every y is −1.

x	y	
0	−2	y-intercept
1	−1	
2 −	0	= 2
3 −	1	= 2

Change in x: 1; change in y: 1

$y = 1x - 2$

$x - y = 2$ Every x − every y is 2.

Facilitation Notes

This version of the problem string lists short notes for important teacher moves during the string. After you've done the string yourself and studied the relationships involved, you might make similar notes for the things you want a reminder of or deem important.

x	y
–1	1
0	0
1	–1
2	–2

What is an equation of the line for these points?
How did you find it? What form is your equation in?
Did anyone notice any patterns in the table?
Between the x's and y's?
What is the equation in standard form?
See any connections?

x	y
–2	3
–1	2
0	1
1	0

What is an equation of the line for these points.
How did you find it? What form is your equation in?
Are the forms equivalent?
Do they represent the same line?
Do the x's and y's add to the same amount for each point?
How does that relate to the standard form of the equation of a line?

x	y
–2	1
–1	0
0	–1
1	–2

Repeat. Quicker.
What have the slopes been so far?
Do the x's and y's have an additive relationship?
How does that fit in with the slope?

x	y
0	–2
1	–1
2	0
3	1

Would patterns help here?
Do the x's and y's add to the same value again?
Or is there another relationship?
How does having a positive slope affect the equation?
What about these data made using patterns possible?
Were the rates helpful? Specific points given?

PROBLEM STRINGS
LESSON

2.0 Solving for y

At a Glance

$2x - 4y = 16$

$-3x + 9y = -15$

$8x - 2y = 10$

$\frac{2}{5}x - \frac{1}{10}y = \frac{1}{4}$

Objectives

The goal of this problem string is to help students build strategies to solve for one variable in the equation of a line in standard form. This will prepare them to solve systems of linear equations by substitution.

Placement

This string could come before or at the start of work on solving systems of linear equations.

You could deliver this string before chapter 2 in the textbook or right along with Lesson 2.1 Linear Systems or Lesson 2.2 Substitution and Elimination.

Guiding the Problem String

When students solve for one variable in similar equations, they sometimes either try to perform the exact same steps no matter what the equations are or guess wildly about the direction to take. With such experience, students could develop instincts about keeping the y-term positive, dividing first to clear fractions, or isolating the y-term first. In this string, you will actively help students develop intuition for directions of attack. Use a different color to highlight each strategy. The first, third, and fourth problems have negative y-terms. All of the equations involve fractions, either in the solution or the original equation. Take the longest on the first equation, ferreting out student ideas, and after the last equation, bringing it all together by describing and comparing strategies.

About the Mathematics

The problems are all given in standard form, $ax + by = c$. By solving for y, students end up with the slope y-intercept form of the line. Both forms can be useful for the different information they offer. The standard form can be useful to see the x- and y-intercepts, with a little bit of work. The slope y-intercept form shows the slope and the y-intercept and is necessary for some graphing calculators to graph the line.

Since all of these problems are solving for y, we refer to the y-term in the following strategy descriptions.

Sometimes it can be helpful to think of getting the y-term positive by either leaving it on the side of the equation it's on if the y-term is already positive or adding it to both sides. With the y-term positive, the coefficient you divide by to isolate y is positive. This can prevent some integer sign errors. More importantly, it can be a strategy students might employ when solving inequalities that helps with "reversing the inequality symbol when multiplying or dividing by a negative number."

Sometimes it can be helpful to divide by the coefficient of y first and sometimes it can be helpful to isolate the y-term first and then divide by the coefficient.

(continued)

Sample Interactions

Use the following as you plan how to elicit and model student strategies. This is not meant as a script, but as a view into the relationships involved and the intent of the problem string.

Teacher: *In this next chapter, it will be handy to solve equations for one of the variables. So, let's get our brains warmed up today by solving for y when I give you an equation like our first problem,* $2x-4y=16$*. If this equation is true, what does just y equal?* Students work briefly and the teacher circulates, looking for students who solve for the y-term first, who add $4y$ to both sides to get the y-term positive, or who divide the whole equation (each term) by 4 or −4, and noting any other strategies.	$2x-4y=16$
Teacher: *I saw some of you keep the −4y and deal with the 2x first. Who can tell us about that?* *When you told me that you divided everything by −4, I can write that in a few different ways.*	$-4y=16-2x$ $\frac{-4y}{-4}=\frac{16}{-4}-\frac{2x}{-4}$ $y=-4+\frac{1}{2}x$ $\frac{-4y}{-4}=\frac{16-2x}{-4}$ $\quad \frac{-4y=16-2x}{-4}$
Teacher: *I saw some of you add 4y to both sides. Tell us about that.* *Let's compare these 2 strategies. How are they different?* *What was handy about getting the y-term positive? You don't have to divide by a negative? Yes, that can be nice!*	$2x=16+4y$ $2x-16=4y$ $\frac{2x}{4}+\frac{-16}{4}=\frac{4y}{4}$ $\frac{1}{2}x-4=y$
Teacher: *Some of you decided to divide right off the bat.* *Just like before, I can write that division a few different ways. From now on, I'll probably choose to write it as dividing each term by −4 because that helps me keep track of all of the terms, so I don't miss any.*	$\frac{2x}{-4}-\frac{4y}{-4}=\frac{16}{-4}$ $\quad \frac{2x-4y}{-4}=\frac{16}{-4}$ $\quad \frac{2x-4y=16}{-4}$
Teacher: *So, you decided to divide everything by −4, keep going.*	$\frac{2x}{-4}-\frac{4y}{-4}=\frac{16}{-4}$ $\quad \frac{2x}{-4}+\frac{-4y}{-4}=\frac{16}{-4}$ $-\frac{1}{2}x+y=-4$ $y=\frac{1}{2}x-4$

Teacher: *Now that we have solved for y, I have a question. Are the original equation and this "answer" the same line? Let's check that out quickly by graphing them both.*

Graph both the original and the slope-*y*-intercept form at the same time, toggling back and forth to show the same line but in different colors.

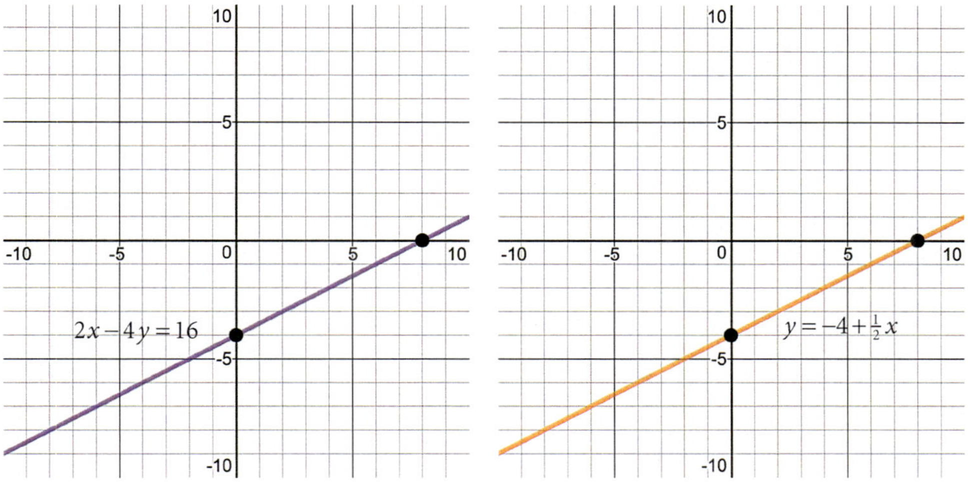

Teacher: *What do you think? Yep, same line.*

Teacher: *Great! Now we have several ways of looking at solving for y in that equation. Turn and talk about these ways. Do you understand them all? Which one do you gravitate toward?*

I wonder if those strategies will influence how you solve for y in this equation? Our next problem is to solve for y in $-3x + 9y = -15$

$-3x + 9y = -15$

(continued)

Teacher: *What did you get when you solved for y?*

Tell us about your thinking and I'll model it on the board.

Why didn't anyone talk about keeping the y-term positive? Ahhh, it already is!

Let's take a look at the graphs. Yes, they look equivalent.

$-3x+9y=-15$

$y=-\frac{5}{3}+\frac{1}{3}x$

$9y=-15+3x$

$\frac{9y}{9}=\frac{-15}{9}+\frac{3x}{9}$

$y=-\frac{5}{3}+\frac{1}{3}x$

$\frac{-3x}{9}+\frac{9y}{9}=\frac{-15}{9}$

$-\frac{1}{3}x+y=-\frac{5}{3}$

$y=-\frac{5}{3}+\frac{1}{3}x$

y-term is already positive!

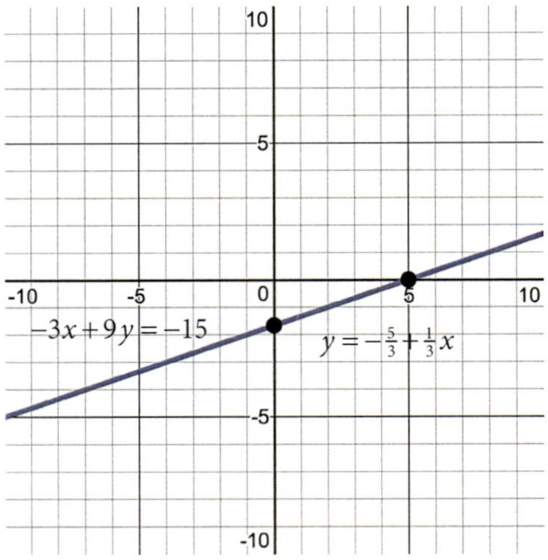

Teacher: *How would you describe the blue strategies? What are these people thinking about?*

How would you describe the red strategy? What's going on there?

Where did fractions show up in these first two problems? Why?

What information can you quickly see in the standard form? If you let x = 0 or y = 0, notice how the x-intercepts pop out.

What information can you see quickly in the slope-y-intercept form?

Teacher: *Before you start solving for y for our next problem, I want you to predict where the fractions will show up this time, if at all. Here is the next problem,* $8x-2y=10$. Students turn and talk, while the teacher listens in.	$8x-2y=10$
Teacher: *Do you think fractions will show up in the slope-y-intercept form? Why or why not?* *Go ahead and solve for y.* *What did you find?*	$y=4x-5$

Ask for and model student strategies, including adding $2y$ to both sides to get the y-term positive. Quickly graph both lines. Look for connections in the graph and equations.

$8x-2y=10$

$y=4x-5$

$\frac{8x}{-2}+\frac{-2y}{-2}=\frac{10}{-2}$

$-4x+y=-5$

$y=4x-5$

$-2y=10-8x$

$\frac{-2y}{-2}=\frac{10}{-2}+\frac{-8x}{-2}$

$y=-5+4x$

$8x=2y+10$

$2y=8x-10$

$\frac{2y}{2}=\frac{8x}{2}+\frac{-10}{2}$

$y=4x-5$

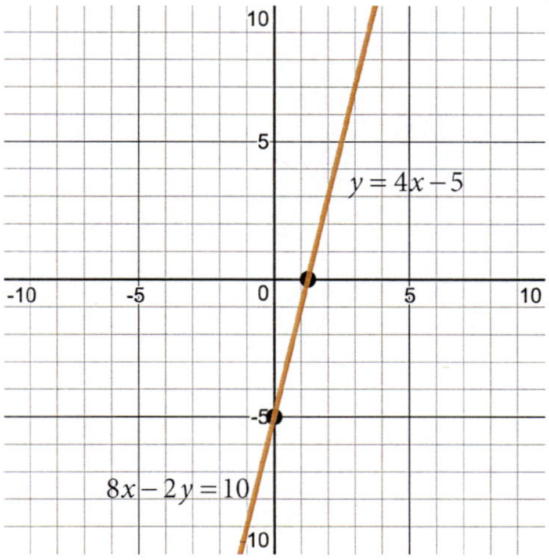

Teacher: *Great work. The last problem of the day is* $\frac{2}{5}x-\frac{1}{10}y=\frac{1}{4}$.

Now we're talking fractions! Before you get started, tell me, what do you think about fractions when you solve for y this time?

What do you think about the rate of this line? Any ideas? Steep? Shallow? Why?

Which strategy do you think you'll try to find y? Why?

$\frac{2}{5}x-\frac{1}{10}y=\frac{1}{4}$

Teacher: *Go ahead and find y. Let's see how this one plays out.*

Repeat asking for and modeling strategies, comparing them, and quickly graphing the lines.

$\frac{2}{5}x-\frac{1}{10}y=\frac{1}{4}$

$y=4x-\frac{2}{5}$

$\frac{2}{5}x-\frac{1}{4}=\frac{1}{10}y$

$10\left(\frac{2}{5}x\right)-10\left(\frac{1}{4}\right)=10\left(\frac{1}{10}y\right)$

$4x-\frac{2}{5}=y$

$-\frac{1}{10}y=-\frac{2}{5}x+\frac{1}{4}$

$-10\left(-\frac{1}{10}y\right)=-10\left(-\frac{2}{5}x\right)+-10\left(\frac{1}{4}\right)$

$y=4x-\frac{2}{5}$

$10\left(\frac{2}{5}x\right)-10\left(\frac{1}{10}y\right)=10\left(\frac{1}{4}\right)$

$4x-y=\frac{2}{5}$

$y=4x-\frac{2}{5}$

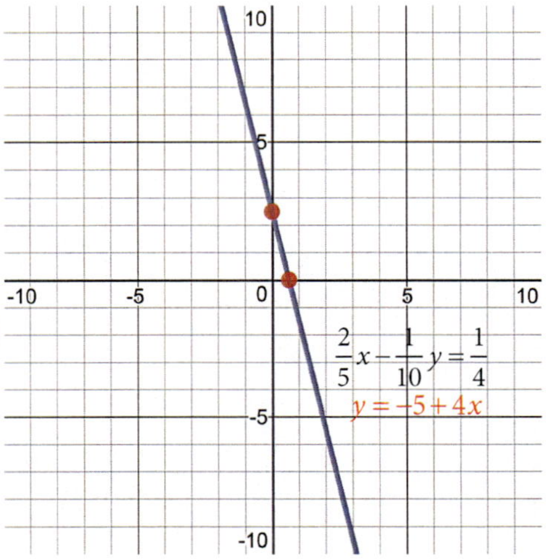

(continued)

Teacher: *Let's talk about the fractions. Look back at all of the equations today. What was true in the standard form if a fraction showed up in the slope-y-intercept form?*

What happens when you start with fractions in the standard form?

Let's describe these strategies that keep coming up.

How do they compare? Which one do you like and why? For which problems?

Teacher: *How would you summarize some of the things that came up in this string today?*

Elicit the following:

- *You can divide first before you move variables around with addition or subtraction, but you have to divide each term.*
- *If the variable you are solving for has a positive coefficient, you don't have to worry about dividing by negative numbers.*
- *Equivalent forms of a linear equation graph the same line.*
- *To solve for y, you can get the y-term positive, divide then solve, or solve then divide.*

Sample Final Display

Your display could look like this at the end of the problem string:

$2x-4y=16$

$y=-4+\frac{1}{2}x$

$-4y=16-2x$

$\frac{-4y}{-4}=\frac{16}{-4}-\frac{2x}{-4}$

$y=-4+\frac{1}{2}x$

$2x=16+4y$

$2x-16=4y$

$\frac{2x}{4}+\frac{-16}{4}=\frac{4y}{4}$

$\frac{1}{2}x-4=y$

$\frac{2x}{-4}-\frac{4y}{-4}=\frac{16}{-4}$

$-\frac{1}{2}x+y=-4$

$y=\frac{1}{2}x-4$

$\frac{2x}{-4}+\frac{-4y}{-4}=\frac{16}{-4}$

$-3x+9y=-15$

$y=-\frac{5}{3}+\frac{1}{3}x$

$9y=-15+3x$

$\frac{9y}{9}=\frac{-15}{9}+\frac{3x}{9}$

$y=-\frac{5}{3}+\frac{1}{3}x$

$\frac{-3x}{9}+\frac{9y}{9}=\frac{-15}{9}$

$-\frac{1}{3}x+y=-\frac{5}{3}$

$y=-\frac{5}{3}+\frac{1}{3}x$

y-term is already positive!

$8x-2y=10$

$y=4x-5$

$\frac{8x}{-2}+\frac{-2y}{-2}=\frac{10}{-2}$

$-4x+y=-5$

$y=4x-5$

$-2y=10-8x$

$\frac{-2y}{-2}=\frac{10}{-2}+\frac{-8x}{-2}$

$y=-5+4x$

$8x=2y+10$

$2y=8x-10$

$\frac{2y}{2}=\frac{8x}{2}+\frac{-10}{2}$

$y=4x-5$

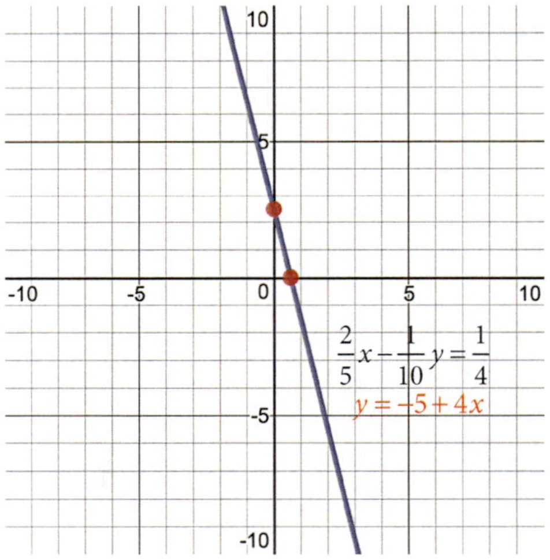

$\frac{2}{5}x-\frac{1}{10}y=\frac{1}{4}$

$y=4x-\frac{2}{5}$

$\frac{2}{5}x-\frac{1}{4}=\frac{1}{10}y$

$10(\frac{2}{5}x)-10(\frac{1}{4})=10(\frac{1}{10}y)$

$4x-\frac{2}{5}=y$

$-\frac{1}{10}y=-\frac{2}{5}x+\frac{1}{4}$

$-10(-\frac{1}{10}y)=-10(-\frac{2}{5}x)+-10(\frac{1}{4})$

$y=4x-\frac{2}{5}$

$10(\frac{2}{5}x)-10(\frac{1}{10}y)=10(\frac{1}{4})$

$4x-y=\frac{2}{5}$

$y=4x-\frac{2}{5}$

Facilitation Notes

This version of the problem string lists short notes for important teacher moves during the string. After you've done the string yourself and studied the relationships involved, you might make similar notes for the things you want a reminder of or deem important.

$2x-4y=16$	Solve for y. Who found y alone first? Divided by -4 first? Anyone get the y term positive first? Different notations for dividing by -4. When dividing, how did you deal with the -4y, + (-4y)/-4 or -(4y/-4)? Graph original and y=. Same graph!
$-3x+9y=-15$	Repeat. Quicker. Start describing strategies: divide then solve or solve then divide. Where did fractions show up? Why? Graph. What connections do you see in original, and y= forms, and graphs?
$8x-2y=10$	Predict: Will there be fractions this time? How do you know? Describe strategies: get y term positive, divide then solve, or solve then divide Where did fractions show up? Why? Graph. Connections?
$\frac{2}{5}x-\frac{1}{10}y=\frac{1}{4}$	Check out these fractions! Predict the rate of this line, steep? Shallow? Which strategy do you think you'll try? Why? Where did fractions show up? Why?

PROBLEM STRINGS LESSON

2.1 Solving Systems of Linear Equations

At a Glance	
$5x - y = -2$ $3x + y = -6$ $5x - y = -2$ $2x - y = 1$ $5x - y = -2$ $10x - 2y = -4$ $5x - y = -2$ $x - 2y = 5$ $5x - y = -2$ $5x - y = 0$	**Objectives** The goal of this problem string is to help students solidify the notion that solutions to systems of linear equations can be a point, the equivalent lines themselves, or no intersection point because the lines are parallel. **Placement** This string could come after students have been introduced to solving systems of linear equations. Students need some sense of what a system is and how to solve one. You could deliver this string any time near the beginning of textbook chapter 2. **Guiding the Problem String** This string is structured to be systems of linear equations where one of the lines is the same for each problem. Let students notice this fact as they solve each problem. Wonder aloud why the string might be that way. As students solve each system, add the graphs to a composite graph of all of the systems. Emphasize that a solution is an intersection point, a solution to both equations. Although the main point of the string isn't strategy, it will come up as students choose how to solve each system. You might start keeping track as a class about student insights as they discuss strategies and coefficients. **About the Mathematics** The first two systems result in a single solution each, where the lines intersect. The third system is a pair of equivalent equations and therefore has a set of infinite solutions. The last system is a pair of parallel lines, where there is no intersection point of the lines and therefore, no solution to the system. A system that has a solution (a point or points of intersection) is called consistent, a system that has no solution is called inconsistent. A system that has infinitely many solutions is called dependent. A system that has exactly one solution is call independent. The equations in the systems are given in standard form which may suggest using elimination as a strategy.

Sample Interactions

Use the following as you plan how to elicit and model student strategies. This is not meant as a script, but as a view into the relationships involved and the intent of the problem string.

Teacher: *Today's problem string starts out with a system of linear equations to solve. Here is the system. What does that mean, to solve a system?*	$5x - y = -2$ $3x + y = -6$

Student: *You find x and y.*

Student: *It's where they intersect.*

Teacher: *You find the x- and y-coordinates of the point where the lines intersect? Will the two lines always have exactly one point of intersection? Are there other possibilities?*

Student: *They could be parallel, so no intersection point. And I think there's another option...*

Teacher: *Maybe that will come up in our work. Okay, go ahead and solve this system.*

Students work. The teacher circulates, looking for students who are using elimination and substitution and noting any other strategies.

Teacher: *First, what solution, x- and y-values, did you find?*

Student: *x is −1.*

Student: *And y is −3.*

Teacher: *So, we can write the point (−1, −3). Let's share some strategies. I saw some of you adding the two equations together. Tell us about that.*

Student: *I added them together and got 8x is −8, so x is −1.*

Teacher: *Can everyone follow what he did?*

$$\begin{array}{r} 5x - y = -2 \\ \underline{+(3x + y = -6)} \\ 8x = -8 \\ x = -1 \end{array}$$

Teacher: *Okay, someone else pick it up from here.*

Student: *Then I plugged in −1 and got −3.*

Teacher: *Which equation did you use?*

Student: *The first one, the 5x.*

The student talks through substituting −1 into the first equation.

Student: *I substituted −1 into the second equation. And I also got −3.*

$$\begin{array}{ll} 5(-1) - y = -2 & 3(-1) + y = -6 \\ -5 - y = -2 & -3 + y = -6 \\ -3 = y & y = -3 \end{array}$$

Teacher: *So, once you've found the value of one variable, we can substitute it into either equation? Why does that make sense?*

Student: *Well, if the point is going to be on both lines, it needs to work in both equations.*

Teacher: *Let's take a look at that. Let's graph both lines. What do you see?*

Student: *They intersect right at (−1, −3).*

Teacher: *So the point (−1, −3) is on both of those lines and is therefore the solution to the system. Nice work. I noticed that some of you started using substitution to solve. What are you thinking about this elimination strategy?*

Student: *Well, when you add the two equations together, it's just so easy, because the y's add to 0. That's pretty efficient.*

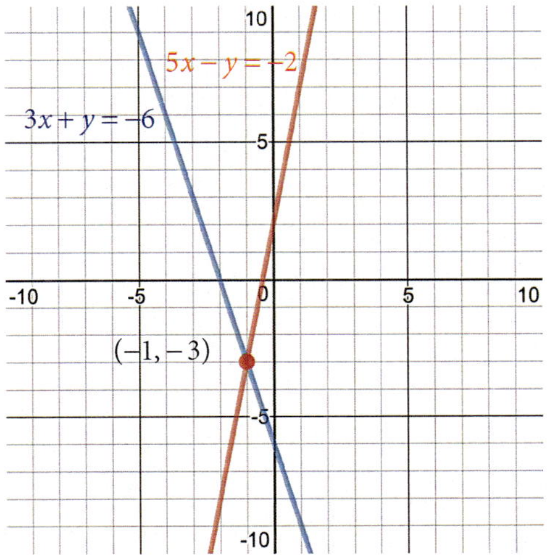

Teacher: *Here is the second system to solve today.*

Students work. The teacher circulates, looking for students who are using elimination and substitution and noting any other strategies.

$$\begin{array}{r} 5x - y = -2 \\ 2x - y = 1 \end{array}$$

(continued)

Teacher: *This time we were a bit more split; lots of people trying different things. Let's start with someone who used substitution. Tell us about that.*

The student describes substituting $5x + 2$ for y in the second equation.

Teacher: *I noticed that when you distributed the* -1*, you got* $-5x - 2$*. Did everyone distribute that the same way? Ahhh, a couple of you are noticing your mistake. Great. That's one of the reasons we share strategies, it gives us a chance to refine our thinking.*

$y = 5x + 2$

$2x - (5x + 2) = 1$

$-3x - 2 = 1$

$-3x = 3, \ x = -1$

Teacher: *Okay, we're not done, right? We only have one variable. I see some smiles. What are you thinking?*

Student: *Well, since the first equation is the same as in the first problem, we already know the y-value.*

Teacher: *But the second equation is different. Doesn't that mean that the solution should be totally different? What's going on here? Turn and talk about what you are thinking? Do we still need to figure out the y-coordinate?*

Students turn and talk while the teacher listens in.

Teacher: *Tell us what you were talking about please.*

Student: *Well, we know that we have to plug in the x, into either equation like we talked about, to find the y, because the solution is the whole point, not just one of the variables. We did that and got* -3*.*

Student: *Yeah, and that got us thinking. We noticed that both systems have the same solution point. But we're not sure why.*

Student: *We thought about that too. One thing is that if* $(-1, -3)$ *is on the line* $5x - y = -2$*, then as soon as you know you're looking for a point with x of* -1*, it's the only point it can be.*

Teacher: *That seems like an important idea. Who can say it in their own words?*

Student: *There is only one point on the line that has an x of* -1*. There can only be that point. For this line, it's* $(-1, -3)$*.*

Teacher: *Did anyone take a look at the graphs? You guys did?*

Student: *Yes, and sure enough, these 2 lines have the same intersection point as the first two.*

Student: *Since the two questions are asking about the red line, it's like asking where does the blue one intersect and then where does the green one intersect? And they all happen to be at the same point.*

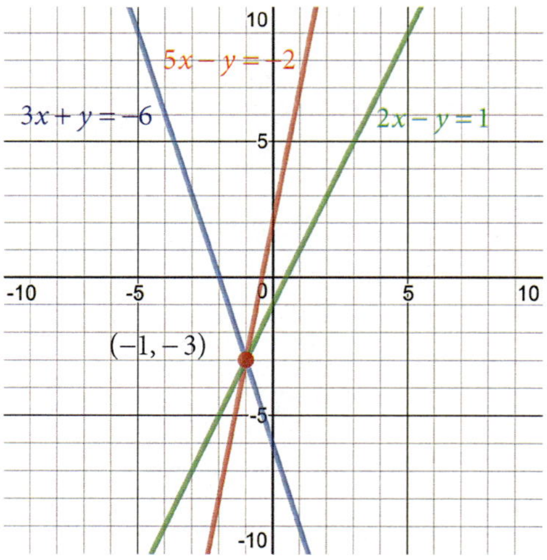

Student: *I also solved for y, but I solved both equations for y and made them equal.*

Teacher: *Talk me through that.*

The student describes solving both equations for y and setting them equal to each other and solving for x.

$5x + 2 = 2x - 1$

$3x = -3$

$x = -1$

Teacher: *Nice work. What about elimination? I saw some of you doing that. Who can tell us about that?* A student talks about using elimination. **Teacher:** *Why did you choose to multiply the second equation by −1 instead of trying to eliminate the x? Another way of saying that is you found the opposite of all of the terms in the second equation. Why?* **Student:** *We would have had to multiply both equations by something so that the x would have the same coefficient. This way we only had to mess with one equation.*	$5x - y = -2 \quad 5x - y = -2$ $-1(2x - y = 1) \quad \underline{-2x + y = -1}$ $3x = -3$ $x = -1$
Teacher: *Let's talk strategy. Some of you used elimination, some substitution. Which do you wish your brain would be inclined to do next time you have a similar problem?* **Student:** *Honestly, I had started doing substitution on the first problem, so since the first equation is the same as in the first problem, I had already solved for y. In the future, if I hadn't already solved for y, I probably would've used elimination.* **Student:** *Oh, so yeah, that makes sense why you'd use it if you already had it.* **Student:** *I think that for this system, both strategies are pretty easy, especially if you'd already solved for y. But since the coefficient of y is just −1, it's pretty easy to solve for it.*	
Teacher: *The next problem is this system. Solve away!* Students work and the teacher circulates, looking for students who are on the verge of understanding that the two equations are equivalent. Bringing these students into the conversation has the potential to help all students make sense of the situation.	$5x - y = -2$ $10x - 2y = -4$
Teacher: *I'm seeing some interesting looks. What's going on here?* **Student:** *We decided to eliminate, so we multiplied the first equation by −2. But then everything went away!* **Teacher:** *What do you mean, went away?* **Student:** *It was zero.*	$-2(5x - y = -2) \quad -10x + 2y = 4$ $10x - 2y = -4 \quad \underline{10x - 2y = -4}$ $0 \;\; + 0 = 0$
Teacher: *Did anyone else find that? I'm seeing some nods. But some of you tried substitution. What did you find?* **Student:** *It all just equaled each other.* **Student:** *And zero equals zero.*	$y = 5x + 2$ $10x - 2(5x + 2) = -4$ $10x - 10x - 4 = -4$

(continued)

Teacher: *Other thoughts?*

Student: *We noticed that the second equation is just twice the first. So, does that mean they are the same line?*

Student: *We wondered that too, so we graphed both of them and they sure look like the same line.*

Teacher: *So what is the solution to a system that is actually two different ways of writing the same line? How many points will be on both of those lines?*

Student: *All of them!*

Student: *Yeah, this is the one that the solution is an infinite set of points.*

Teacher: *Yes, and we call this kind of system dependent. What are you guys smiling about?*

Student: *Well, we were laughing because at first we just assumed that the solution to this would be (−1, −3) again. Then we were sure that it wasn't. Now we are thinking it is, but it's only one of the infinite number of points. Ha!*

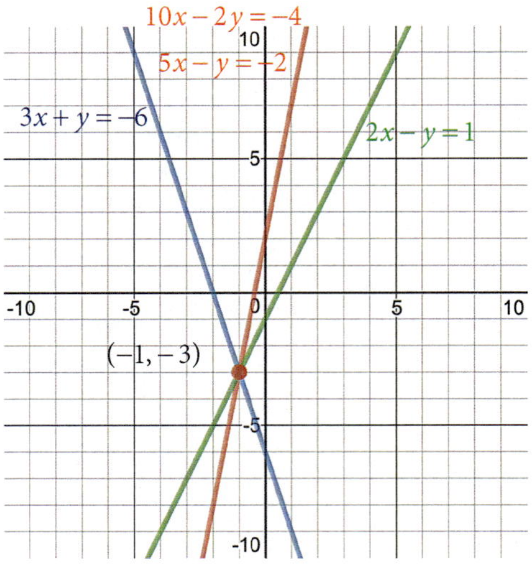

Teacher: *Here is the next system to solve. Yes, it still has the same first line. I wonder how you'll choose to solve this one?*

Students work and the teacher circulates, looking for elimination, substitution, and taking note of other strategies.

$$5x - y = -2$$
$$x - 2y = 5$$

Teacher: *Let's share some thinking about this system.*

Student: *Well, first of all, you did it again! The solution is the same as the first two systems.*

Teacher: *Just the x-value or the y-value?*

Student: *Both, it's the same point. I bet the lines all intersect in the same point.*

Teacher: *We'll take a look at that in a bit. First, I see that some of you used elimination. Are you glad you did? Tell us about that.*

Student: *It wasn't too bad. I decided to multiply the bottom one by −5 to get rid of the x's.*

Student: *And I multiplied the top equation by −2 so that the y's would cancel.*

Teacher: *When you say cancel, do you mean add to 0? Okay, so let's put both of those up.*

$$\begin{array}{ll} 5x - y = -2 & 5x - y = -2 \\ -5(x - 2y = 5) & \underline{-5x + 10y = -25} \\ & 9y = -27 \\ & y = -3 \end{array} \qquad \begin{array}{ll} -2(5x - y = -2) & -10x + 2y = 4 \\ x - 2y = 5 & \underline{x - 2y = 5} \\ & -9x = 9 \\ & x = -1 \end{array}$$

Teacher: *Do either of these strike you now as more efficient?*

Student: *Not really. With both of them you start by multiplying by a negative and you end up with pretty easy last equations.*

Teacher: *Is everyone okay with that? That you could've chosen either and been equally efficient?*

Teacher: *I see that some of you used substitution. Are you glad you did? Tell us about that.* **Student:** *Yes, because even if we hadn't already solved for y for the first equation, you could easily do that.* The student details the substitution while the teacher models it on the board.	$y = 5x + 2$ $x - 2(5x + 2) = 5$ $x - 10x - 4 = 5$ $-9x = 9, \quad x = -1$
Teacher: *Anyone wish they would've chosen the other strategy?* **Student:** *Maybe the substitution has fewer steps, though the math is easy enough either way that maybe for this problem, you could just choose whatever comes naturally.* **Teacher:** *Let's look at the graph now. What do you think it will look like when we add this line? Predict. Turn and comment to your partner while I get it up here.* **Teacher:** *Is that similar to what you thought? By the way, we call a system that has one solution an independent system. Okay, we have one more problem. Looking at the graph, where do you think it might be? Just guess.*	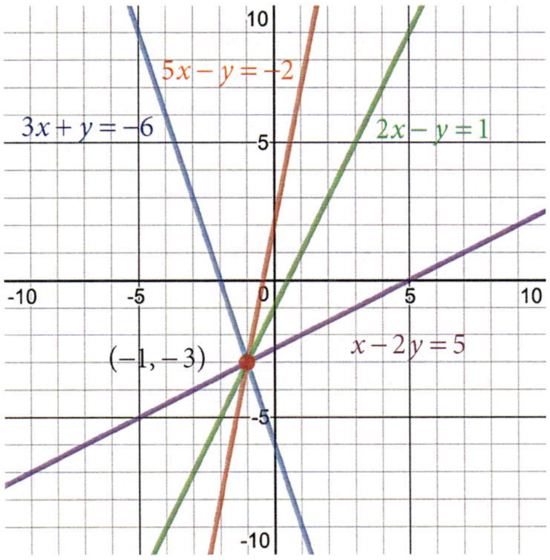
Teacher: *Well, the next problem is this system. Let me know when you're ready to talk about the solution.* Students work and the teacher circulates, looking for students who are thinking about graphs and noting other strategies.	$5x - y = -2$ $5x - y = 0$
Teacher: *I noticed that some of you tried elimination. Tell us about that.* **Student:** *Yeah, we multiplied the bottom equation by −1 and when we added them together, we got weirdness, zero doesn't equal −2?*	$5x - y = -2$ $\underline{-1(5x - y = 0)}$ $0 \quad + 0 = -2$
Teacher: *I noticed some of you were thinking about the graphs of these two lines. Does that help you make sense of their weirdness?* **Student:** *We solved them both for y. Well, we already had the first one. But anyway, if you think about the graphs, the lines will be parallel.*	$y = 5x + 2$ $y = 5x$

(continued)

Student: *Oh! They have the same slope. They won't ever intersect.*

Student: *Is that why we got 0 = −2? Is that what it means, that there's no solution because the lines don't intersect?*

Teacher: *Let's take a look at the graphs.*

Student: *Right, it is parallel. So, no solution.*

Student: *So, no points work in both equations.*

Teacher: *And we call that kind of system inconsistent. Nice connections and reasoning everyone!*

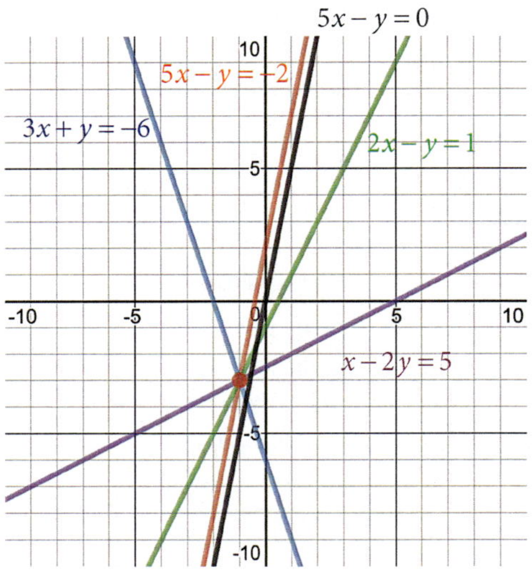

Teacher: *How would you summarize some of the things that came up in this string today?*

Elicit the following:

- *Thinking about the graphs can help make sense of the results.*
- *A solution to a system can be the point where the lines intersect.*
- *If the lines are the same, the solution is infinitely many points—all of the points on the line.*
- *If the lines are parallel, there is no intersection point, therefore there is no solution.*

Sample Final Display

Your display could look like this at the end of the problem string:

$5x-y=-2$ ✓
$3x+y=-6$ ✓
$(-1,-3)$

$$\begin{array}{r} 5x-y=-2 \\ +(3x+y=-6) \\ \hline 8x=-8 \\ x=-1 \end{array}$$

$$\begin{array}{l} 5(-1)-y=-2 \\ -5-y=-2 \\ -3=y \end{array}$$

$$\begin{array}{l} 3(-1)+y=-6 \\ -3+y=-6 \\ y=-3 \end{array}$$

$5x-y=-2$ ✓
$2x-y=1$ ✓
$(-1,-3)$

$$\begin{array}{l} y=5x+2 \\ 2x-(5x+2)=1 \\ -3x-2=1 \\ -3x=3,\ x=-1 \end{array}$$

$$\begin{array}{l} 5x+2=2x-1 \\ 3x=-3 \\ x=-1 \end{array}$$

$$\begin{array}{l} 5x-y=-2 \\ -1(2x-y=1) \end{array}$$

$$\begin{array}{r} 5x-y=-2 \\ -2x+y=-1 \\ \hline 3x=-3 \\ x=-1 \end{array}$$

$5x-y=-2$ ✓
$10x-2y=-4$ ✓
dependent
equations are equivalent
infinite solutions

$$\begin{array}{l} -2(5x-y=-2) \\ 10x-2y=-4 \end{array}$$

$$\begin{array}{r} -10x+2y=4 \\ 10x-2y=-4 \\ \hline 0\ \ +0=0 \end{array}$$

$$\begin{array}{l} y=5x+2 \\ 10x-2(5x+2)=-4 \\ 10x-10x-4=-4 \\ 0=0 \end{array}$$

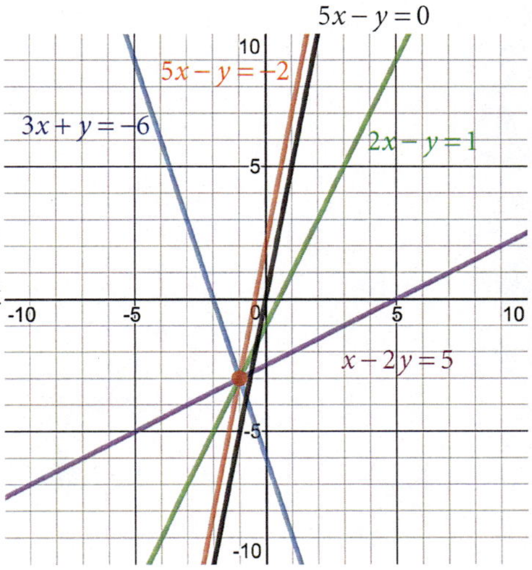

$5x-y=-2$ ✓
$x-2y=5$ ✓
$(-1,-3)$

$$\begin{array}{l} 5x-y=-2 \\ -5(x-2y=5) \end{array}$$

$$\begin{array}{r} 5x-y=-2 \\ -5x+10y=-25 \\ \hline 9y=-27 \\ y=-3 \end{array}$$

$$\begin{array}{l} -2(5x-y=-2) \\ x-2y=5 \end{array}$$

$$\begin{array}{r} -10x+2y=4 \\ x-2y=5 \\ \hline -9x=9 \\ x=-1 \end{array}$$

$$\begin{array}{l} y=5x+2 \\ x-2(5x+2)=5 \\ x-10x-4=5 \\ -9x=9,\quad x=-1 \end{array}$$

$5x-y=-2$ ✓
$5x-y=0$ ✓
no solution
parallel lines
inconsistent

$$\begin{array}{r} 5x-y=-2 \\ -1(5x-y=0) \\ \hline 0\ \ +0=-2 \end{array}$$

$$\begin{array}{l} y=5x+2 \\ y=5x \end{array}$$

(continued)

Facilitation Notes

This version of the problem string lists short notes for important teacher moves during the string. After you've done the string yourself and studied the relationships involved, you might make similar notes for the things you want a reminder of or deem important.

$5x-y=-2$ $3x+y=-6$	*What does it mean to solve a system? Find solution.* *What is solution? Just an x? Just a y? Either? A point?* *Model elimination, substitution.* *Once you've found one variable, you can substitute it into either? Why?* *Graph.*
$5x-y=-2$ $2x-y=1$	*Why are you smiling? Same solution? Did you need to find the other variable? Why not?* *Model elimination, substitution. Graph.* *Which strategy do you wish your brain would be inclined to in a similar prob?*
$5x-y=-2$ $10x-2y=-4$	*Same line. Look for students on the verge of understanding. Share.* *Model elimination, substitution. Graph. Same line!* *Called dependent system.*
$5x-y=-2$ $x-2y=5$	*Look at the previous graphs. Where might this next one be?* *Are you glad you chose the strategy you did? Why?* *Predict the graph. Graph. When there's one solution, called independent.*
$5x-y=-2$ $5x-y=0$	*Share elimination. Graph. Parallel lines!* *Called inconsistent (no solution, no intersection)*

PROBLEM STRINGS LESSON 2.2

Substitution or Elimination?

At a Glance	
$x = \frac{3}{2}y - 4$ $y = -\frac{3}{2}x + 7$ $2x - 3y = -8$ $3x + 2y = 14$ $16x + 2y = -60$ $-16x - 5y = 54$ $-3x + y = -17$ $y = 2x - 11$ $y = -2x - 2$ $-2x + 4y = 12$	**Objectives** The goal of this problem string is to develop students' intuition for choosing a smart strategy—substitution or elimination—when solving a system of linear equations. **Placement** This string could come any time after students have learned about solving systems of linear equations using substitution and elimination. You could deliver this string any time during textbook chapter 2. Since textbook lesson 2.1 reviews what a system of equations is and why you solve them, it works well right after. **Guiding the Problem String** The first two systems are equivalent because the first equations are equivalent, as are the second equations. Thus, students can begin to discuss that the form of the equations can suggest a strategy, but that the form does not have to dictate the strategy. Spend time on these two systems solving and comparing the strategies, discussing the equivalent forms of the equations, and emphasizing the nature of the solution, the intersection of the two lines, and the point that satisfies both equations. The next two systems are set up to be more suited to elimination and substitution, respectively. The last system is structured to be easy to argue that either strategy would be just as efficient. You might choose to spend time on the last three systems just discussing what students' plan of attack would be and why, instead of having students solve them. All of this discussion and reasoning about which strategy to choose helps students develop the habit of taking time to think before they start and helps them understand what to look for to make decisions. **About the Mathematics** Even when students can successfully use either elimination or substitution to solve a system of linear equations they can get caught up in the procedural "doing" of it all and lose sight of strategy. Some students can use either strategy when it's dictated but have trouble choosing between them. Use this string to bring these issues to light and help students work to keep meaning and relationships at the forefront.

Sample Interactions

Use the following as you plan how to elicit and model student strategies. This is not meant as a script, but as a view into the relationships involved and the intent of the problem string.

Teacher: *Let's get started today solving some systems of linear equations. The first problem today is this system. Before you start doing anything, what does it mean to solve a system of equations?* *What does the solution mean in the equations?* *What does the solution mean in the graph?*	$x = \frac{3}{2}y - 4$ $y = -\frac{3}{2}x + 7$

(continued)

Teacher: *Some of you substituted for y in the first equation. Tell us about that and I'll represent that on the board.* *Some fun work with fractions! Whew! Good work.* *Now that we have the x-value, we're done, no? How did you solve for y? Substituted 2? I'm not going to record that right now, but I do want to ask, what is the solution? An ordered pair? A point? Yes!*	$x=\frac{3}{2}(-\frac{3}{2}x+7)-4$ $x=-\frac{9}{4}x+\frac{21}{2}-4$ $\frac{13}{4}x=\frac{13}{2}$ $x=\frac{13}{2}\cdot\frac{4}{13}=2$
Teacher: *Some of you substituted for x in the second equation. Let's get that thinking up here. Please tell us about that.* *And once you found that y is 4, then how did you find x? Again, we won't record that part. But it's important to remember to solve for both coordinates.* *So, with either of these strategies, you were substituting to solve. Why? What about these equations nudged you to use substitution?*	$y=-\frac{3}{2}(\frac{3}{2}y-4)+7$ $y=-\frac{9}{4}y+6+7$ $\frac{13}{4}y=13$ $y=4$
Teacher: *The next problem in our string is to solve this system. Go!*	$2x-3y=-8$ $3x+2y=14$
Teacher: *Okay, good work. What is the solution to that system?* *The same as the first system? How odd. Are you sure?* *Let's look at your work. I saw some of you eliminate the x-term first. Tell us about that.* *Now that we have that y is 4, did you find the x also? I won't take the time to put up that work, but why is it important to find both x and y?*	$3(2x-3y=-8)$ $6x-9y=-24$ $-2(3x+2y=14)$ $+-6x-4y=-28$ $-13y=-52$ $y=4$
Teacher: *Some of you eliminated the y-term first. Please explain that thinking.* *Sure enough, using elimination either way found the same x- and y-values. Interesting. Is that a coincidence? Do you see any relationship between the previous and this system?* *The systems are equivalent? How do you know?*	$2(2x-3y=-8)$ $4x-6y=-16$ $3(3x+2y=14)$ $+9x+6y=42$ $13x=26$ $x=2$

Teacher: *Do you think that equivalent systems will always have the same solution? Why? What does it have to do with the graphs? The equations?*

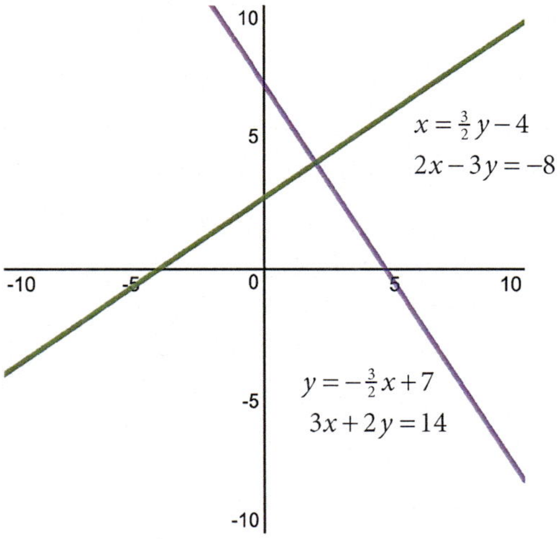

Teacher: *Let's look back at these two problems. You're telling me that they are equivalent. Why did you feel like substituting to solve the first system and eliminating to solve the second system?*

So the structure of the equations might help you decide what to do?

Some of you were really not liking all of the fraction work in the first system. Would it have been possible to try to find an equivalent form to see if the work might be more to your liking? Would it be worth the trouble?

It seems like we are developing some important ideas about how the structure might help you determine how to solve a system. I wonder how those ideas will play out with our next system.

Teacher: *Here is the next system. Now before you do anything, let's engage our strategy brains. Predict, without actually doing anything, what might be a good plan of attack to solve this system.*

$$16x+2y=-60$$
$$-16x-5y=54$$

Teacher: *What do you think would be a way to solve this system? Why? What about the structure helps?*

$$16x+2y=-60$$
$$-16x-5y=54$$

Elimination, easy to eliminate *x*.

$$16x+2y=-60$$
$$\underline{-16x-5y=54}$$
$$-3y=-6$$

Teacher: *We might not solve the next problem either. Let's really be thinking about how we might solve it.*

$$-3x+y=-17$$
$$y=2x-11$$

Teacher: *What do you think would be a nice way to solve this system? Why? What about the structure helps?*

Why is the coefficient of 1 nice?

coefficient 1

$$-3x+y=-17$$
$$y=2x-11$$

already solved for *y*

Substitution $\quad -3x+(2x-11)=-17$

(continued)

Teacher: *Here is the last problem in the string for us to think about. How might you go about solving this system?*	$y = -2x - 2$ $-2x + 4y = 12$
Teacher: *I hear some of you talking about substitution. Tell us about that.* *Who agrees? Disagrees? Why do you think elimination? Tell us about that.* *What about the structure of this system, of these equations, makes you think the way you do?* *What makes you think it's just as efficient either way?* $y = -2x - 2$ $-2x + 4y = 12$ Substitution $-2x + 4(-2x - 2) = 12$ Elimination $2x + y = -2$ $-2x + 4y = 12$ $5y = 10$ or $y = -2x - 2$ $4y = 2x + 12$ $5y = 10$	
Teacher: *Let's get some of these ideas on an anchor chart.*	

Sample Anchor Chart

Your display could look like this at the end of the problem string.

Eliminate or Substitute?	
Look at the structure!	
If one equation is already solved for a variable, try substitution.	$-3x + y = -17$ $y = 2x - 11$
If the coefficients of a variable are the same in both equations, try to eliminate that variable.	$16x + 2y = -60$ $-16x - 5y = 54$

Sample Final Display

Your display could look like this at the end of the problem string:

$x=\frac{3}{2}y-4$
$y=-\frac{3}{2}x+7$

Substitution

$x=\frac{3}{2}(-\frac{3}{2}x+7)-4$
$x=-\frac{9}{4}x+\frac{21}{2}-4$
$\frac{13}{4}x=\frac{13}{2}$
$x=\frac{13}{2}\cdot\frac{4}{13}=2$

$y=-\frac{3}{2}(\frac{3}{2}y-4)+7$
$y=-\frac{9}{4}y+6+7$
$\frac{13}{4}y=13$
$y=4$

$2x-3y=-8$
$3x+2y=14$

Elimination

$3(2x-3y=-8)$
$-2(3x+2y=14)$

$6x-9y=-24$
$+-6x-4y=-28$
$-13y=-52$
$y=4$

$2(2x-3y=-8)$
$3(3x+2y=14)$

$4x-6y=-16$
$+9x+6y=42$
$13x=26$
$x=2$

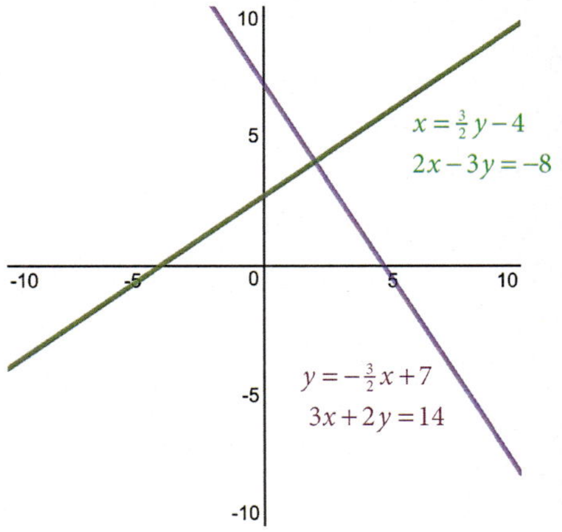

$16x+2y=-60$
$-16x-5y=54$

Elimination, easy to eliminate x.

$16x+2y=-60$
$-16x-5y=54$
$-3y=-6$

coefficient 1

$-3x+y=-17$
$y=2x-11$

already solved for y

Substitution

$-3x+(2x-11)=-17$

$y=-2x-2$
$-2x+4y=12$

Substitution

$-2x+4(-2x-2)=12$

Elimination

$2x+y=-2$
$-2x+4y=12$
$5y=10$

or

$y=-2x-2$
$4y=2x+12$
$5y=10$

(continued)

Facilitation Notes

This version of the problem string lists short notes for important teacher moves during the string. After you've done the string yourself and studied the relationships involved, you might make similar notes for the things you want a reminder of or deem important.

$x=\frac{3}{2}y-4$ $y=-\frac{3}{2}x+7$	*First, what does it mean to solve a system? Equations? Graphs?* *Who substituted for y? How?* *Who substituted for x? How?* *What is the solution? x, y or ordered pair?* *Why do you think you used substitution for this system?*
$2x-3y=-8$ $3x+2y=14$	*What is the solution to this system?* *Who eliminated x first? Tell us about that.* *Who eliminated y first? Tell us about that.* *So you can eliminate either? Interesting.* *Any relationship to the first system? Equivalent? How do you know? Graph both.* *Might you even consider changing structure?*
$16x+2y=-60$ $-16x-5y=54$	*Predict—how would you solve? Why?* *What about structure nudges you to eliminate?*
$-3x+y=-17$ $y=2x-11$	*Predict—how would you solve? Why?* *What about structure nudges you to substitute?*
$y=-2x-2$ $-2x+4y=12$	*Predict—how would you solve? Why?* *What about structure allows for either strategy?* *Let's get these important ideas on an anchor chart.*

PROBLEM STRINGS
LESSON

2.2 Think Before You Eliminate

At a Glance	
$56x - 10y = 25$ $-56x + 5y = -20$ $-3x - 10y = 5$ $7x + 5y = -20$ $36x + 45y = 412$ $-18x + 90y = 21$ $2x - y = 5$ $-6x + 7y = 60$ $-4x + 5y = 12$ $7x - 8y = 5$	**Objectives** The goal of this problem string is to help students develop intuition for first steps in using elimination to solve a system of linear equations. By looking at the relationships in the problems, students can choose efficient first steps. **Placement** This string could come during your work on solving systems of linear equations using elimination. You could deliver this string any time during textbook chapter 2, but it might be handy to help students decide how they will use elimination in textbook Lesson 2.2 Substitution and Elimination. **Guiding the Problem String** Each of the problems in the string has a structure based on relationships that beg for a different first step in using elimination to solve. Your goal is to pull from students what they notice about each system and how the relationships between the coefficients of the variables interact with the constant terms to nudge them toward one strategy or another. Encourage students to look, notice, evaluate. Do not solve the systems. **About the Mathematics** These systems of linear equations were created to bring to the surface some underlying relationships that can influence students' first moves. With each of these systems, you could choose to eliminate either variable, but with the exception of the last system, the others each have something in their structure that makes a certain first move more efficient. When both sets of coefficients are relatively prime (the only common factor is 1), you need to multiply both equations in order to eliminate a variable. Deciding which variable to eliminate can be helped by considering which multiplication, if either, is easier for the coefficients or the constant terms. Considering the signs of the coefficients may play a role as well.

(continued)

Important Questions

Use the following as you plan how to elicit and model student strategies.

- *In this problem string, we will be solving systems of equations by elimination. You do not need to solve this system of equations. Study it and decide what you think a good plan of attack would be. What would be a good first move?*
- *Why would you choose that first move?*
- *What relationship are you using to decide? How does that help?*
- *How do the coefficients influence your strategy?*
- *How do the signs of the coefficients influence your strategy?*
- *What happens when the coefficient of a variable is a factor of the other coefficient of that variable? For example when the system has* $-10y$ *and* $5y$*?*
- *What happens when the coefficients of a variable are relatively prime?*
- *How do the constant terms influence your strategy?*

Teacher: *How would you summarize some of the things that came up in this string today?*

Elicit the following:

- *If a coefficient is* 1, *just multiply that equation by whatever you need to eliminate that variable.*
- *If one set of coefficients are opposite signs but the same absolute value, just add the equations to eliminate that variable.*
- *If one variable's coefficients have opposite signs, you can take advantage of that.*
- *If the coefficient of a variable is a factor of the coefficient of that variable in the other equation, you only have to scale one equation. You may have to consider the signs as well.*

Sample Final Display

Your display could look like this at the end of the problem string:

$56x-10y=25$ $-56x+5y=-20$		$56x-10y=25$ $-56x+5y=-20$ $0x$				Opposite equal coefficients add to 0.
$-3x-10y=5$ $7x+5y=-20$	$2(7x+5y=-20)$	$-3x-10y=5$ $14x+10y=-40$ $0y$				Only have to multiply one equation. 5 is a factor of 10.
$36x+45y=412$ $-18x+90y=21$	$2(-18x+90y=21)$	$36x+45y=412$ $-36x+180y=42$ $0x$	or	$-2(36x+45y=412)$	$-72x-90y=-824$ $-18x+90y=21$ $0y$	Opposite signs are nice.
$2x-y=5$ $-6x+7y=60$	$7(2x-y=5)$	$14x-7y=35$ $-6x+7y=60$ $0y$				Look for a coefficient of 1 or −1.
$-4x+5y=12$ $7x-8y=5$	$7(-4x+5y=12)$ $4(7x-8y=5)$		or	$8(-4x+5y=12)$ $5(7x-8y=5)$		When both sets of coefficients are relatively prime, you have to multiply both, but maybe look at the constants to see which is easier for you.

Facilitation Notes

This version of the problem string lists short notes for important teacher moves during the string. After you've done the string yourself and studied the relationships involved, you might make similar notes for the things you want a reminder of or deem important.

$56x-10y=25$ $-56x+5y=-20$	We are not solving these systems, just helping each other analyze them. How would you use the relationships in this system to solve? Why? What could we say about that? Let's keep track of these ideas.
$-3x-10y=5$ $7x+5y=-20$	What strikes you about this system? What could you do? What is more efficient?
$36x+45y=412$ $-18x+90y=21$	Some of you want to scale the top equation by −2? Others the bottom equation by 2? Does the constant term influence your choice?
$2x-y=5$ $-6x+7y=60$	What's nice about this one? How does the coefficient of −1 influence your choice?
$-4x+5y=12$ $7x-8y=5$	What do you do when the coefficients are relatively prime? What did you notice about the structure of all of these systems?

PROBLEM STRINGS LESSON

2.3 How Many Solutions?

At a Glance	
How many possible solutions are there for systems of equations that form: • a line and a line • a line and a parabola • a line and an exponential function • a line and a cubic equation • a line and a circle • a line and a hyperbola • a parabola and a parabola • a parabola and a hyperbola	**Objectives** The goal of this problem string is to help students develop notions about the possible number of solutions for systems of linear and non-linear equations.
	Placement This string could come as you are working to solve systems of non-linear equations to support your work by helping students reason about the possible number of solutions. You could deliver this string any time as you work on textbook Lesson 2.3 Linear and Non-linear Systems of Equations.
	Guiding the Problem String The first problem should be a quick review, the possible number of solutions for a system of 2 linear equations. For the rest of the problems, you may have to help students with the long run behavior of the non-linear functions. You can do this by graphing a few examples of parabolas, including $x = y^2$ for the second problem, a few examples of exponential functions for the third problem, and so forth. You might need to have several examples of cubic equations, including $x = y^3$ and $y = x^3 - 2x$. Ask students to draw or act out each case to support their reasoning. You might need to stipulate that these systems are coplanar equations.
	About the Mathematics The long run behavior of each parent function is important to consider. For this problem string, supply the long run behavior if necessary. Don't make students guess. You might consider quickly putting up a calculator graph so that students can get a feel for any unknown parent functions. This string should not be a test of prior parent knowledge, but rather a harbinger of future work. Some cubic functions have a "dip" that allows for two intersections with some lines. The asymptotes of hyperbolas allow for zero intersections with those lines.

Important Questions

Use the following as you plan for ideas on how to elicit and model student thinking.

- *How many solutions are possible in a system of equations where both equations are lines?*
- *How do you know?*
- *Is it possible for there to be no solutions? Exactly one solution? Exactly two solutions? More solutions?*
- *Draw each case or use "math aerobics" (have students use their bodies to act out) to physically demonstrate each number of solutions.*
- *What is the long run behavior of each of the functions?*
- *How does the long run of the function or equation force the number of possible solutions?*

Teacher: *How would you summarize some of the things that came up in this string today?*

Elicit the following:

- *Systems of equations can have different numbers of solutions.*
- *Knowing about the shape of the graph can help you find the number of possible solutions.*
- *Using the long run behavior of a function can help you decide the possible number of solutions.*
- *Horizontal and vertical placement, as well as slope contribute to the number of possible solutions. Stretch or compression in parabolas should be considered too.*

(continued)

Sample Final Display

Your display could look like this at the end of the problem string:

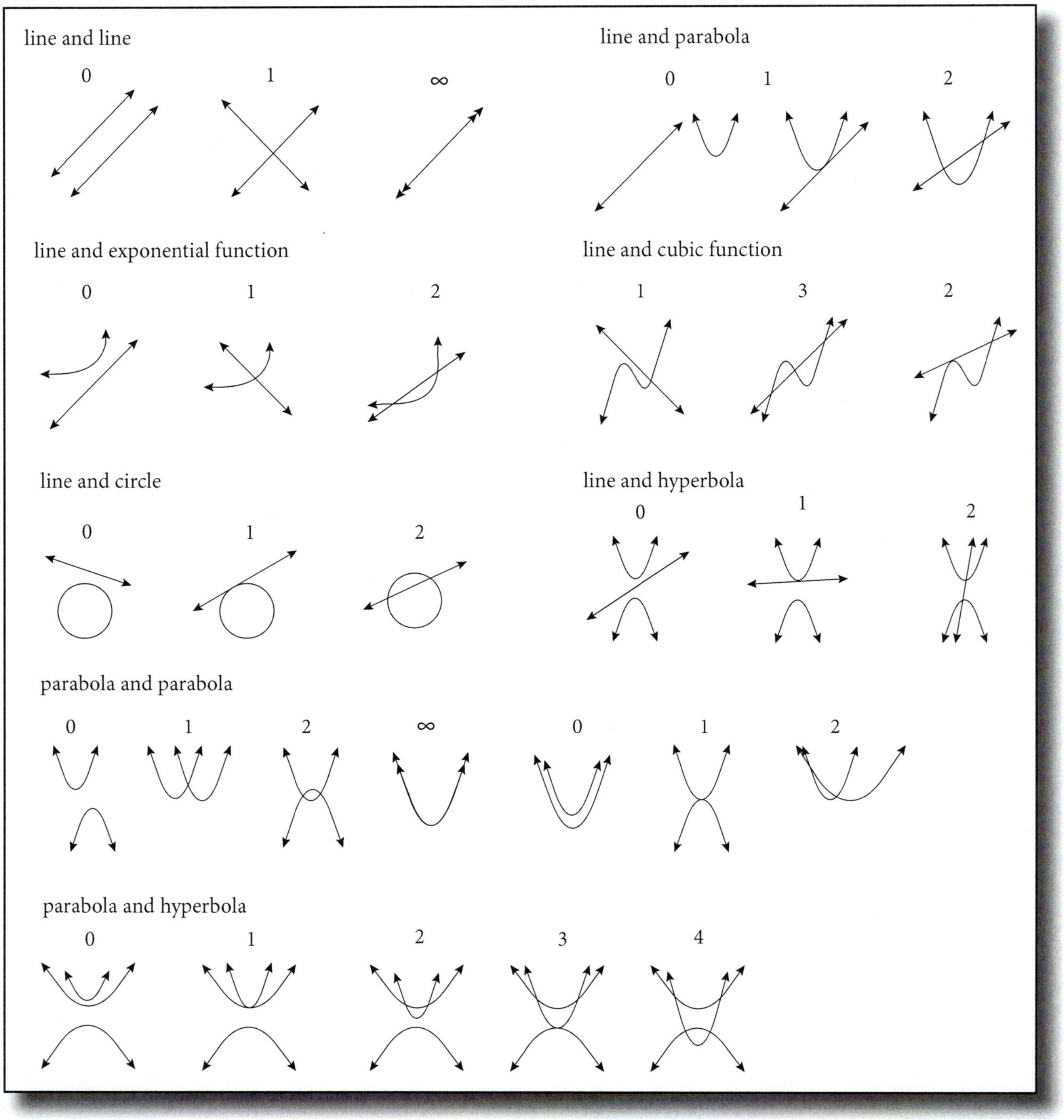

Facilitation Notes

This version of the problem string lists short notes for important teacher moves during the string. After you've done the string yourself and studied the relationships involved, you might make similar notes for the things you want a reminder of or deem important.

a line and a line	*How many intersections are possible? How do you know? Sketch.*
a line and parabola	*Now how many? Is it possible for no solutions? Exactly one solution? Tangent? More? Math aerobics (stand up and "show")*
a line and an exponential function	*What is the long run behavior? How does that help?*
a line and a cubic equation	*What is the long run behavior? What about $y=x^3-2x$? Can there be 2 solutions? Tangent?*
a line and a circle	*Tangent?*
a line and a hyperbola	*What is a hyperbola? Look at $y=1/x$, $y=1/x^2$. Asymptotes?*
a parabola and a parabola	*Now we're having fun! Don't forget they can be tangent.*
a parabola and a hyperbola	*Tangent?*

PROBLEM STRINGS
LESSON

2.4 What's Your Solution?

At a Glance	
	Objectives The goal of this problem string it to support students' understanding of the different solution possibilities of different kinds of equations and inequalities. This string provides an opportunity for students to further their understanding that functions and inequalities differ from one-variable equations because they have sets of solutions rather than one solution.
$x = -4$ $x \leq -4$ $x > -4$ $2x - 4 = 0$ $y = 2x - 4$ $y < 2x - 4$ $y \geq 2x - 4$ $y > 2x - 4$ $x > -4$	**Placement** This string could come after students have worked on graphing systems of linear equations and as students begin working with linear inequalities. You could use this string before or during your work with textbook Lesson 2.4 Systems of Inequalities.
	Guiding the Problem String The first three problems should go quickly and set the routine of identifying solutions of equations compared to inequalities. Model these three problems horizontally in a row on separate number lines. The second set of three problems are in two dimensions and should be represented on a coordinate grid; their graphs should be underneath their corresponding one-dimensional model to juxtapose their meanings and foster conversation. In each case, ask students to identify values that make the statements true, tag these as "solutions." Suggest a non-solution and a solution for students to evaluate. The next problem, $2x - 4 = 0$, where the solution, $x = 2$, is modeled on a number line, while $y = 2x - 4$, where the solution is a set of ordered pairs that form the line, is modeled on the coordinate grid.
	About the Mathematics The number line is a useful model to represent solutions of one variable equations and inequalities. Single solutions of equations are represented with a point, while inequalities have an infinite number of solutions and are represented by a ray, line, or line segment. The coordinate grid is a model for two-variable linear equations and functions; these have an infinite number of solutions and are also represented by lines. Two-variable linear inequalities also have an infinite numbers of solutions and are represented as parts of a plane by shaded areas of the graph. The special cases of vertical and horizontal lines have only one variable in the equation and are represented by $x = c$ and $y = c$, respectively, but have solutions with both an x and a y component. Solutions to equations and inequalities are values which make the statement true. Systems of inequalities have solutions that satisfy both (or all) of the individual inequalities.

Sample Interactions

Use the following as you plan how to elicit and model student strategies. This is not meant as a script, but as a view into the relationships involved and the intent of the problem string.

Teacher: *The first few problems today are really quick. If we start with the equation $x = -4$, an algebraic representation, how can we represent that on a number line?* **Student:** *You just draw a number line and put a dot on −4.* **Teacher:** *Yep, too easy, right?*	$x = -4$ 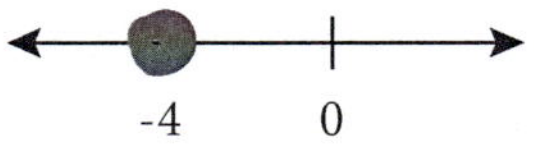
Teacher: *Next up is $x \le -4$. What are some values of x that would make this a true statement?* **Students:** *x is −4 or −10. x could be −7.* **Teacher:** *Okay, I'm going to represent what you said on a new number line.* Teacher puts points on the values students suggested. **Teacher:** *How does that look? Did we represent all the values of x that make this a true statement?* **Student:** *No, there are a lot more numbers that can work for x. Shade the number line from the −4 to the left and put an arrow on it.* **Teacher:** *Why would we shade part of the number line instead of a point like in the first problem?* **Student:** *A line means all the points, even the points in between the integers. So −4.5 works too.*	$x \le -4$ 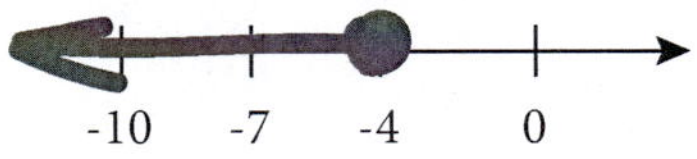
Teacher: *Nice, the next problem is $x > -4$. What are some values that make that statement true? What will that look like?* **Students:** *Zero. 3. Draw a circle on −4 and draw a line to the right.* **Teacher:** *So you are saying that zero is greater than −4. What about −6? Talk to us about true statements?* **Student:** *That is to the left, so it is smaller, −6 is greater than −4 isn't true.*	$x > -4$ 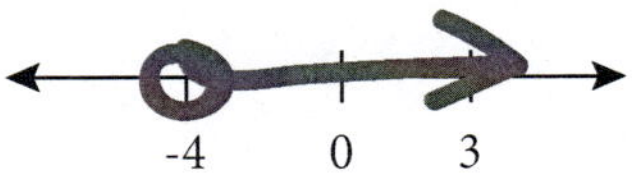

(continued)

Teacher: *That was a nice review. Now let's look at each of those statements in two dimensions. How can we model them on a coordinate grid? Let's start with $x = -4$. If x is -4, what is y? What are some points that make it a true statement? Would a table help us think about it?*

Student: *So, x has to be -4. It doesn't tell us what y is, so I guess y could be anything.*

Students: *Yeah, like $(-4, 0)$ or $(-4, 2)$.*

Teacher: *Let me put those on the graph. Is that all the points that will work? Will the point $(2, -4)$ work? Does that make a true statement?*

Student: *No, because x has to be -4.*

Student: *But all the y values will work, as long as x is -4. So draw a vertical line where $x = -4$.*

$x = -4$

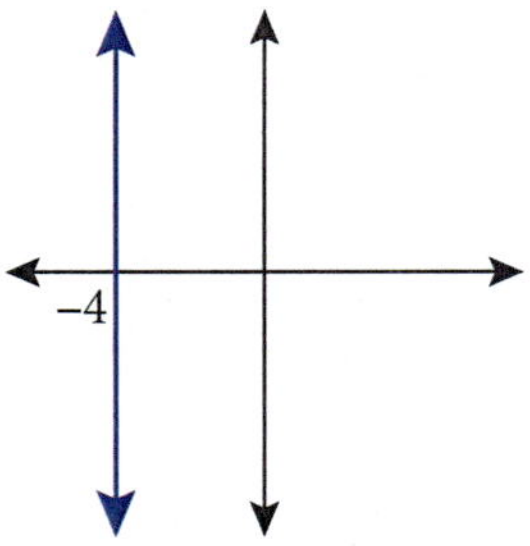

Teacher: *The next one is $x \leq -4$. What are some values that would make this a true statement?*

Student: *Well, x has to be less than or equal to -4, so the point $(-5, 2)$ will work.*

Several other students offer points while the teacher records them on a table and a coordinate grid. If no one offers a point with an x value of -4, the teacher can ask students to consider one.

Teacher: *Will $(-4, 3)$ satisfy the statement?*

Student: *Oh yeah, x can be -4 because of the less than or equal to sign.*

Teacher: *Once again, did we show all the possible solutions?*

Student: *No, there are tons of points that will work. We need a solid line on $x = -4$ and shade in the left side of the graph.*

The teacher repeats this conversation with the next problem, $x > -4$, making note of the dashed line separating the solution set from the rest of the grid.

$x \leq -4$

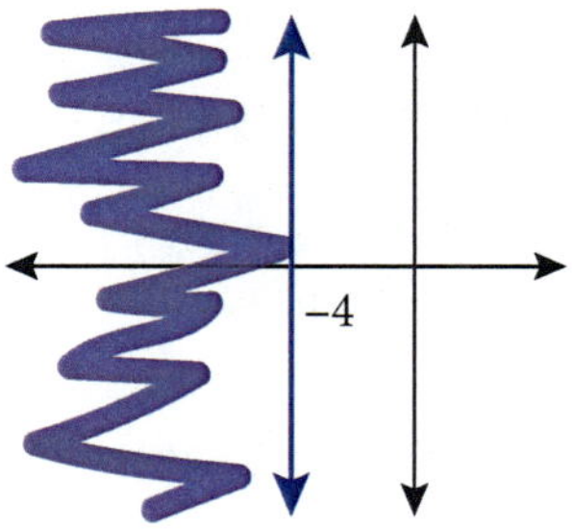

$x > -4$

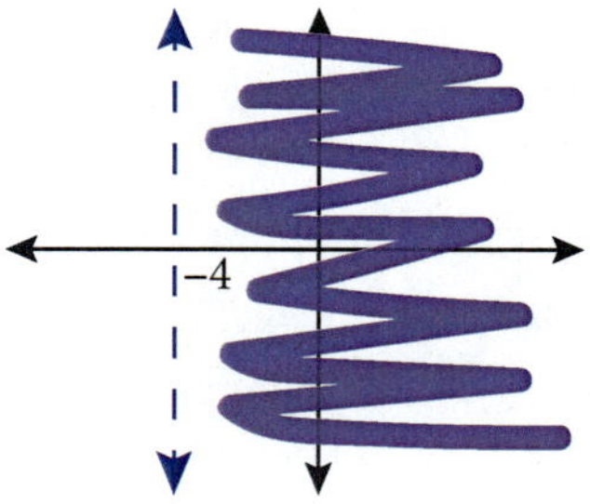

Teacher: *The next one is $2x - 4 = 0$. What values make this a true statement?*

Student: *I solved it and got $x = 2$.*

Teacher: *So we got just one answer here when before we were getting lots of solutions. Interesting.*

$2x - 4 = 0$

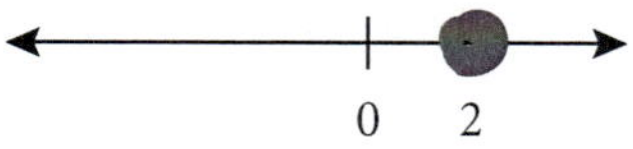

Teacher: *The next problem is $y = 2x - 4$. What changes from the previous? What values work for x and y here? I am going to use a table to record your ideas and put them on a graph.*

Students: *The y intercept is -4. If x is 1, then y is -2. (2, 0).*

Student: *There are going to be an infinite number of answers. Just draw a line to connect the points.*

Teacher: *So are you saying that any point on this line will be a solution to this function?*

Students: *Yeah, even the points in between the integers and on to infinity.*

Teacher: *How many solutions are there? And how is this different from the previous problem?*

Student: *There wasn't a y in the other one, so we just got one answer. But when there is an x and a y there are lots of combinations that work.*

Student: *And when it was in two dimensions or an inequality we got lots of points that worked.*

$y = 2x - 4$

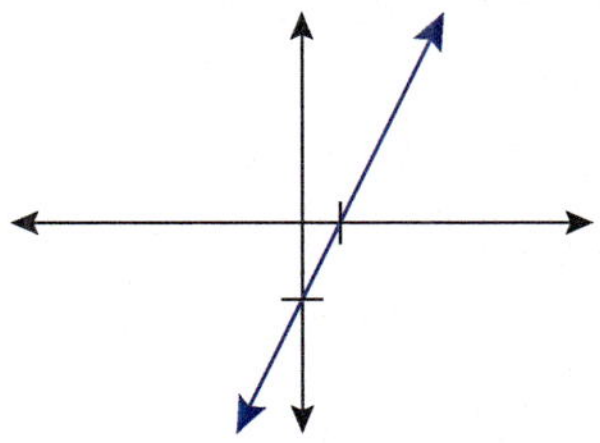

x	y
0	-4
1	-2
2	0

Teacher: *Next, tell us about the solution to $y < 2x - 4$. What does it look like?*

Student: *I graphed $y < 2x - 4$ to look like the line $y = 2x - 4$, but shaded below the line and used a dotted line.*

Teacher: *Someone else, explain why she chose a dotted line?*

Student: *Because it's not equal, it's just less than.*

Teacher: *How did you know whether to shade above or below? Turn and talk about this.*

Students talk.

Teacher: *What did you decide?*

Student: *We tried some points and the ones that work are all below the line.*

Student: *Since it says that y is less than the line, I graphed points underneath the line.*

Teacher: *Does that make sense?*

$y < 2x - 4$

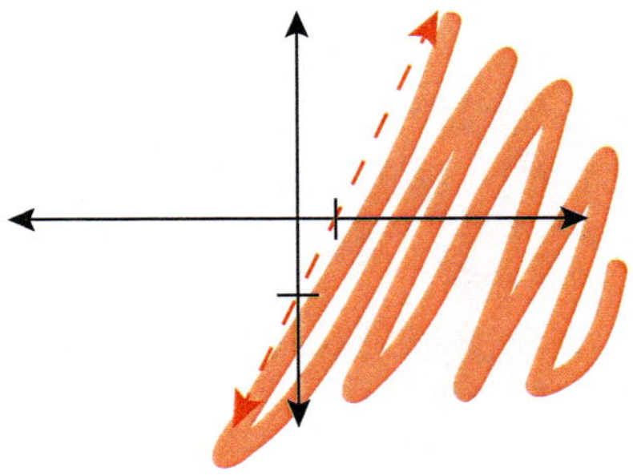

(continued)

Teacher: *What if it changes just in the tiniest bit, and there is a greater than or equal to symbol now?*

Students: *Now the line is not dotted, it's solid. And you shade the other half of the plane, the part above the line.*

Teacher: *So now we shade the other part of the plane? Great work.*

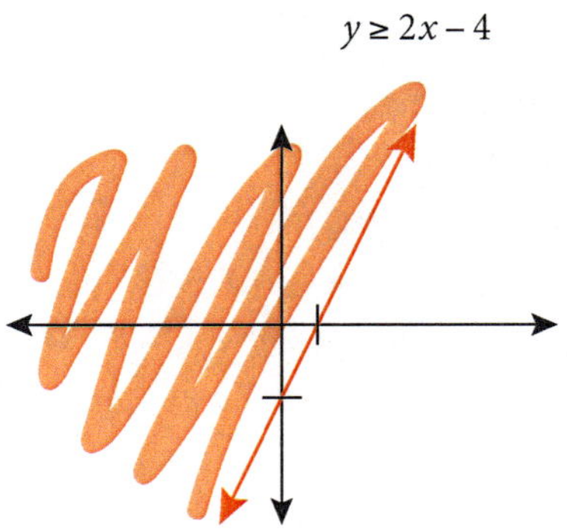

Teacher: *All right, the last problem is a system of two inequalities: $y > 2x - 4$ and $x > -4$. What points will work in both those inequalities? Will it be better to list our solutions in a table or represent them with a graph?*

Student: *There should be an infinite number of solutions, just like in the other inequalities, so listing them doesn't make sense. We should graph the solutions.*

Teacher: *Okay, work on this one and represent your solutions on a graph.*

Students work while the teacher circulates.

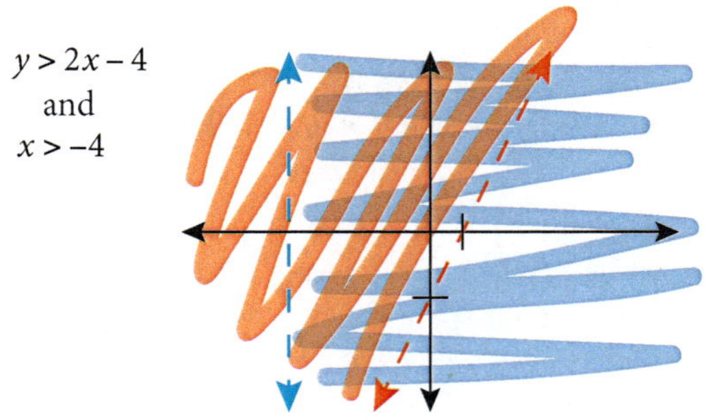

Teacher: *Tell us what your graph looks like.*

Student: *I graphed $y > 2x - 4$ to look like the line $y = 2x - 4$, but shaded in the left side and used a dotted line. Then I graphed $x = -4$, but with a dotted line and shaded in the right side.*

Teacher: *Okay, I've got that sketched up here. What are the solutions to the system? Can you name some points that work in both inequalities? How do you know they will work?*

Student: *I tested (0, 0) in both inequalities and it works in both of them.*

Teacher: *I am going to mark that point on our graph. What are some other points that are solutions?*

Student: *Well, (1, 1) is in that same overlap place. It will work in both inequalities. All those points in the overlap should work.*

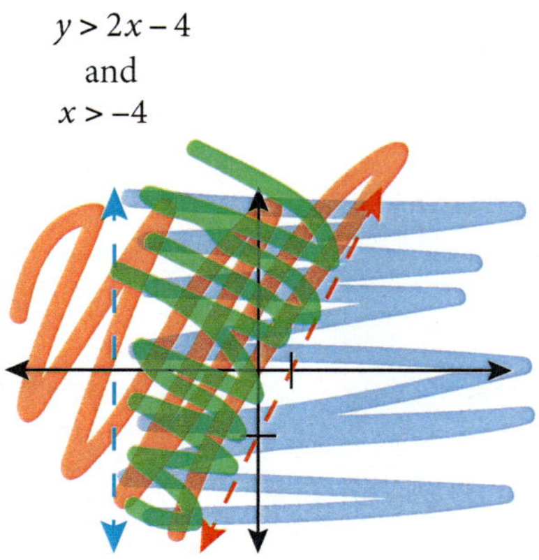

x	y
0	0
1	1

Teacher: *Create a table and list some points you think will work in both inequalities. When you have identified at least three, turn yo your partner and compare them.*

Students work and partners share.

Teacher: *What did you find?*

Student: *As long as our points were in the shaded region, they worked in both inequalities. That's kind of what the overlapping area means anyway.*

Teacher: *Nice insights. Great work today. Let's summarize what was important from today's discussion.*

Teacher: *How would you summarize some of the things that came up in this string today?*

Elicit the following:

- *Locations on a number line can represent values that make one-variable equations true.*
- *Rays represent values that make one-variable inequalities true.*
- *Lines represent values that make two-variable or two-dimensional equations true.*
- *Shaded areas (sections of planes) represent values that make two-variable or two-dimensional inequalities true.*
- *Overlapping shaded areas (sections of planes) represent values that make systems of inequalities true.*

Sample Final Display

Your display could look like this at the end of the problem string:

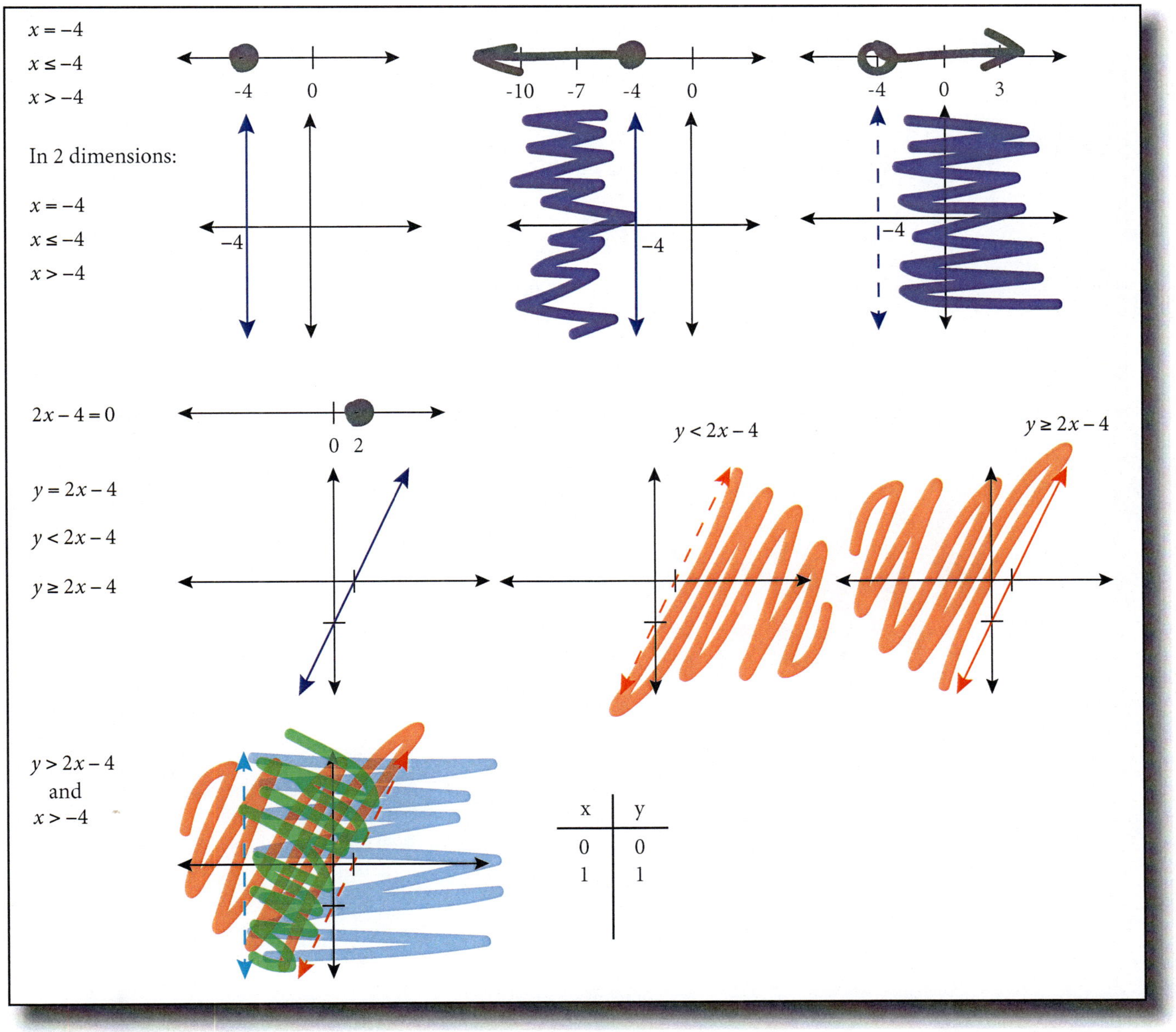

(continued)

Facilitation Notes

This version of the problem string lists short notes for important teacher moves during the string. After you've done the string yourself and studied the relationships involved, you might make similar notes for the things you want a reminder of or deem important.

$x = -4$ $x \leq -4$ $x > -4$	The first three are quick. Graph on number lines horizontally in a row. What would these look like? Rational numbers? Points vs. rays?
$x = -4$ $x \leq -4$ $x > -4$	Now, in 2 dimensions. If x is 4, what is y? What are some points that make it a true statement? Suggest non-solution point.
$2x - 4 = 0$	Number line again. Solution is a point on the number line.
$y = 2x - 4$	What changes? Graph, ask for solutions, table. Solution is the whole line.
$y < 2x - 4$	What changes? Solution is a part of the plane, does not include the line.
$y \geq 2x - 4$	What changes? Solution is a part of a plane and includes the line.
$y > 2x - 4$ $x > -4$	Overlaps, pair share some solutions. Solution is section of the plane.

PROBLEM STRINGS LESSON

3.1 Function Notation

At a Glance	
$(-2, 5)$ $f(2) = -3$ $f(5) = -9$ $f(1) =$ ___ $f($___$) = 0$ $f(x) =$ _________	**Objectives** The goal of this problem string is to help students learn the meaning and construct the use of function notation.
	Placement This problem string could be used to introduce or remind students about function notation. Students should have prior experience writing equations of lines to be successful with the last three problems of the string. You can use this problem string to begin textbook Chapter 3: Functions and Relations, before the chapter, during Lesson 3.1 Interpreting Graphs or right before Lesson 3.2 Function Notation.
	Guiding the Problem String The first three problems should go quickly. Listen for students who are noticing that the three points might be colinear. If no one says anything, ask what they think or comment about it yourself. Take longer to let students wrestle with the fourth and fifth problems. Refer back to the previous points if needed to help ground them in the function notation. Even though it may appear that the point of the string is to write the equation of the line that contains the points, the emphasis is on the meaning and use of function notation. Keep asking students to clarify what the function notation for each question means. The last question is meant to help students realize that function notation can represent a point, but it can also represent sets of points that follow the rule. If things like the vertical line test come up, address them quickly, but keep the emphasis on the meaning and use of function notation.
	About the Mathematics Function notation is a social construct. By convention, the community of mathematicians have chosen to write function notation using potentially confusing notation, with parentheses that otherwise mean a grouping symbol and multiplication. The use of and facility with function notation is logico-mathematical and must be constructed.

(continued)

Sample Interactions

Use the following as you plan how to elicit and model student strategies. This is not meant as a script, but as a view into the relationships involved and the intent of the problem string.

Teacher: *To get us started today, put your pencils down and get your brain ready to visualize. What are some ways that mathematicians might represent* $(-2, 5)$*?*

$(-2, 5)$

Student: *It's a point, so you could graph it.*

Student: *Put it in a table of x's and y's.*

The teacher graphs the point and puts it in a table. Since no one mentions function notation, the teacher does.

Teacher: *Does anyone remember anything about function notation? How could we represent this point using f and parentheses? Nothing? Does this help you remember? Mathematicians can represent the point* $(-2, 5)$ *by saying that for a function where the input, the x-value is* -2*, the function value or y-value is* 5*. We say this, "f of negative* 2 *is* 5 *or equals* 5*."*

$(-2, 5)$

$f(-2) = 5$

x	y
-2	5

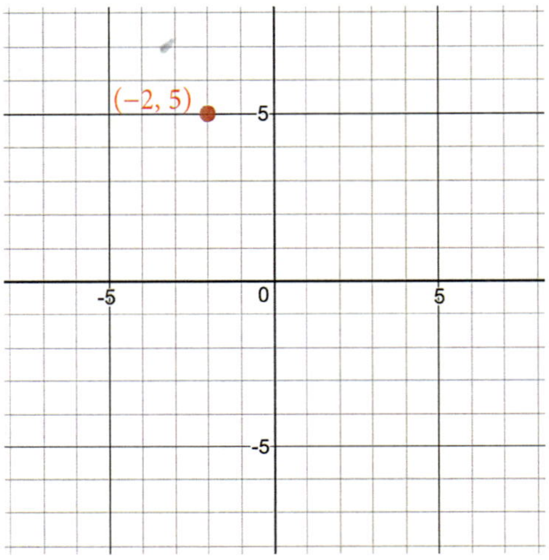

Teacher: *Let's go the other way. For the second problem, we have the function notation. So for the same function, f, f of* 2 *is* -3*. How might mathematicians represent that? Can you picture it?*

With student input, the teacher writes the point, puts it in the table, plots it on the graph, and repeats quickly for the next problem, $f(5) = -9$. The teacher listens for anyone who notices that the points might be colinear.

$f(2) = -3$

$f(5) = -9$

$(-2, 5)$ $f(-2) = 5$

$f(2) = -3$ $(2, -3)$

$f(5) = -9$ $(5, -9)$

x	y
−2	5
2	−3
5	−9

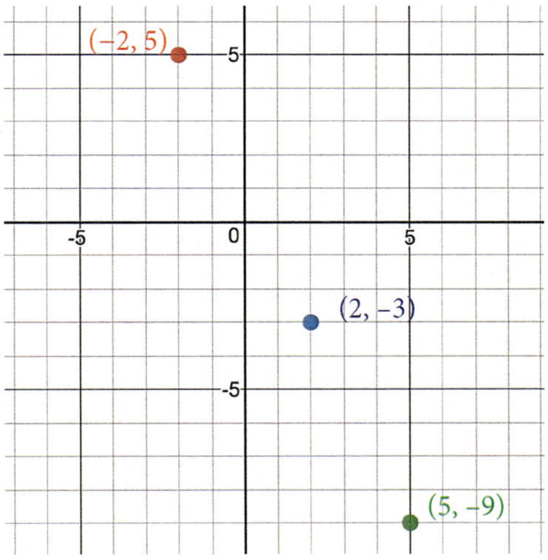

Teacher: *Those almost look linear, huh? Okay, next problem. For the same function, what do you think f of 1 is? If the input is 1, what is the output? If the x-value is 1, what is the y-value?*

$f(1) =$ ___

Student: *What do you mean?*

Teacher: *Yeah, good question. Can anyone help us here?*

Student: *So, I'm thinking that you've given us the x-value of a point, and we need to find the y-value.*

Student: *I think those points are on the same line.*

Teacher: *If they are, how would that help?*

Student: *We could find the point on the line with the x of 1.*

Student: *I think they are on the same line. When I kind of try to draw a line between them, they look like they are.*

Student: *They are on the same line because I looked at the slopes between the three points. The slope is −2.*

Teacher: *Can someone pick that up? Are the slopes all −2?*

(continued)

The teacher records students' thoughts as they talk about the ratios of the distances between the points and sketches the line between the points.

Teacher: *So, we think the points are colinear, all on this line. This line, this function, we have been calling f. I'm going to label it that. How does that help us find f(1)?*

Student: *Looking at the line, it seems like it's the point (1, −1).*

Teacher: *How can we be sure?*

Student: *Since we know the slope is −2, I think we can find the equation of that line. Then we can put in 1 to see what we get out.*

Teacher: *What is the equation of the line?*

Student: *I think the y-intercept is 1, so it would be the line y = −2x + 1.*

Student: *I agree. So now we just plug in x.*

Teacher: *This is getting busy, I am going to erase these blue lines.*

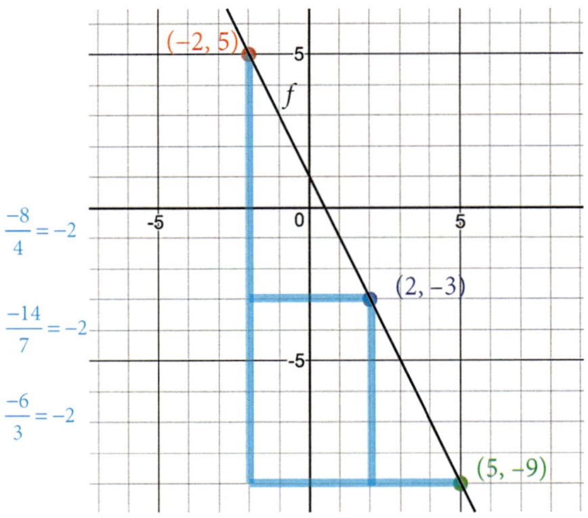

Teacher: *Okay, I'll write down the equation of the line you said. What is this function, this line, at 1? I will write that as f(1).*

Student: *That is −2 times 1 plus 1. So that's −2 and 1 is −1.*

Student: *Right, and the point is (1, −1).*

The teacher records the thinking with the function and plots the points (0, 1) and (1, −1).

(−2, 5) — $f(-2) = 5$

$f(2) = -3$ — (2, −3)

$f(5) = -9$ — (5, −9)

$f(1) = -1$ — (1, −1)

$y = -2x + 1$

(0, 1)

$f(1) = -2\,(1) + 1 = -2 + 1 = -1$

x	y
−2	5
2	−3
5	−9
1	−1

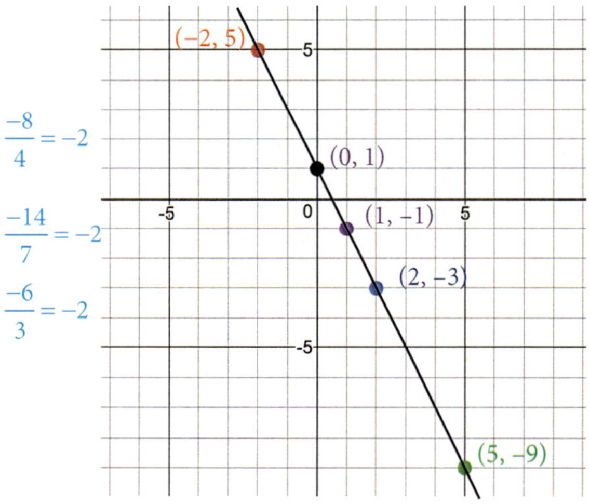

Teacher: *The next problem is kind of like the previous and kind of not. If I write f(___) = 0, what does that mean?*

Student: *Do we plug in 0?*

Student: *I don't think so. What you plug in goes inside the parentheses.*

Student: *Yeah, I think we need to find a y-value of 0.*

Teacher: *What about a y-value of 0?*

Student: *What is the x-value when the y-value is 0. It looks to me like it's in between 0 and 1.*

Teacher: *Right here? Nice. How could we get an exact answer?*

Student: *We know the slope is −2 but it seems like there must be an easier way.*

$f(\_\_\_) = 0$

Student: *Can we put the equation equals 0 and solve?*

Teacher: *Walk us through that and I'll write it up here.*

The student talks about solving for *x* as the teacher represents it on the board. They plot the point and write in the 0.5 in the function notation.

(−2, 5) $f(-2) = 5$

$f(2) = -3$ (2, −3)

$f(5) = -9$ (5, −9)

$f(1) = -1$ (1, −1)

$y = -2x + 1$

(0, 1)

$f(1) = -2\,(1) + 1 = -2 + 1 = -1$

$f(\underline{0.5}) = 0$ (0.5, 0)

$f(?) = -2x + 1 = 0$

$1 = 2x$

$x = 0.5$

x	*y*
−2	5
2	−3
5	−9
1	−1
0.5	0

$\frac{8}{4} = 2$

$\frac{14}{7} = 2$

$\frac{6}{3} = 2$

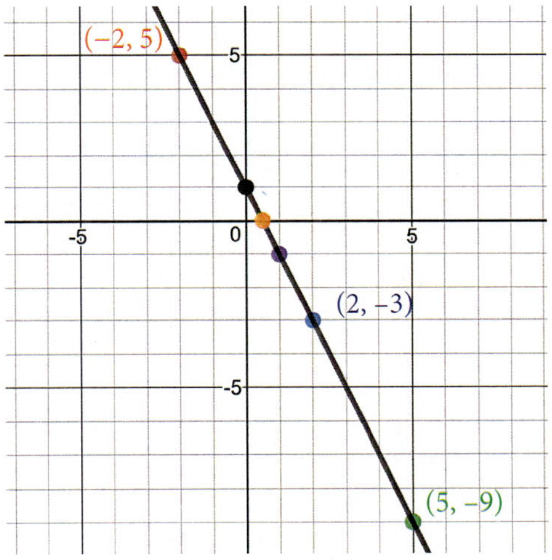

Teacher: *The last problem today is what is f of x? And what does that mean?*

Student: *I don't know. How do we plug an x in? What does that mean? It's not a number.*

Student: *I think it means that we do exactly that, plug an x in. Write the equation of the line.*

Student: *That line stands for all of the values of x and y that are on that line, so yeah, f(x) = −2x + 1.*

Teacher: *Turn and talk about what you are thinking about f of x.*

Students turn and talk.

$f(x) = \_\_\_\_\_\_\_\_\_\_$

(continued)

Teacher: *I heard lots of conversation about x's and y's and f. The function that we have been talking about this whole problem string is called f. And for that function, the y-value is −2 times any x-value plus 1. So, I will add that to the table and the graph.*

$(-2, 5)$ $f(-2) = 5$

$f(2) = -3$ $(2, -3)$

$f(5) = -9$ $(5, -9)$

$f(1) = -1$ $(1, -1)$

$y = -2x + 1$ $f(1) = -2(1) + 1 = -2 + 1 = -1$

$(0, 1)$

$f(\underline{0.5}) = 0$ $(0.5, 0)$ $f(?) = -2x + 1 = 0$

$1 = 2x$

$x = 0.5$

$f(x) = \underline{-2x + 1}$ $(x, -2x + 1)$

x	$y = -2x + 1$
−2	5
2	−3
5	−9
1	−1
0.5	0
x	$-2x + 1$

$\frac{-8}{4} = -2$

$\frac{-14}{7} = -2$

$\frac{-6}{3} = -2$

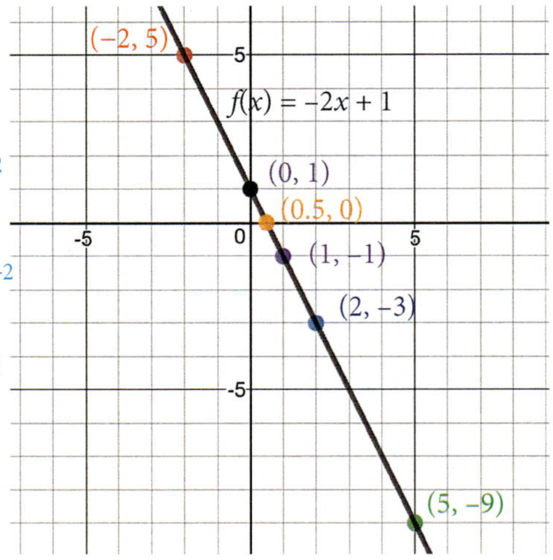

Teacher: *How would you summarize some of the things that came up in this string today?*

Elicit the following:

- *The name of this function is f.*
- *A function can have a lot of different points that have the same relationship.*
- *f(2) means the y-value when x is 2.*
- *When you write a point in function notation, it's f of the x-value equals the y-value f(x-value) = y-value.*
- *f(x) is another way of referring to y-values.*
- *When you write out the function, f(x) = −2x + 1, that represents all of the (x, y) points that work in the function.*

Sample Final Display

Your display could look like this at the end of the problem string:

(−2, 5)	$f(-2) = 5$		
$f(2) = -3$	(2, −3)		
$f(5) = -9$	(5, −9)		
$f(1) = -1$	(1, −1)		
	$y = -2x + 1$ (0, 1)	$f(1) = -2(1) + 1 = -2 + 1 = -1$	
$f(\underline{0.5}) = 0$	(0.5, 0)	$f(?) = -2x + 1 = 0$ $1 = 2x$ $x = 0.5$	
$f(x) = \underline{-2x + 1}$	$(x, -2x + 1)$		

x	$y = -2x + 1$
−2	5
2	−3
5	−9
1	−1
0.5	0
x	$-2x + 1$

$\frac{-8}{4} = -2$

$\frac{-14}{7} = -2$

$\frac{-6}{3} = -2$

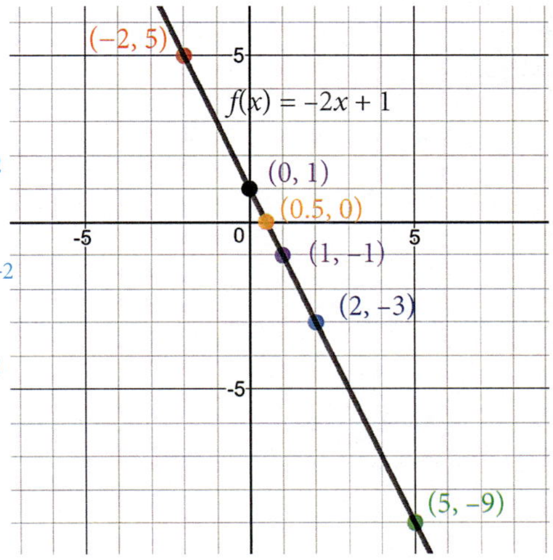

Facilitation Notes

This version of the problem string lists short notes for important teacher moves during the string. After you've done the string yourself and studied the relationships involved, you might make similar notes for the things you want a reminder of or deem important.

(−2, 5)	*What are some ways a mathematician might represent? Graph, table, function notation.*
$f(2) = -3$	*Other direction. How might a mathematican represent? Ordered pair, graph, table.*
$f(5) = 9$	*Repeat. Quick.* *Notice if anyone is thinking the points are colinear.* *If no one, notice aloud that the points look linear.*
$f(1) = \_\_\_$	*What does this mean?* *If the points are colinear, how does that help?* *Same slopes? Can you write the equation of the line?*
$f(\_\_\_) = 0$	*Similar, but different. What does this mean?* *How could we get an exact value? Use -2x+1 = 0, solve for x.*
$f(x) = \_\_\_\_\_\_$	*What does this mean? What are all the y-values if we use x for the independent values?*

PROBLEM STRINGS
LESSON

3.2 Combining Functions

At a Glance

$f(x) = 2x + 3$

$g(x) = 4$

$f(x) + g(x)$

$h(x) = -5$

$f(x) + h(x)$

$^*w(x) = -2x$

$^*f(x) + w(x)$

*optional problems

Objectives

The goal of this problem string is to give students experience reasoning with function notation when combining functions, where students connect algebraic, tabular, and graphic perspectives. This foreshadows transformations of functions.

Placement

This string can happen as students are learning to use function notation to label functions, to combine functions with arithmetic, and before students formally study transformations of functions. It can also be used to shore up understanding of vertically translating functions.

You can use this string during or after textbook Lesson 3.2 Function Notation.

Guiding the Problem String

The first two problems should go quickly, drawing on students' prior knowledge. Take longer with the third problem, probing students for understanding and connection between the representations. Refer to y-values of the function as "function values" so that students get used to the vocabulary. The fourth problem should again go quickly; take longer to work on the fifth problem which is a different combination problem. If you have time, the last two problems offer a look at combining two lines with a potentially surprising result, a horizontal line. Make sure students know that this is a special case—not all combinations of non-horizontal lines result in a horizontal line. Use the algebraic result, the addition of values in the table, and distances on the graph to substantiate the results.

About the Mathematics

The combination of linear functions using addition and subtraction underpin the larger concept of both vertical translations of functions (a function plus a constant function) and polynomials (the sum of power functions).

Vertical translations of functions, $f(x) + k$, can be thought of as the sum of a function and a constant function, $f(x) + g(x)$ where $g(x) = k$.

Polynomials, $p(x) = a_n x^n + a_{n-1} x^{n-1} + \ldots + a_0$ can be thought of as the sum of power functions. A power function is a single term polynomial, $f(x) = a_m x^m$. This understanding can be helpful when determining the long run behavior of polynomials. In this string, this can be seen as the long run behavior of the non-horizontal lines dominating the long run behavior of the constant function, resulting in a non-horizontal line that has been translated. In more general polynomials, the long run behavior of the term of highest degree dominates the long run behavior of the entire polynomial.

Sample Interactions

Use the following as you plan how to elicit and model student strategies. This is not meant as a script, but as a view into the relationships involved and the intent of the problem string.

<table>
<tr>
<td>Teacher: So, we've just looked at labeling functions like f(x), g(x), h(x), where x is the variable used in the function. Let's work with that notation some more. The first problem in today's string is to tell me everything you can about the function $f(x) = 2x + 3$. Pick three important things.</td>
<td>$f(x) = 2x + 3$</td>
</tr>
<tr>
<td>The teacher circulates, looking for students to share: y-intercept, rate of change, graph, and table.
Teacher: I noticed some of you sketched a graph. Tell us about your graph, please.
Student: The y-intercept is 3 and the rate is 2, so start at 3.
Teacher: And when you say start at 3, you mean when x is 0, the y is 3?
Student: Yes, and then go over 1 and up 2 because the rate is 2.
Teacher: So according to that, the function also contains the point (1, 5). I'll mark that.
Student: Then draw the line between those points.
Teacher: Okay, I'll draw the line that contains those points. We have often written the equation for this line with y = , but today let's write it with f(x).</td>
<td>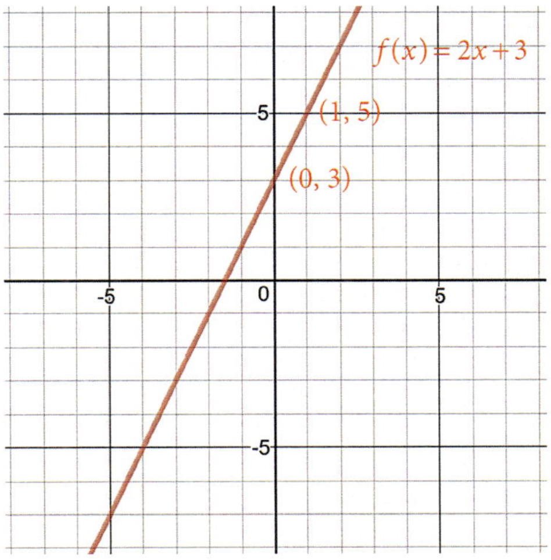
</td>
</tr>
<tr>
<td>Teacher: That might help us remember that we use f(x) to represent the y-values of the function that is written with the variable x. And some of you had these values in a table, but you had at least one more point, right?
Student: Yeah, I had (−1, 1) too.</td>
<td>$f(x) = 2x + 3$
<table><tr><th>x</th><th>y = f(x)</th></tr><tr><td>−1</td><td>1</td></tr><tr><td>0</td><td>3</td></tr><tr><td>1</td><td>5</td></tr></table></td>
</tr>
<tr>
<td>Teacher: The next problem in our string is $g(x) = 4$. Tell me what you know about this function.</td>
<td>$g(x) = 4$</td>
</tr>
</table>

(continued)

Student: *It's a horizontal line, y = 4. Whatever the x-value is, the y-value is always 4.*

Teacher: *Let's get all of that up here.*

$f(x) = 2x + 3$

$g(x) = 4$

x	$y = f(x)$	$y = g(x)$
−1	1	4
0	3	4
1	5	4

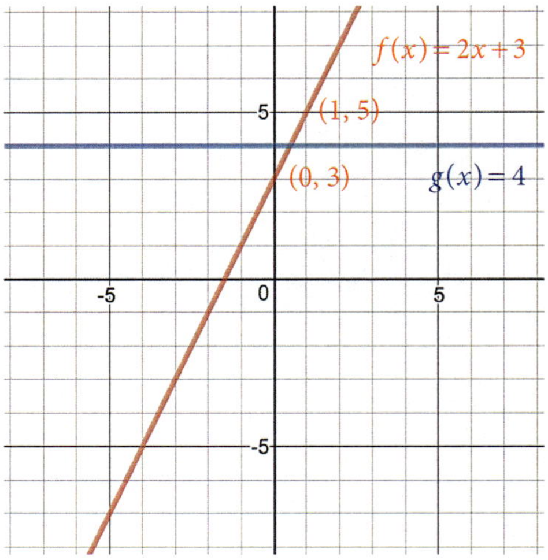

Teacher: *And our next problem is $f(x) + g(x)$, what can you tell me about that? What do you think it means?*

$f(x) + g(x)$

Student: *In the table, you just add the y-values together. Five, seven, nine.*

x	$y = f(x)$	$y = g(x)$	$y = f(x) + g(x)$
−1	1	4	5
0	3	4	7
1	5	4	9

Student: *You can add the equations together. That would be 2x plus 3 plus 4, so 2x plus 7.*

$f(x) + g(x) = 2x + 3 + 4 = 2x + 7$

Teacher: *I wonder how we can make sense of this with the graphs?*

Student: *What do you mean?*

Student: *I think it means that the y-values of the graph are related.*

Teacher: *Can you add the y-values of the red graph to the y-values of the blue graph? What do you get?*

Pause.

Teacher: *Let's take a point on the red line. Choose one.*

Student: *How about (−2, −1)?*

Teacher: *Great. Now what happens if I add the y-value of that point, −1 to the y-value of the blue function which is always 4.*

Student: *You get 3.*

Teacher: *Let's plot that point, (−2, 3). Let's do that again for a couple more points.*

The teacher repeats adding 4 to the *y*-values for two more points.

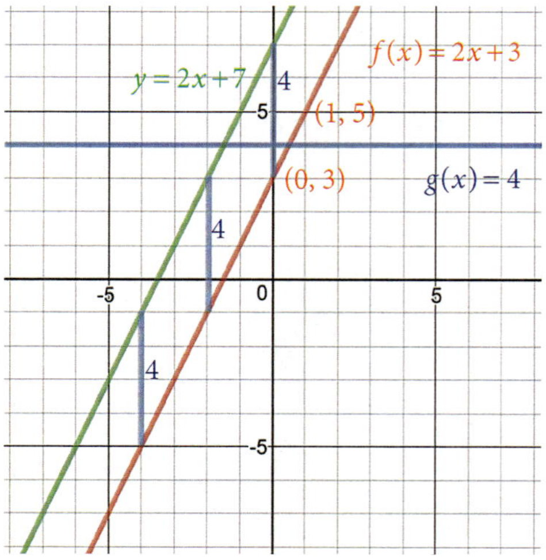

Teacher: *What happens if we keep adding 4 to every y-value of every point on the red line?*

Student: *You get a new line, the red line shifted up 4.*

Teacher: *How do you know it will be a line?*

Student: *If you shift every point up 4, you haven't changed the shape.*

Student: *And also, when we added the functions together, we got a line,* $y = 2x + 7$.

Teacher: *Nice connection! Cool, so a line plus a constant function results in another line.*

$f(x) = 2x + 3$

$g(x) = 4$

x	$y = f(x)$	$y = g(x)$	$y = f(x) + g(x)$
−1	1	4	5
0	3	4	7
1	5	4	9

$f(x) + g(x) \quad = 2x + 3 + 4 = 2x + 7$

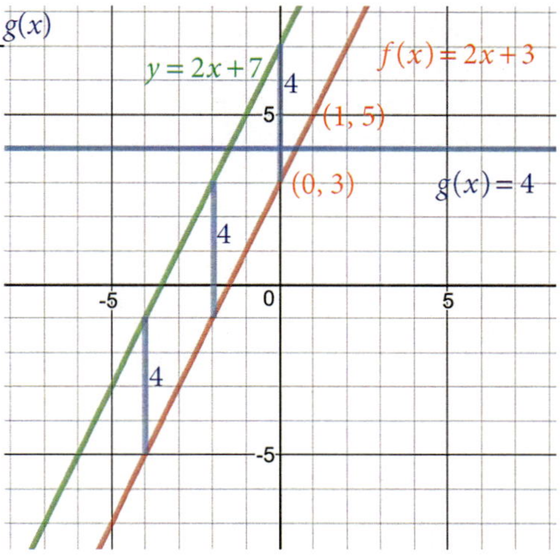

Teacher: *I'm going to erase everything except the red f(x) function. The next question is* $h(x) = -5$. *Tell me all about it.*

$h(x) = -5$

The teacher records students' insights and then asks the next question, $f(x) + h(x)$. The teacher represents the students' thinking about the combination function.

$f(x) = 2x + 3$

$g(x) = 4$

x	$y = f(x)$	$y = h(x)$	$y = f(x) + h(x)$
−1	1	−5	−4
0	3	−5	−2
1	5	−5	0

$f(x) + g(x) \quad = 2x + 3 + 4 = 2x + 7$

$h(x) = -5$

$f(x) + h(x) \quad = 2x + 3 + (-5) = 2x - 2$

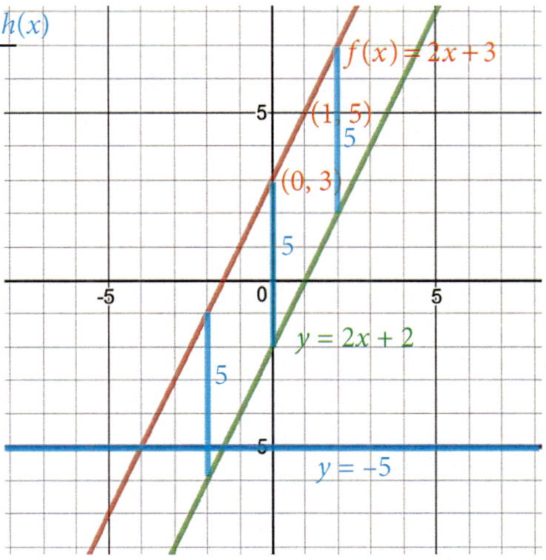

(continued)

Teacher: *Can someone talk about what happened when we added a constant function to a line and what happened when we added a negative constant function, or subtracted a constant function, from a line?*

Student: *When we added the horizontal line, the whole line went up. When we added the negative horizontal line, the whole line went down.*

Student: *It's like every y-value went up or down, depending on whether you added to it or subtracted from it.*

Teacher: *Was this more about the x-values or the y-values?*

Student: *Kind of both. We looked at what was happening at x-values, but we were adding and subtracting y-values.*

Teacher: *That is one reason why you will hear mathematicians refer to the y-values of a function as the function values. They are the value of a function at a certain x-value. So when we are adding functions, we are adding y-values.*

If time permits, the last two problems could look like the following.

$f(x)=2x+3$

$g(x)=4$

$f(x)+g(x) \quad =2x+3+4=2x+7$

$h(x)=-5$

$f(x)+h(x) \quad =2x+3+(-5)=2x-2$

$w(x)=-2x$

$f(x)+w(x) \quad =2x+3+(-2x)=3$

x	$y=f(x)$	$y=w(x)$	$y=f(x)+w(x)$
−1	1	2	3
0	3	0	3
1	5	−2	3

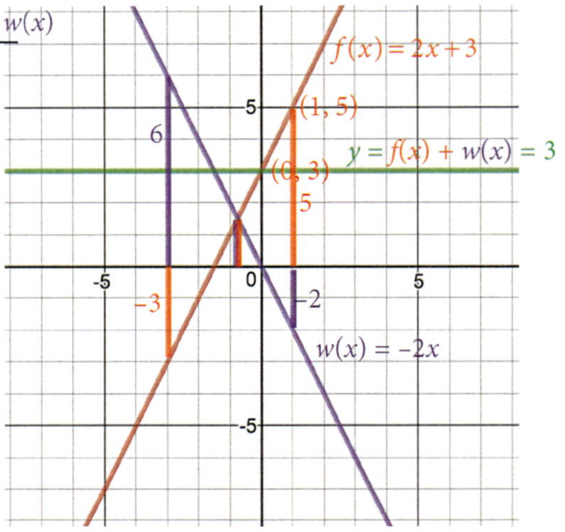

Teacher: *How would you summarize some of the things that came up in this string today?*

Elicit the following:

- *Function notation is a way to think about the y-values of a function.*
- *You can combine functions, algebraically, in a table, and on a graph.*
- *When you combine functions, you think about combining the y-values at x-values.*
- *A line added to a constant function results in a translation of the original line.*

Sample Final Display

Your display could look like this at the end of the problem string:

$f(x)=2x+3$

$g(x)=4$

x	$y=f(x)$	$y=w(x)$	$y=f(x)+w(x)$
-1	1	2	3
0	3	0	3
1	5	-2	3

$f(x)+g(x) \quad =2x+3+4=2x+7$

$h(x)=-5$

$f(x)+h(x) \quad =2x+3+(-5)=2x-2$

$w(x)=-2x$

$f(x)+w(x) \quad =2x+3+(-2x)=3$

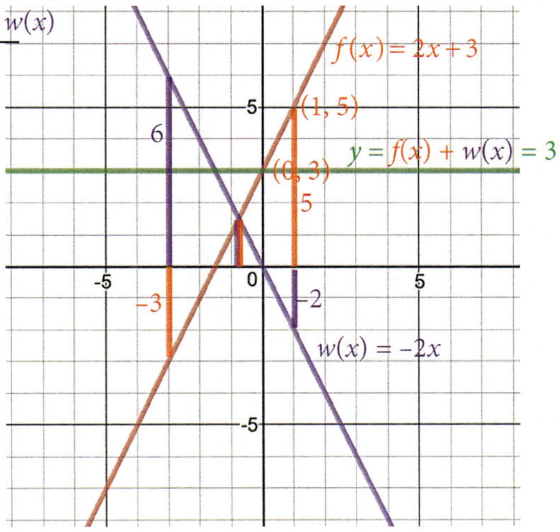

Facilitation Notes

This version of the problem string lists short notes for important teacher moves during the string. After you've done the string yourself and studied the relationships involved, you might make similar notes for the things you want a reminder of or deem important.

$f(x)=2x+3$	We've been learning about functions. Tell me at least 3 important things about this function. Elicit y-intercept, rate, graph, table. Quick.
$g(x)=4$	Very quick. Use color. Graph, table.
$f(x)+g(x)$	What can you tell me about this? Linger. Make sense of table, equations, graph.
$h(x)=-5$	Erase all but first function on table and graph. Very quick. Graph, table.
$f(x)+h(x)$	Linger. Make sense of table, equations, graph. Generalize a line added to a constant function.
*$w(x)=-2x$	Optional. Erase all but first function. Elicit y-intercept, rate, graph, table.
*$f(x)+w(x)$	Linger. Make sense of table, equations, graph.

(continued)

PROBLEM STRINGS
LESSON

3.4 Translations and the Quadratic Family

At a Glance

$y = x^2$

$y = (x - 200)^2$

$y = x^2 - 750$

$y = (x + 1300)^2$

$y = x^2 + 24{,}000$

Objectives

The goal of this problem string is to develop students' facility with translations of functions. To do this students find appropriate viewing windows for translations of the parent function $y = x^2$. Using the power of technology, students can quickly test their assumptions by trying and adjusting. As they are pressed to defend their window choices, students build skill using translations with quadratic functions.

Placement

This is the first in a series of four problem strings that use finding appropriate viewing windows as a vehicle to build and strengthen students' understanding and facility with parent functions and transformations. You could use this problem string as students are learning about vertical and horizontal translations.

This problem string could come during the work of textbook Lesson 3.4 Translations and the Quadratic Family.

Guiding the Problem String

The first problem is intended to ground students in the behavior of the parent function $y = x^2$ and to establish some parameters for appropriate viewing windows. If students are very familiar with quadratic functions, focus this question on what it means to find a good viewing window. The rest of the problems are designed to be well outside of the beginning window for most graphing software. Allow students to swipe or zoom, but encourage them to predict before they just start guessing. Then press them for justification when they do find the function.

- *Why is the function there?*
- *How can you use your knowledge of transformations to defend your choice?*

When you gather the students and collectively find an appropriate window, use the window setting feature and ask students to justify each choice based on transformations.

About the Mathematics

There are infinite possibilities for appropriate viewing windows for each function. It is not important that students find the same windows. It is important that the windows are appropriate (showing the important features without much extra space). It is most important that students begin to use their knowledge of transformations to guide their thinking and justify their choices.

The graphs of these functions represent continuous, infinite relationships. As you work with transformations, do not refer to them as the letter *u* or use language that would suggest that the transformations are moving around a static shape. The parent parabola is not a static shape that eventually goes vertical, but an infinite set of points that keeps increasing as *x* increases and symmetrically increases as *x* decreases. In this string, each of those points are being translated, creating a transformed, continuous function.

Sample Interactions

Use the following as you plan how to elicit and model student strategies. This is not meant as a script, but as a view into the relationships involved and the intent of the problem string.

Teacher: *To start off today, let's visualize the graph of the parent function* $y = x^2$*. Turn and talk about what you picture in your mind.*

Students turn and talk while the teacher listens in.

Teacher: *I'm going to graph it with our display grapher. Is this a good viewing window for this function?*

Student: *Sure, you can see the whole function.*

Student: *Well, not the whole function since it goes on and on forever, but you get the gist of it.*

Teacher: *What do you mean the gist of it?*

Student: *You can see the important parts.*

Teacher: *What are the important parts of a quadratic function?*

Student: *The shape, the x-intercepts, the y-intercepts, and the vertex.*

Teacher: *And we can see all of that so we think it's a good viewing window, or an appropriate viewing window. Good can be subjective. Can we agree that a good viewing window shows the important parts of a function?*

Student: *We don't really need all of that blank area at the bottom.*

$y = x^2$

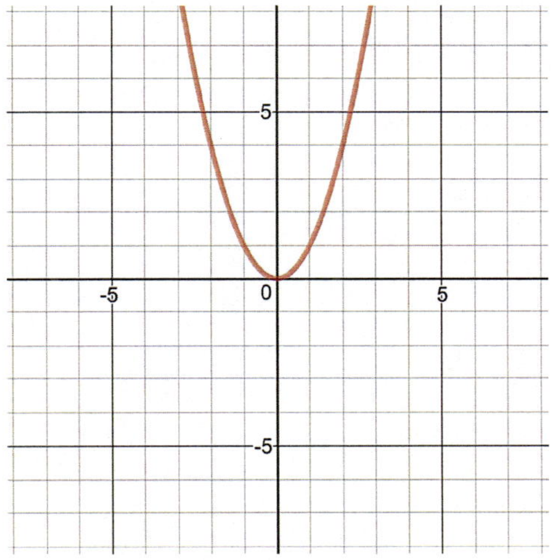

Teacher: *Is anyone bothered by all of the extra space in the bottom of the graph?*

Student: *You could get rid of that and fill the window more with it so you could see more of the function.*

Teacher: *How?*

Student: *Make the y's start at 0 or maybe a little lower, like* -1.

Student: *Now it feels like we could see less of the x's so that it would spread out a bit.*

Student: *What do you mean?*

Student: *There's too much space on the sides. It could spread out a bit.*

Teacher: *How would you do that?*

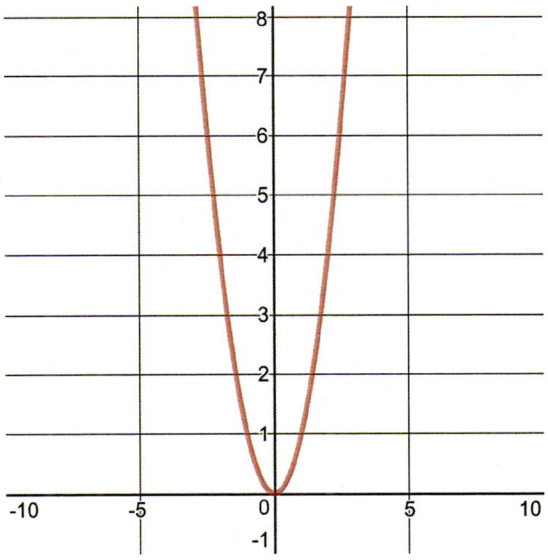

(continued)

Student: *Right now, we only see the parabola in between −4 and 4. You could change the x's to −4 and 4.*

Student: *Yes, that looks better.*

Teacher: *Why do you say that it looks better?*

Student: *We got rid of the extra and the parabola looks good.*

Teacher: *Could we have other windows that we could call good? So, let's just say that there could be lots of good windows and when we find a window, it just needs to be good enough, showing the important parts, but not a lot of empty space.*

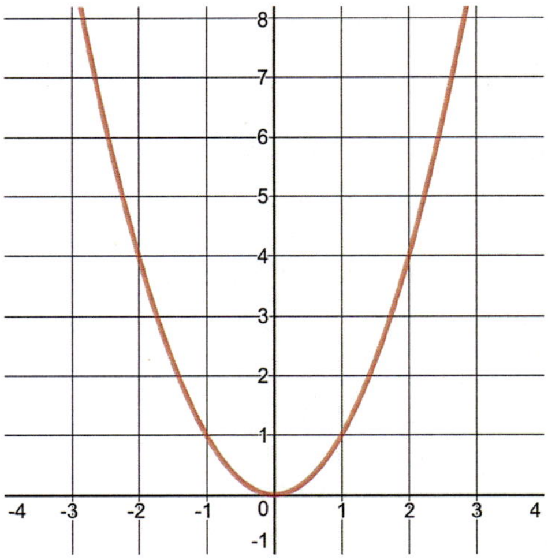

Teacher: *Okay, the next problem in today's string is to find a good viewing window for* $y = (x - 200)^2$. *You can use your devices to find it, but try to reason about it, rather than just randomly guessing.*

The teacher circulates and looks for students who are reasoning and students who are swiping or zooming in and out.

If the students do not have devices on which to graph, ask students to visualize where they could find the graph. Then work together as a class to find an appropriate window.

Teacher: *Who thinks they've found a good window?*

Student: *I think mine's fine. It's just over to the right so I can see 200.*

Teacher: *How did you find it?*

Student: *I just swiped over and there it was.*

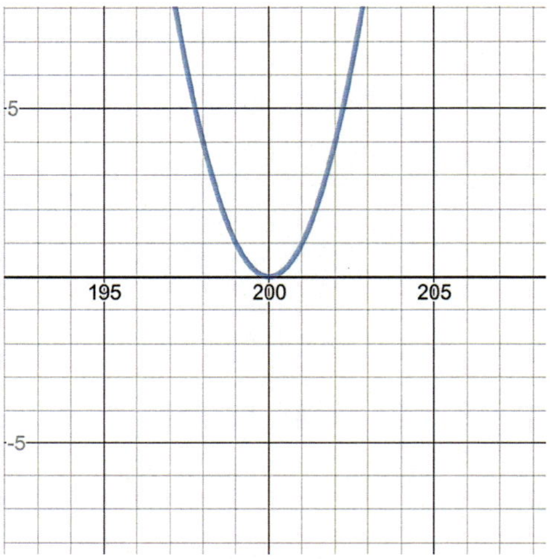

Teacher: *So, with a bit of luck, you ran into it? Can we make sense of why we can see the function in this window?*

Student: *You've got x − 200 in the parentheses.*

Teacher: *Right, we've replaced x with x − 200.*

Student: *Yeah, so the function has to be 200 to the right of where the parent function was. But, I would lose the bottom part again. We don't need that extra space.*

Student: *And while you're at it, close in on the x's too. You've got too much horizontal space, just use from 195 to 205.*

Teacher: *Like this? What do you all think? I see lots of nodding. Why look to the right?*

Student: *When you replace x with x – 200, the graph shifts right.*

Teacher: *Let's make a note of that.*

$y = x^2$

$y = (x - 200)^2$ Look right!

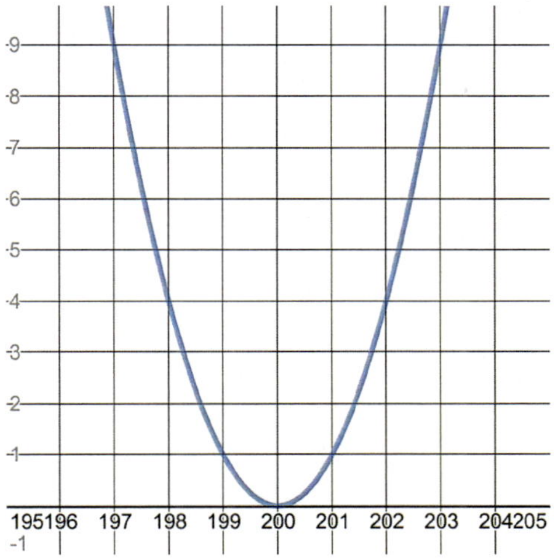

Teacher: *For this next problem, I want you to think and predict first, before you just start swiping or zooming. The next problem is to find an appropriate viewing window for* $y = x^2 - 750$ *in as few moves as possible. Be ready to explain why you are looking where you are looking.*

The teacher circulates, asking students to defend their choices or justify a window if they just happen on it by swiping.

Teacher: *When you are ready, turn and share your window and defend your choices.*

Students turn and talk and the teacher circulates.

(continued)

Teacher: *Let's come back together as a group and find a viewing window that we can agree is good enough. I noticed that you two were swiping and then you gave up on that. Would you tell us about that please?*

Student: *We thought it might be down but after swiping down we still didn't see it, so we finally just changed the window by putting in numbers.*

Teacher: *What window did you decide on?*

The teacher starts with their window and then with student input finds a suitable window.

Teacher: *What could we make a note of here, with this function and why?*

Student: *Look down! Because you have the original x^2 but down 750.*

$y = x^2$

$y = (x - 200)^2$ Look right!

$y = x^2 - 750$ Look down!

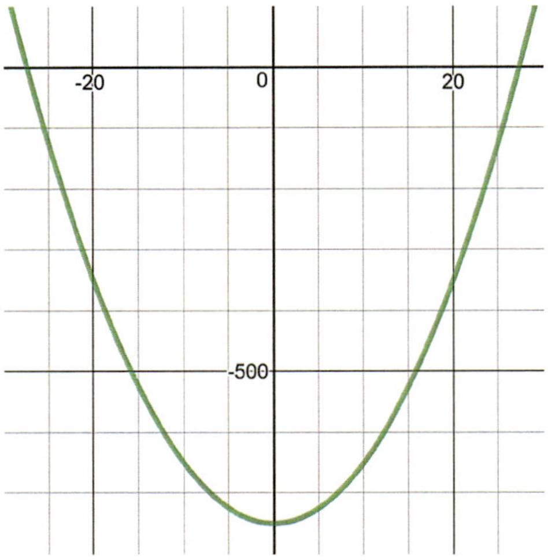

This is only one possible appropriate viewing window for this function. It is less important that students find the same windows. It is important that the windows are appropriate (showing the important features with not a lot of extra space.)

Teacher: *For the next problem, I want you to work with your partner. One of you uses the grapher. The other partner makes suggestions. You have to agree before the graphing partner can input the ideas. So, there should be lots of discussing and agreeing before the partner making the changes actually makes any changes. Got it? Find an appropriate viewing window for $y = (x + 1300)^2$. Remember, talk and agree first. Then make the changes.*

The teacher circulates as students work, encouraging good partner work and pressing for justification. Then the teacher calls the class together and they decide on an appropriate viewing window such as the following. The class suggests adding the note to “Look left!”

$y = x^2$

$y = (x - 200)^2$ Look right!

$y = x^2 - 750$ Look down!

$y = (x + 1300)^2$ Look left!

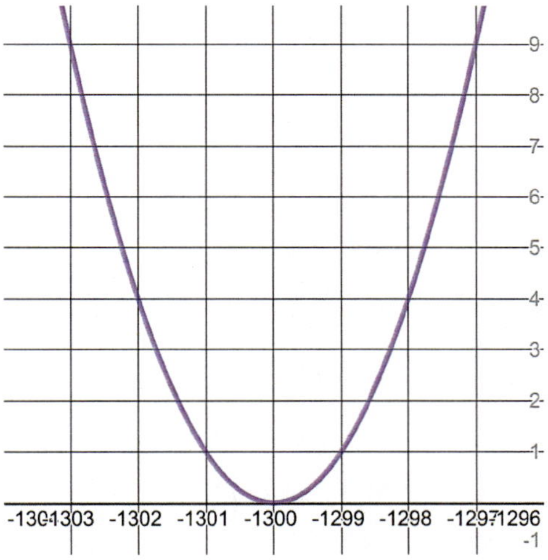

Teacher: *The last problem today is to find a suitable viewing window for* $y = x^2 + 24{,}000$. *What’s going on with this function? How can that help you find a good viewing window? This time, change who is holding the grapher. Remember, you have to agree before that person makes any changes. Go!*

See the Sample Final Display for a possible viewing window and notes for this problem.

Teacher: *How would you summarize some of the things that came up in this string today?*

Elicit the following:

- *Translations really do shift functions around! Using our knowledge of translations, we can find the function that has been translated.*
- *If you add to or subtract from the whole function, the y-values change.*
- *If you replace x with x plus or minus something, the function shifts left or right.*
- *There can be lots of different appropriate viewing windows for the same function, but they should all include the important features of the function and not extra space.*

(continued)

Sample Final Display

Your display could look like this at the end of the problem string:

$y = x^2$	
$y = (x - 200)^2$	Look right!
$y = x^2 - 750$	Look down!
$y = (x + 1300)^2$	Look left!
$y = x^2 + 24{,}000$	Look up!

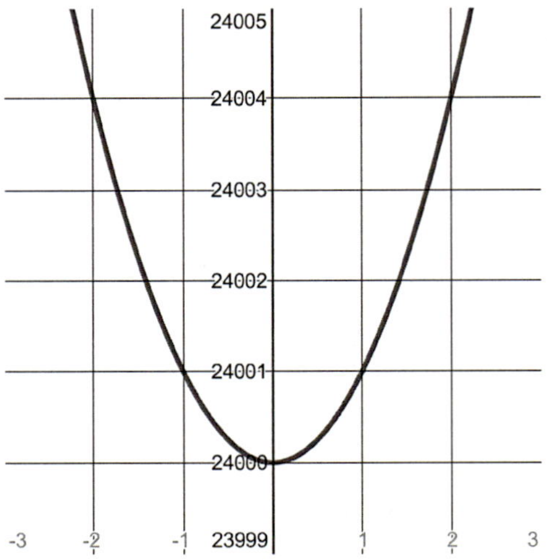

Facilitation Notes

This version of the problem string lists short notes for important teacher moves during the string. After you've done the string yourself and studied the relationships involved, you might make similar notes for the things you want a reminder of or deem important.

$y = x^2$	*Visualize: what does the graph look like?* *Display with display grapher.* *What makes a good, appropriate viewing window?* *Shows important features, no extra space.*
$y = (x - 200)^2$	*Find a good viewing window.* *Work together as a class. "Replace x with x-200".* *Make a note "Look right".*
$y = x^2 - 750$	*Predict! Find a good viewing window in as few moves as possible.* *Turn and defend choices.* *Find an appropriate viewing window as a class.* *Press for justification. Note "Look down".*
$y = (x + 1300)^2$	*Partner up. One holds grapher. Only make changes when both agree.* *Talk, defend, change, adjust.* *Find an appropriate viewing window as a class.* *Press for justification. Note "Look left".*
$y = x^2 + 24{,}000$	*Repeat.* *What do we know about translations?*

PROBLEM STRINGS
LESSON

3.5 Reflections and the Square Root Family

At a Glance

$y = \sqrt{x}$

$y = -\sqrt{x}$

$y = \sqrt{-x}$

$y = -\sqrt{-x}$

$y = -\sqrt{x - 15}$

$y = -\sqrt{x + 300}$

$y = \sqrt{-x} - 15$

$y = \sqrt{-x} + 300$

*$y = \sqrt{-x + 25}$

*optional problem

Objectives

The goal of this problem string is to develop students' facility with reflections of functions. To do this students find appropriate viewing windows for reflections of the parent function $y = \sqrt{x}$. Using the power of technology, students can quickly test their assumptions by trying and adjusting. As they are pressed to defend their window choices, students build skill using reflections with square root functions.

Placement

This is the second in a series of four problem strings that use finding appropriate viewing windows as a vehicle to build and strengthen students' understanding and facility with parent functions and transformations. You could use this problem string as students are learning about reflections.

This problem string could come during the work of textbook Lesson 3.5 Reflections and the Square Root Family.

Guiding the Problem String

Use the first question to establish the behavior of the parent function $y = \sqrt{x}$. The next three problems play with reflecting the square root function over the y-axis, the x-axis and then both axes. These should go fairly quickly, but take the time to watch a few points being reflected to reinforce that you are transforming a whole set of relationships, not a static shape. The remaining problems combine translations and reflections with the square root function. You can change up the rhythm of the problem string by asking students to predict first, having students work with a partner where they must agree before changing anything, having students work individually and then comparing windows with a partner, or finding an appropriate window together as a class. The emphasis should always be on pressing for justification of the viewing window based on the transformations of the parent function.

See problem string 3.4 for a sample dialogue of a problem string with a similar structure.

About the Mathematics

The graphs of these functions represent continuous, infinite relationships. As you work with transformations, do not refer to the functions using language that would suggest that the transformations are moving around a static shape. The parent square root function is not a static shape that eventually goes horizontal, but an infinite set of points that keeps increasing as x increases. In this string, each of those points are being reflected across an axis, creating a transformed, continuous function.

(continued)

Important Questions

Use the following as you plan how to elicit and model student strategies.

- *What is the general behavior of* $y = \sqrt{x}$ *?*
- *What happens to the y-values of f(x) in the transformation −f(x)?*
- *What happens to the function f(x) when you replace x with −x, f(−x)?*
- *What changes, the x- or y-values, when you reflect a function over the x-axis? How can we represent that using function notation?*
- *What changes, the x- or y-values, when you reflect a function over the y-axis? How can we represent that using function notation?*

Teacher: *How would you summarize some of the things that came up in this string today?*
Elicit the following:

- *Reflections are transformations where you reflect the function over a line.*
- *When a function f(x) reflects over the x-axis, all of the y-values take on the opposite sign. The y-values that were once positive are now negative. The y-values that were once negative are now positive. This is represented by −f(x).*
- *When a function f(x) is reflected over the y-axis, all of the y-values that used to correspond to x, now correspond to −x. This is represented by f(−x).*
- *We can combined reflections and translations. It makes sense to reflect over the axis first then shift.*

Sample Final Display

Your display could look like this at the end of the problem string:

$y = \sqrt{x}$ 1st quadrant

$y = -\sqrt{x}$ 4th quadrant

$y = \sqrt{-x}$ 2nd quadrant

$y = -\sqrt{-x}$ 3rd quadrant

$y = -\sqrt{x - 15}$ reflection over x, look right

$y = -\sqrt{x + 300}$ reflection over x, look left

$y = \sqrt{-x} - 15$ reflection over y, look down

$y = \sqrt{-x} + 300$ reflection over y, look up

$y = \sqrt{-x + 25}$ $= \sqrt{-(x - 25)}$ reflection over y, look right

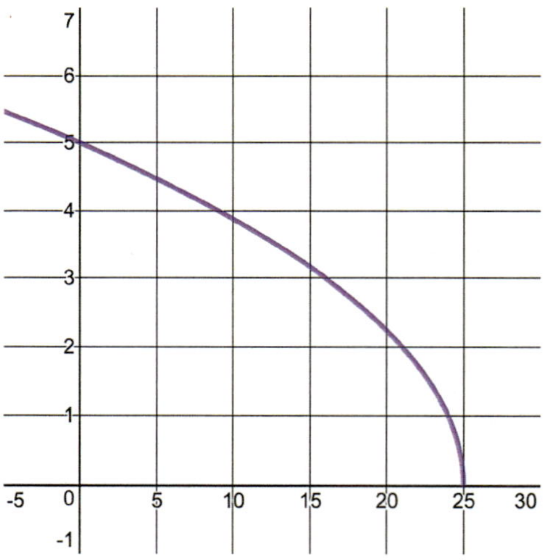

Facilitation Notes

This version of the problem string lists short notes for important teacher moves during the string. After you've done the string yourself and studied the relationships involved, you might make similar notes for the things you want a reminder of or deem important.

$y=\sqrt{x}$	Visualize: what does the graph look like? Display with display grapher. Review appropriate viewing windows.
$y=-\sqrt{x}$	Find a good viewing window. Note "4th quadrant".
$y=\sqrt{-x}$	Replace x with -x, the opposite of x. Repeat. What do we know about reflections? Note "2nd quadrant".
$y=-\sqrt{-x}$	Predict! How can you use what you know? Note "3rd quadrant".
$y=-\sqrt{x-15}$	Partner up. One holds grapher. Only make changes when both agree. Talk, defend, change, adjust. Find an appropriate viewing window as a class. Press for justification. Note "Reflection over x, look right."
$y=-\sqrt{x+300}$	Repeat. Note "Reflection over x, look left."
$y=\sqrt{-x}-15$	Repeat. Note "Reflection over y, look down."
$y=\sqrt{-x}+300$	Repeat. Note "Reflection over y, look up."
$y=\sqrt{-x+25}$	Why is the graph there? What is going on? Factor -1. Note "Reflection over y, look right."

PROBLEM STRINGS LESSON 3.6

Dilations and the Absolute-Value Family

At a Glance

$y = |x|$

$y = 0.01|x|$

$y = 50|x|$

$y = -|x|$

$y = |-x|$

$y = 2|x+15| - 60$

$y = -\frac{1}{2}|x-28| + 1$

$y = 10|x+12| + 300$

Objectives

The goal of this problem string is to develop students' facility with dilations of functions. To do this students find appropriate viewing windows for vertical stretches and compressions of the parent function $y = |x|$. Using the power of technology, students can quickly test their assumptions by trying and adjusting. As they are pressed to defend their window choices, students build skill using dilations with absolute-value functions.

Placement

This is the third in a series of four problem strings that use finding appropriate viewing windows as a vehicle to build and strengthen students' understanding and facility with parent functions and transformations. You could use this problem string as students are learning about vertical dilations.

This problem string could come during the work of textbook Lesson 3.6 Dilations and the Absolute-Value Family.

Guiding the Problem String

Use the first question to establish the behavior of the parent function $y = |x|$. The next two problems deal with first vertically compressing the absolute-value function by a scale factor of 0.01 and then vertically stretching the absolute-value function by a scale factor of 50. Take some time to follow a few points being dilated to reinforce that you are transforming a whole set of relationships, not a static shape. The next two problems deal with reflecting the absolute-value function which, since it is even, will have no visible change when reflecting across the y-axis in the problem $y = |-x|$.The remaining problems combine translations, reflections, and dilations with the absolute-value function. You can change up the rhythm of the problem string by asking students to predict first, having students work with a partner where they must agree before changing anything, having students work individually and then comparing windows with a partner, or finding an appropriate window together as a class. The emphasis should always be on pressing for justification of the viewing window based on the transformations of the parent function.

About the Mathematics

The graphs of these functions represent continuous, infinite relationships. As you work with transformations, do not refer to the function as the letter *v* or use language that would suggest that the transformations are squishing a static shape. Words like skinnier, taller, fatter, and smaller all connote a static shape. The absolute-value parent function is not a static shaped *v* that "ends," but an infinite set of points that keep increasing as x increases and decreases. In this string, each of those points are being dilated, creating a transformed, continuous function. Use words like, "the y-values are getting bigger faster" to describe vertical stretches or "the y-values are getting bigger more slowly" to describe vertical compressions.

Sample Interactions

Use the following as you plan how to elicit and model student strategies. This is not meant as a script, but as a view into the relationships involved and the intent of the problem string.

Teacher: *We are going to find some more viewing windows today. The first problem is the absolute value of x. Let's talk about how to enter that in your graphers.*	$y=\|x\|$

The teacher explains the way to enter the absolute-value function into the students' graphers.

Teacher: *Find an appropriate viewing window. That didn't take too long. I'm seeing two choices over and over. I will sketch them both on the board.*

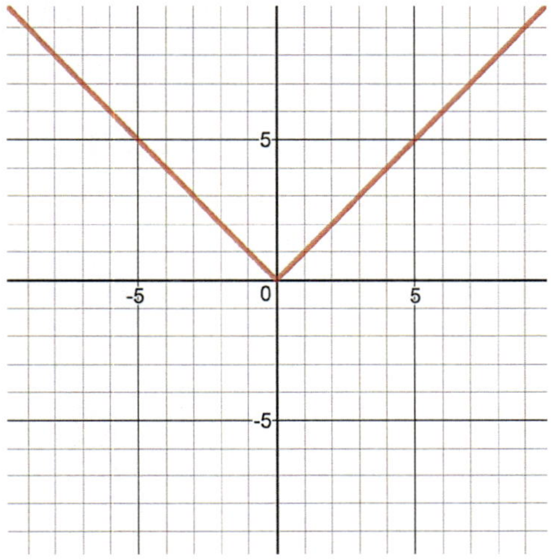

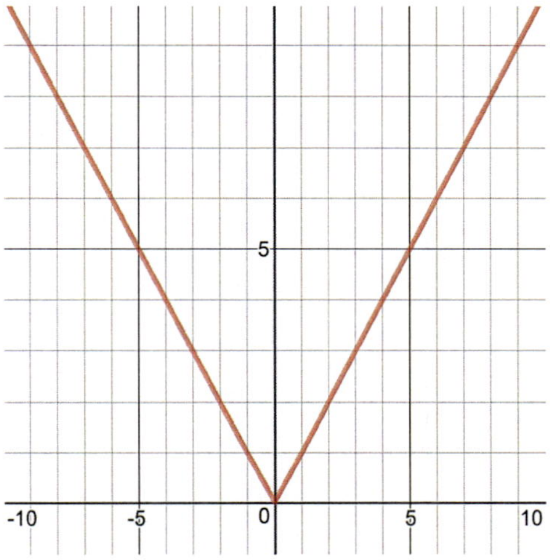

Teacher: *I wonder why you chose each of those?*

Student: *I chose the one on the right because it shows less empty space. The graph fills the window.*

Teacher: *And we've said that was one of our goals, right? I wonder why someone might choose the one on the left? The one with the extra space in the third and fourth quadrants showing?*

Student: *I was just being lazy.*

Student: *One reason you might actually choose it is because the window is square so the lines are actually at a 45 degree angle. The one on the right, look at the grid lines. They are not square.*

Student: *Right, so the lines look steeper than they actually are.*

Teacher: *That idea of steep and not steep graphs might come up more today. That might be helpful to keep in mind.*

(continued)

Teacher: *The next problem is to use our knowledge of transformations to find an appropriate viewing window for* $y = 0.01|x|$. *Let's work with this together. I'll put the function in the display grapher in the same window we had with the parent function. What do you think is happening?*

Student: *That's weird.*

Student: *It looks really flat.*

Teacher: *What does* 0.01 *mean?*

Student: *One one-hundredth. Like a penny. Small.*

$y = 0.01|x|$

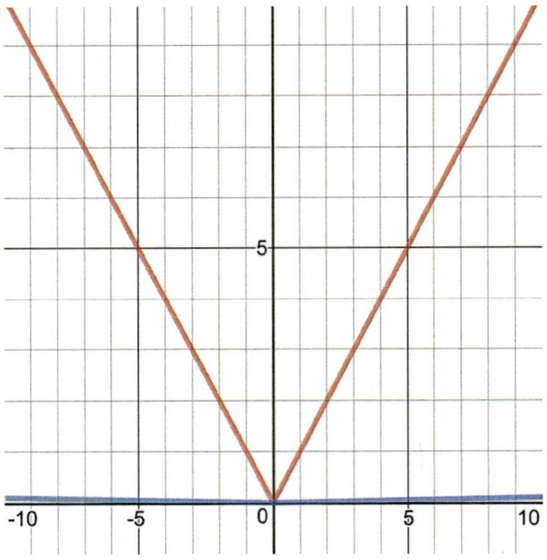

Teacher: *Let's take a look at what happened to a couple of points. When x is* 1, *what is y on the parent function,* $y = |x|$?

Student: *At one.*

Teacher: *So, here is the point* (1, 1). *What happens to that point in the new function,* $y = 0.01|x|$?

Student: *That would be* 1 *times* 0.01 *and that's* 0.01.

Teacher: *I'll plot that way down here. What about another easy point, like x is* 5.

Student: *That is the point* (5, 5).

Teacher: *What is the y-value for that x-value of* 5 *on the transformed function?*

Student: *That is five times one-hundredth, or five hundredths. That's not too much higher!*

Student: *Ahhh, that's why it's so flat.*

Teacher: *Yes, that is why the y-values are increasing so much slower for* $y = 0.01|x|$ *than for* $y = |x|$. *This viewing window is giving us a good view of how the two functions relate. Could we find a viewing window for just* $y = 0.01|x|$? *I'll turn off the parent function. Find a window and be ready to share.*

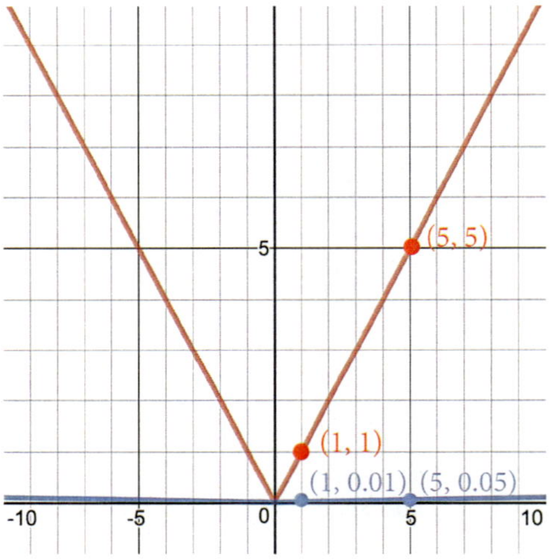

Teacher: *It seems to me like some of you took drastically different approaches. Let's start with your group. Tell me what window you had and I'll put it in the display grapher.*

Student: *We went from −10 to 10 on the x's and −0.01 to 0.11 on the y's.*

Teacher: *Got it. Did anyone else take this approach? I see some nods. What were you thinking?*

Student: *I could see that the y-values were getting big slowly, so I decided to see less y-values, kind of zoom in on the y's. When I made them small enough, I could see the function pretty well.*

Teacher: *Tell me about the x's in your window.*

Student: *We just left them alone. They are the same as they were.*

Teacher: *So you concentrated on changing the y-values. Does that make sense to everyone?*

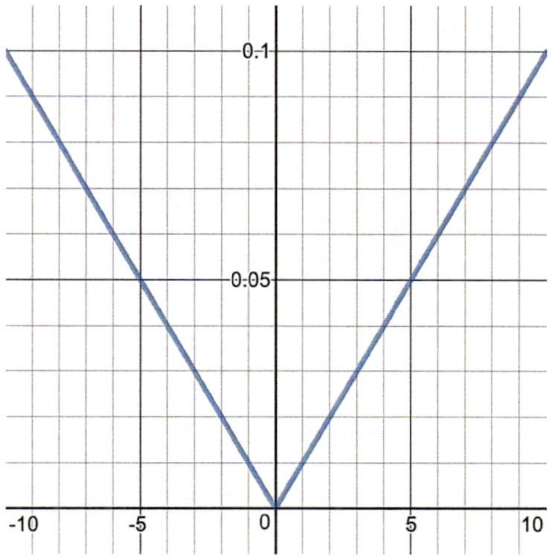

Teacher: *But that is not what some of you did. Tell us about that.*

Student: *We did just the opposite. When it was so flat, we thought about going way far out so that it could grow.*

Teacher: *What do you mean, way far out? Tell me your window.*

Student: *We kept the y's the same, but went −1,000 to 1,000 on the x's.*

Student: *Whoa.*

Teacher: *Why do you say that?*

Student: *The windows are so different but the graph ends up looking the same. That's interesting.*

Student: *One group changed the x's and the other the y's.*

Student: *I am not sure I am clear on all of this.*

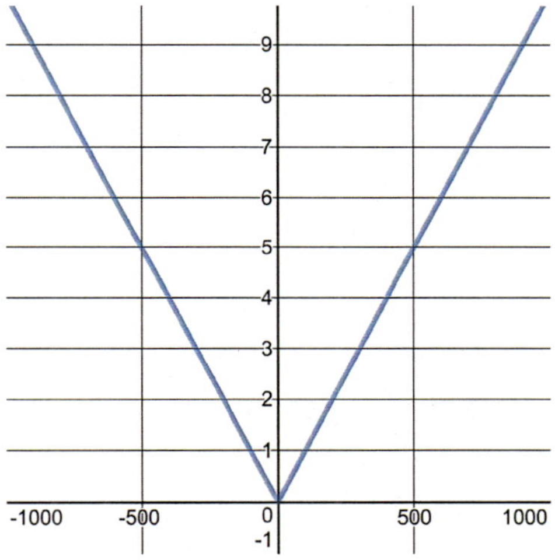

(continued)

Teacher: *Can someone help us understand how you went from that short, squatty graph to the nice looking absolute value that filled the screen.*

Student: *It's like they zoomed in really close on the y's and then stretched the y's up to fill the window.*

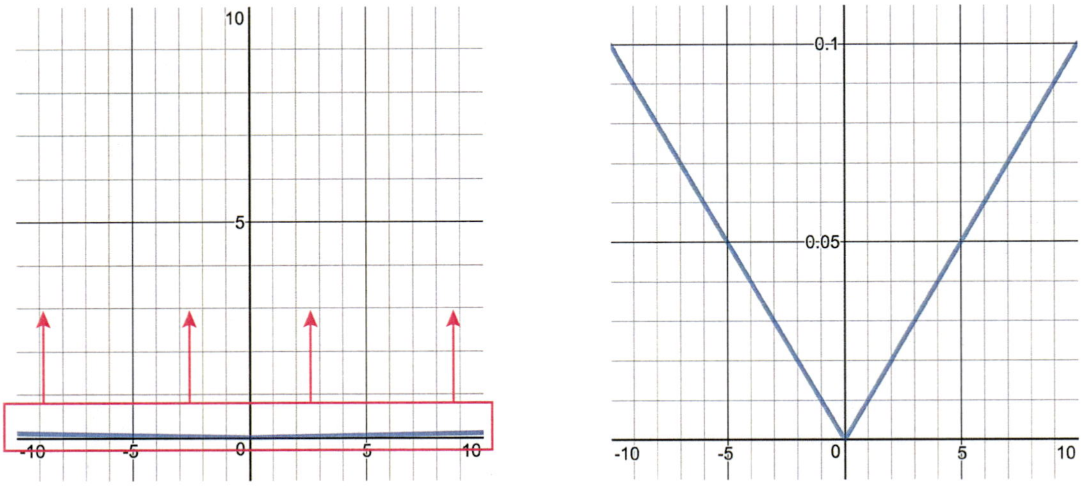

Student: *And this group went way outside the x's and then squished the whole thing in to fit into the window.*

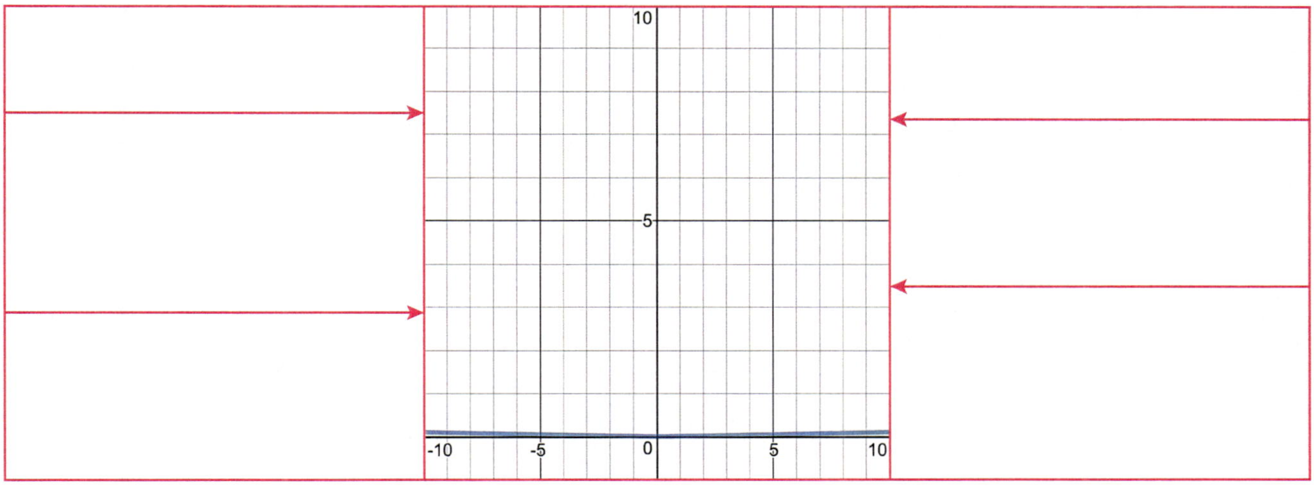

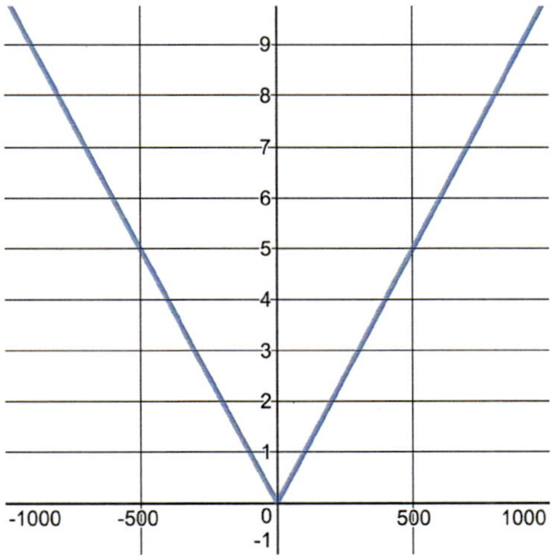

Teacher: *Great work. I'm going to put the parent function back up in the graph and our next problem,* $y = 50\|x\|$*, in the viewing window that we decided was good for the parent. Tell me about this graph.*	$y = 50\|x\|$

(continued)

Student: *That looks really skinny.*

Teacher: *Let's talk about why it looks that way. What's going on?*

Student: *Well, it's 50 times the y-values.*

Teacher: *Who can add on to that? No one? I wonder if looking at a couple of points might help. What's going on at* $x = 1$*?*

Student: *In the parent function, that's the point* $(1, 1)$*. In the new one, that's the point* $(1, 50)$*. Ah, that's way up there. It's off our graph.*

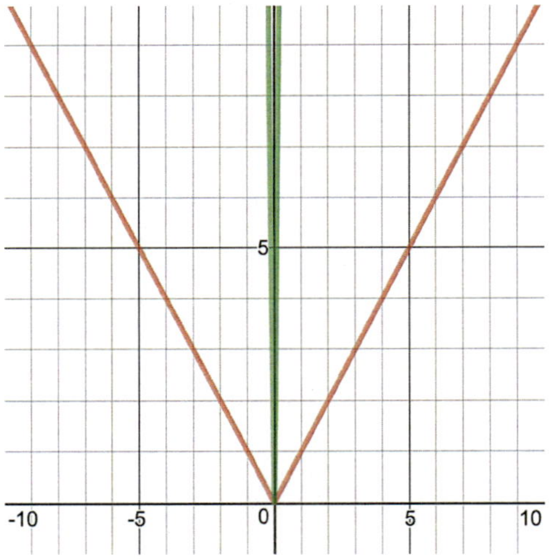

Teacher: *So, on the parent function, when you go over 1, you go up 1. But on the new function, when you go over 1, you go up 50? So, those y-values have really been stretched, haven't they. Go ahead and work to find a good viewing window for this one.*

Students work. The teacher chooses students to share who have either zoomed in on the x's or the y's and facilitates a conversation around their choices.

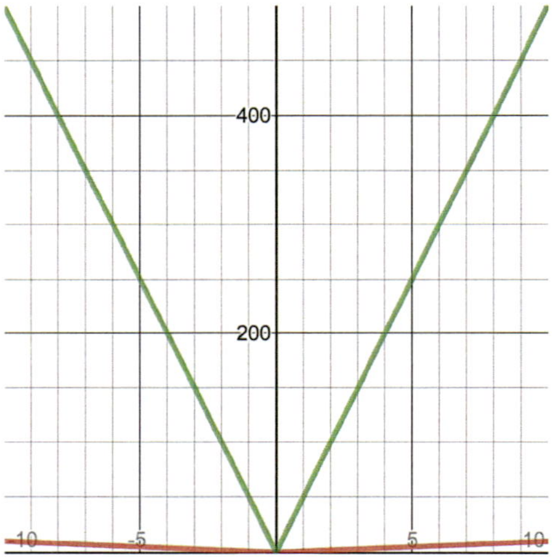

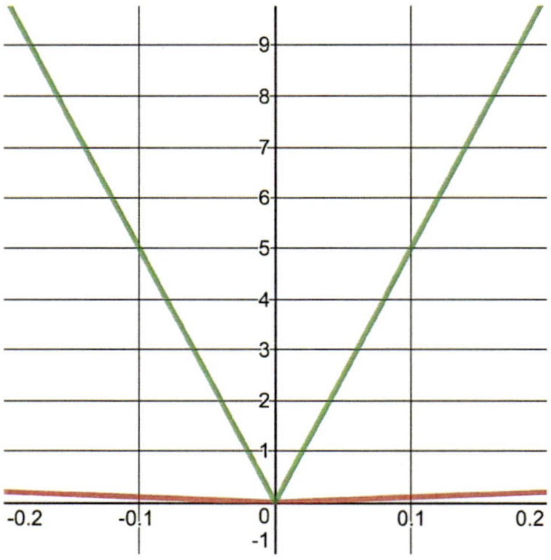

Teacher: *Thinking about the last two graphs, why did one vertically compress and the other vertically stretch?*

Student: *I think it has to do with fractions. If it's bigger or smaller than 1.*

Student: *The very first graph was being multiplied by 1, so it makes sense that multiplying it by something less than one would give y-values smaller than the parent function. That's a compression.*

Student: *Less than 1 but still positive?*

Teacher: *I think you are talking about values between zero and one, those fractions would produce a vertical compression.*

Student: *Yeah, and bigger than one makes a vertical stretch.*

The teacher gives students the next two problems in the string one at a time, $y = -|x|$ and $y = |-x|$, and facilitates a conversation around reflecting the absolute-value function.

$y = -|x|$

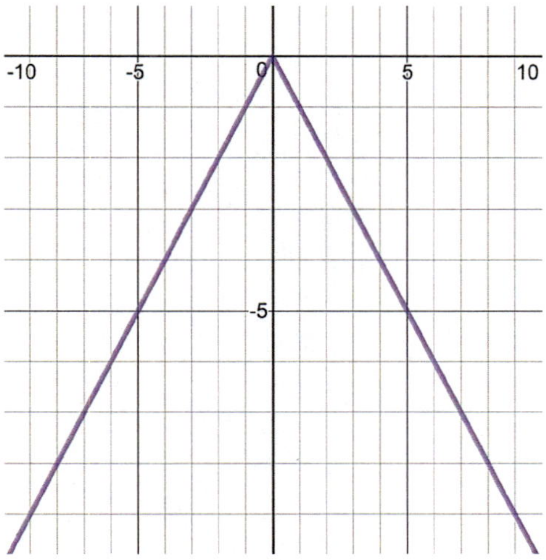

$y = |-x|$

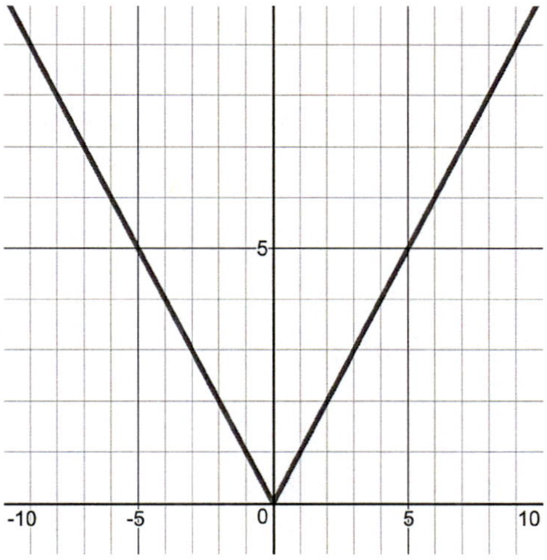

Given time, the teacher gives students the next three problems in the string one at a time, $y = 2|x+15| - 60$, $y = -\frac{1}{2}|x-28| + 1$, and $y = 10|x+12| + 300$ (for a sample graph, see the Final Display) and facilitates a conversation around combining transformations.

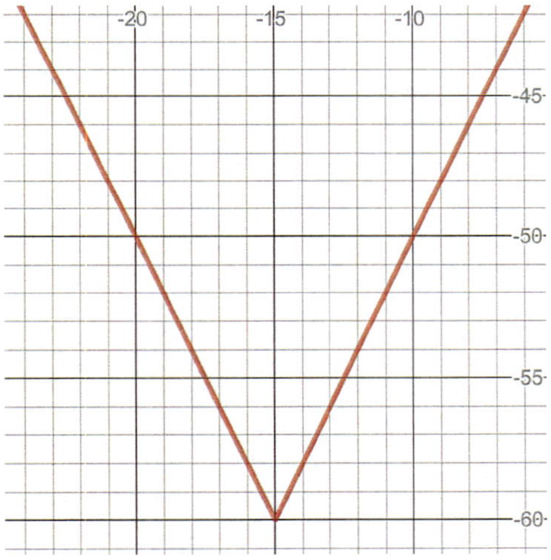

$y = 2|x+15| - 60$

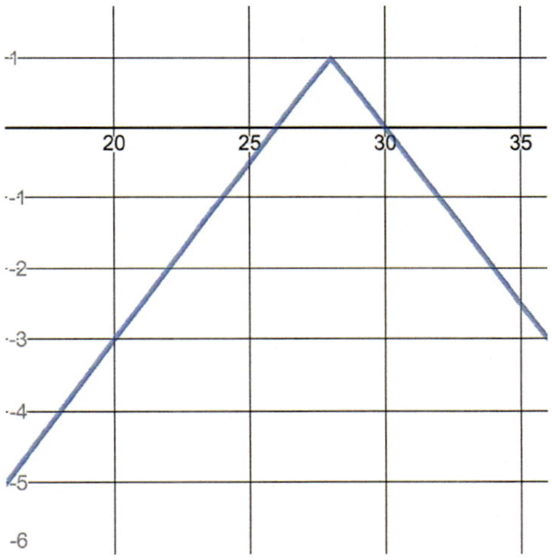

$y = -\frac{1}{2}|x-28| + 1$

Teacher: *How would you summarize some of the things that came up in this string today?*

Elicit the following:

- *When you multiply a function by a value, it's multiplying all of the y-values, scaling the y-values.*
- *For af(x), when the scale factor a is greater than* 1*, the result is a vertical stretch because the y-values are getting bigger faster.*
- *For af(x), when the scale factor a is between* 0 *and* 1*, the result is a vertical compression because the y-values are getting bigger more slowly.*

(continued)

Sample Final Display

Your display could look like this at the end of the problem string:

$y=\|x\|$	related to $y = x$
$y=0.01\|x\|$	As x is happening, the y-values are only 1⁄100 of what they were.
$y=50\|x\|$	As x is happening, the y-values are 50 times what they were.
$y=-\|x\|$	opposite of original y-values
$y=\|-x\|$	opposite of original x-values
$y=2\|x+15\|-60$	vertical stretch, look left and down
$y=-\frac{1}{2}\|x-28\|+1$	reflection over y-axis, vertical compression, look right and up
$y=10\|x+12\|+300$	vertical stretch, look left and up

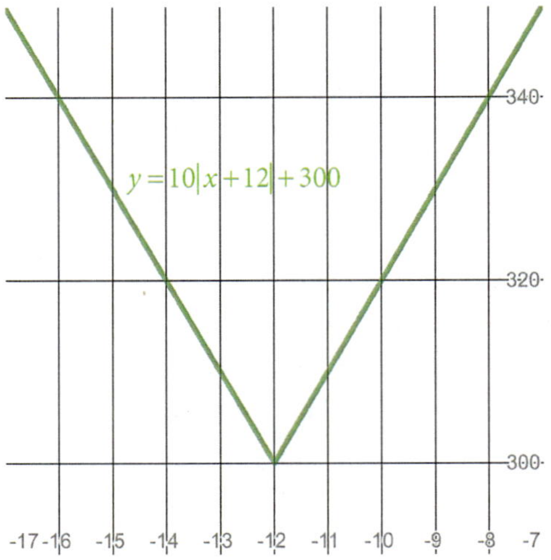

Facilitation Notes

This version of the problem string lists short notes for important teacher moves during the string. After you've done the string yourself and studied the relationships involved, you might make similar notes for the things you want a reminder of or deem important.

$y=\|x\|$	Visualize: what does the graph look like? Display with display grapher. Home window and just quadrants 1,2.
$y=0.01\|x\|$	What does 0.01 mean? What's happening to the y-values? Compare to parent when x=1, 5. Compare windows, zooming in on x's or on y's. Note "As x is happening, the y-values are only 1/100th of what they were".
$y=50\|x\|$	What's happening to the y-values? Compare to parent when x=1 Compare windows, zooming in on x's or on y's. Note "As x is happening, the y-values are 50 times what they were".
$y=-\|x\|$	Predict! How can you use what you know? Note "opposite of y-values".
$y=\|-x\|$	Predict! How can you use what you know? Note "opposite of x-values".
$y=2\|x+15\|-60$	Partner up. One holds grapher. Only make changes when both agree. Talk, defend, change, adjust. Find an appropriate viewing window as a class. Press for justification. Note "Vertical stretch, look left and down".
$y=-\frac{1}{2}\|x-28\|+1$	Repeat. Note "Reflection over y, vertical compression, look right and up".
$y=10\|x+12\|+300$	Repeat. Note "vertical stretch, look left and up".

PROBLEM STRINGS
LESSON

3.7 Transformations

At a Glance	
$y = -\frac{1}{3}(x-12)^2 - 12$ $y = 1.5\lvert x+30\rvert + 15$ $y = 30\sqrt{-x} - 100$	**Objectives** The goal of this problem string is to develop students' facility with combining the transformations of translations, reflections, and dilations of functions. To do this students find appropriate viewing windows for the transformations of the functions $y = x^2$, $y = \sqrt{x}$, and $y = \lvert x\rvert$. Using the power of technology, students can quickly test their assumptions by trying and adjusting. As they are pressed to defend their window choices, students build skill transforming functions.
	Placement This is the fourth in a series of four problem strings that use finding appropriate viewing windows as a vehicle to build and strengthen students' understanding and facility with parent functions and transformations. You could use this problem string after students have learned about translation, reflections, and vertical dilations. This problem string could come during the work of textbook Lesson 3.7 Transformations and the Circle Family.
	Guiding the Problem String These problems are meant to bring together the three functions and combinations of transformations. See problem string 3.4 for a sample dialogue of a problem string with a similar structure.
	About the Mathematics These problems combine translations, reflections, and dilations with quadratic, square root, and absolute-value functions. You can change up the rhythm of the problem string by asking students to predict first, having students work with a partner where they must agree before changing anything, having students work individually and then comparing windows with a partner, or finding an appropriate window together as a class. The emphasis should always be on pressing for justification of the viewing window based on the transformations of the parent function.

Important Questions

Use the following as you plan how to elicit and model student strategies.

- *What effect is f(x) + b on the graph of f(x)? Why? How does that help you find an appropriate viewing window?*
- *What effect is f(x + b) on the graph of f(x)? Why? How does that help you find an appropriate viewing window?*
- *What effect is –f(x) on the graph of f(x)? Why? How does that help you find an appropriate viewing window?*
- *What effect is f(–x) on the graph of f(x)? Why? How does that help you find an appropriate viewing window?*
- *What effect is af(x) on the graph of f(x) when* $a > 1$*? Why? How does that help you find an appropriate viewing window?*
- *What effect is af(x) on the graph of f(x) when* $0 < a < 1$*? Why? How does that help you find an appropriate viewing window?*

(continued)

Teacher: *How would you summarize some of the things that came up in today's string?*

Elicit the following:

- *Translations are transformations f(x) + b where you shift the function f(x) right for b < 0 and left for b > 0.*
- *Reflections are transformations where you reflect the function f(x): –f(x) over the x-axis or f(–x) over the y-axis.*
- *Dilations are transformations af(x)where you vertically stretch f(x) by a scale factor of a when a > 1, and vertically compress f(x) by a scale factor of a when 0 < a < 1.*

Sample Final Display

Your display could look like this at the end of the problem string:

$y=-\frac{1}{3}(x-12)^2-12$	parabola, vertically compressed, reflected over the *x*-axis, look right and down
$y=1.5\lvert x+30\rvert+15$	absolute value, vertically stretched, look left and up
$y=30\sqrt{-x}-100$	square root, vertically stretched, reflected over the *y*-axis, look down

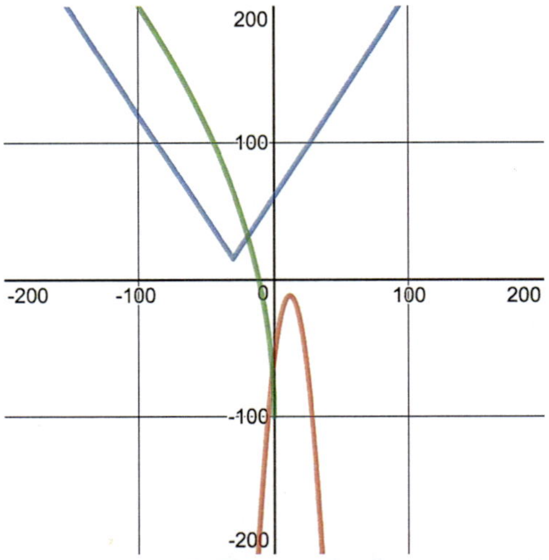

Facilitation Notes

This version of the problem string lists short notes for important teacher moves during the string. After you've done the string yourself and studied the relationships involved, you might make similar notes for the things you want a reminder of or deem important.

$y=-\frac{1}{3}(x-12)^2-12$	Partner up. One holds grapher. Only make changes when both agree. Talk, defend, change, adjust. Find an appropriate viewing window as a class. Press for justification. Note "parabola, vertically compressed, reflected over the x-axis, look right and down."
$y=1.5\lvert x+30\rvert+15$	Repeat. Note "absolute value, vertically stretched, look left and up"
$y=30\sqrt{-x}-100$	Repeat. Note square root, vertically stretched, reflected over the y-axis, look down" For fun, find a window to see all three at once!

PROBLEM STRINGS
LESSON

4.0 Exponent Relationships 1

At a Glance

$2^3 \cdot 2^2 = 2^6$

$2^a \cdot 2^b = 2^{ab}$

$\frac{2^6}{2^2} = 2^3$

$\frac{2^a}{2^b} = 2^{\frac{a}{b}}$

$\frac{1}{3^2} = \frac{1}{9}$

$\frac{2^2}{2^5} = 2^{-3}$

$\frac{1}{2^a} = 2^{-a}$

$\left(2^3\right)^2 = 2^6$

$\left(2^a\right)^b = 2^{ab}$

Objectives

The goal of this problem string is to refresh students' knowledge of many of the relationships of exponents, including the product property of exponents, the quotient property of exponents, the definition of negative exponents, and the power of a power property of exponents.

Placement

This is the first in a series of four problems strings that work with exponent relationships. This string serves to refresh many of the important relationships that students learned before. Use this problem string to get students minds warmed up to work with exponents and exponential functions. The second and third strings in the series work with rational exponents and the fourth string pulls it all together.

You can use this problem string before your work in the textbook Chapter 4 Exponential, Power, and Logarithmic Functions, or at least before Lesson 4.2 Properties of Exponents and Power Functions.

Guiding the Problem String

This problem string is in the format of equations that are either true or false. As you present each problem, ask students to envision the relationships without paper and pencil. If any of your students have gotten into a rut of mindless following of procedure with exponents, this can help them focus on thinking and reasoning about relationships. As students decide if the statement is true or false, press them to justify their thinking.

The first two problems get students thinking about the product property of exponents. The next five are about the quotient property of exponents and the definition of negative exponents. The last two problems bring up the power of a power property of exponents. The first problem in each set has a specific example of the property. Use that first problem to compare student strategies of writing out all of the 2s involved and then computing versus figuring out each power and then computing. This prompts students to reason about the equivalence. Then use the last problem of each set to ask students to generalize.

About the Mathematics

Note that the problems in this string are all focused on bases of 2 or 3. This is purposeful as a precursor before students abstract to any base.

The equations in this problem string are statements that are either true or false. One non-example is enough to prove a statement false. The reasoning that students make to justify that a statement is true are not often analytic proofs of the statement, but they should provide enough generalization to support the assertion.

In the relationships of $a^b = c$ a is the base, b is the exponent and c is the power. Sometimes teachers say "a raised to the power of b" but we will use "a raised the exponent of b" to differentiate it from using "power" to describe c. For example, "Two raised to the fourth power is 16, so 16 is a power of 2."

(continued)

Sample Interactions

Use the following as you plan how to elicit and model student strategies. This is not meant as a script, but as a view into the relationships involved and the intent of the problem string.

<table>
<tr>
<td>Teacher: Let's kick off today by warming up our brains with a problem string. You will not need paper and pencil for most of this string, maybe all of it. The format for this string is a bit different than many we've done. I'm going to give you a statement, an equation. You decide if the statement is true—if the expressions are actually equivalent. For example, I could say that we live in an active volcano. You would respond that my statement is not true. Which is a good thing. Okay, our first problem is the equation $2^3 \cdot 2^2 = 2^6$. Is this a true statement? Are the expressions equivalent?
Brief think time.</td>
<td>True or False?

$2^3 \cdot 2^2 = 2^6$</td>
</tr>
<tr>
<td colspan="2">Teacher: What do we think?
Student: Yes, it's true.
Student: No, it's not true.
Student: What do you mean?
Teacher: Is 2 cubed times 2 squared equivalent to 2 to the sixth? Justify your claim please.</td>
</tr>
<tr>
<td>Teacher: Did anyone think about lots of 2s?
Student: I thought about all of the 2s on the left. Two raised to the third means there's three 2s and then 2 more 2s is five 2s, so false, it's not six 2s.
Teacher: Let me write what you just said. You're saying it's not six 2s multiplied together?
Student: I thought about it differently. Two raised to the third is 8 and 2 squared is 4, so 8 times 4 is 32. And 32 is 2 to the fifth.
Teacher: So you both agree that the equation is false. The two cubed times two squared is actually 32 which is two to the fifth? Yes? Nice justification.</td>
<td>$(2\cdot2\cdot2)(2\cdot2) = 2^5 \neq 2^6$
False

$8 \cdot 4 = 32 = 2^5$</td>
</tr>
<tr>
<td>Teacher: Then what do you think about this more general statement, that 2 raised to some exponent, a, times 2 raised to some exponent, b, is equivalent to 2 raised to a times b?
Student: That's false. We just showed that wasn't true.
Teacher: When you have one non-example, that's enough? Yes? What relationship could we write? A true statement?
Student: It should be two raised to the a plus b.
Teacher: Like this? Do you think it will always be true? Convince us.
Student: Yes, if you have that many 2s, a 2s and b 2s all times each other, that's the same as 2 raised to a plus b.
Teacher: Are we convinced?</td>
<td>$2^a \cdot 2^b = 2^{a \cdot b}$
False

$2^a \cdot 2^b = 2^{a+b}$
a 2s times each other; b 2s times each other; (a+b) 2s times each other</td>
</tr>
</table>

Teacher: *The next question is another true/false. Is this equation true or false?* **Student:** *I don't remember the rule.* **Teacher:** *I wonder if you know something about two raised to the sixth and 2 squared and division to figure it out?* Brief think time.	$\frac{2^6}{2^2} = 2^3$
Teacher: *Who has an idea they want to put forth?* **Student:** *If you think about six 2s on the top and two 2s on the bottom, then two 2s cancel.* **Teacher:** *Cancel is not really a mathematical action. Let me write what you said, six 2s added in the numerator?* **Student:** *No, two to the sixth means six 2s times each other.* **Teacher:** *Right, and ...?* **Student:** *Two 2s times each other in the denominator and so they cancel.* **Teacher:** *These two 2s each divide with the two 2s in the denominator? So now it's all equivalent to?* **Student:** *Four 2s times each other in the numerator, 2 raised to the 4 and that's 16. And since 16 is not 8, the equation is wrong.* **Teacher:** *So you're saying that the left side is equivalent to* 2^4 *which is 16 and the right side is equivalent to 8, so the equation is not true?* **Student:** *Yes.*	$\frac{2\cdot2\cdot2\cdot2\cdot2}{2\cdot2} = \frac{\overset{1}{\cancel{2}}\cdot\overset{1}{\cancel{2}}\cdot2\cdot2\cdot2\cdot2}{\underset{1}{\cancel{2}}\cdot\underset{1}{\cancel{2}}} = \frac{2\cdot2\cdot2\cdot2}{1} = 2^4 = 16$ $2^3 = 8 \neq 16$ False
Teacher: *Are we convinced? Did anyone think about what* 2^6 *is first?* **Student:** *I did. I got 64 divided by 4 and that's 16. But the equation said it should be 8, so yeah, I agree that it's false.*	$\frac{2^6}{2^2} = \frac{64}{4} = 16 = 2^4 \neq 2^3$
Teacher: *So, let's get a little general again. Is 2 raised to some exponent, a, divided by 2 raised to some exponent, b, equivalent to 2 raised to a divided by b?* **Student:** *No, it's not. That's false.*	$\frac{2^a}{2^b} = 2^{\frac{a}{b}}$ False

(continued)

Student: *It's not raised to a divided by b, it's a minus b.* **Teacher:** *Like this? Why?* **Student:** *It's like there are a bunch of 2s in the numerator and a bunch of 2s in the denominator, so when they all go away...* **Teacher:** *When they divide out, each 2 in the numerator divided by a 2 in the denominator is equivalent to 1...* **Student:** *Yeah, then you have the left over 2s left.* **Teacher:** *How many left over 2s?* **Student:** *There will be a minus b left over 2s.* **Teacher:** *All multiplied together and we write that as exponentiation, 2 raised to the a minus b.*	$\frac{2^a}{2^b} = 2^{a-b}$ $\frac{2^a}{2^b} = 2^{a-b}$ 2^a ← *a* 2s times each other 2^b ← *b* 2s times each other 2^{a-b} → (*a* – *b*) 2s times each other; *a* – *b* (the difference between *a* and *b*)
Teacher: *This next problem is another true/false. Is 1 divided by 3 squared equivalent to one-ninth or 1 divided by 9? I know this isn't a hard one.* **Student:** *Yes.*	$\frac{1}{3^2} = \frac{1}{9}$ True
Teacher: *Right, that might be helpful later on. Here's the next question. True or false, is 2 squared divided by 2 to the fifth equivalent to 2 raised to negative 3, or opposite of 3?* **Student:** *Well, according to what we did before, yes, because 2 minus 5 is negative 3.* **Student:** *Yes, but also if you think about two 2s times each other divided by five 2s, you would be left with three 2s in the denominator. Right?*	$\frac{2^2}{2^5} = 2^{-3}$ $\frac{2 \cdot 2}{2 \cdot 2 \cdot 2 \cdot 2 \cdot 2} = \frac{\overset{1}{\cancel{2}} \cdot \overset{1}{\cancel{2}}}{\underset{1}{\cancel{2}} \cdot \underset{1}{\cancel{2}} \cdot 2 \cdot 2 \cdot 2} = \frac{1}{2 \cdot 2 \cdot 2} = \frac{1}{2^3} = \frac{1}{8}$
Student: *Oh, that's right, those are equivalent, the 2 to the −3 and the 1 over 2³.* **Teacher:** *These are equivalent?* **Student:** *Yes.* **Teacher:** *How? Why?* **Student:** *Well, if you follow what we said before, that when you're dividing the 2s you subtract the exponents, then that's what you get.* **Student:** *Well, I thought about 4 divided by 32 which is one-eighth. Does that mean one-eighth is 2 to the −3?* **Teacher:** *Good question. What does everyone think?* **Students:** *Looks like it. Maybe. Shouldn't it be negative?*	$\frac{1}{2^3} = 2^{-3}$ $\frac{2^2}{2^5} = 2^{2-5} = 2^{-3}$ $\frac{2^2}{2^5} = \frac{4}{32} = \frac{1}{8} = 2^{-3}$ True

Teacher: *It might be helpful to look at a pattern of raising a number to an exponent and letting the exponents decrease. I'm going to change to the number 3 just so it's not all about the number 2. You all know that 3^3 is 27, 3^2 is 9, and 3 to the first is 3. So, if we follow this pattern, what is 3 raised to the 0? Yes, 1. In fact, any number to the 0 is 1 for the same reason. Then what is 3 raised to the −1? And how about 3 raised to the −2? In fact, what is 3 raised to any negative exponent equivalent to? Turn and talk about that.* Students turn and talk briefly. **Teacher:** *I heard many students remembering working with negative exponents before. What are some of your insights?* **Student:** *It's weird because the negative exponents have nothing to do with the number being positive or negative, just if it's in the numerator or denominator.* **Student:** *And we were looking back at the previous problem, where 1 divided by 3 squared was one-ninth. And that's 3 to the −2. It all fits.*	$3^3 = 27$ $3^2 = 9$ $3^1 = 3$ $3^0 = 1$ $3^{-1} = \frac{1}{3}$ $3^{-2} = \frac{1}{3^2}$ $3^{-p} = \frac{1}{3^p}$
Teacher: *So does this general statement capture what we were just talking about? True or false?* **Student:** *True.*	$\frac{1}{2^a} = 2^{-a}$ True
Teacher: *The next problem in our exponent relationship string is this equation, true or false? Think about it and then tell your partner what you are thinking.* Brief think and partner talk time.	$\left(2^3\right)^2 = 2^6$
Teacher: *What do you think? True or false? How do you know?* **Student:** *I think it's true. I thought about the squared part. That means that 2^3 is multiplied by itself. So then you have six 2s and that's 2^6.* **Teacher:** *You have three 2s multiplied together times three 2s multiplied together, so you end up with six 2s multiplied together?*	$\left(2^3\right)^2 = 2^3 \cdot 2^3 = (2 \cdot 2 \cdot 2)(2 \cdot 2 \cdot 2) = 2^6$
Student: *I did the 2^3 first, that's 8. So then 8^2 is 64, and I figured out that 64 is 2^6. So yes, it's true.*	$\left(2^3\right)^2 = 8^2 = 64 = 2^6$ True
Teacher: *Let's get a little general. For any exponents a and b, is this statement true?* Brief think time.	$\left(2^a\right)^b = 2^{a \cdot b}$

(continued)

Teacher: *So, how are you making sense of this?*

Student: *We think it's true because that's what we just did with the 2^3 squared.*

Teacher: *Can anyone get a little more general than that one example?*

Student: *It's like you have a bunch of 2s raised to the a exponent. You have b of them.*

Teacher: *I will represent your thinking like this. Does this make sense to everyone?*

$$\left(2^a\right)^b = \underbrace{2^a \cdot 2^a \cdot \ldots \cdot 2^a}_{b\,(2^a)\text{s multiplied together}} = 2^{a \cdot b} \longrightarrow a \cdot b \text{ 2s times each other}$$

True

Teacher: *How would you summarize some of the things that came up in this string today?*

Elicit the following:

- *One non-example can prove a statement false. You must be able to reason generally to claim that a statement is true.*
- *It can be helpful to list out the bases multiplying like $2^3 = 2 \cdot 2 \cdot 2$ or you can simplify them like $2^3 = 8$ to think about relationships.*
- *You can make sense out of exponents by reasoning about examples.*

Sample Final Display

Your display could look like this at the end of the problem string:

$2^3 \cdot 2^2 = 2^6$ False $(2 \cdot 2 \cdot 2)(2 \cdot 2) = 2^5 \neq 2^6$ $8 \cdot 4 = 32 = 2^5$

$2^a \cdot 2^b = 2^{a \cdot b}$ False $2^a \cdot 2^b = 2^{a+b}$ (a 2s times each other; b 2s times each other; $(a+b)$ 2s times each other)

$\frac{2^6}{2^2} = 2^3$ False $\frac{2 \cdot 2 \cdot 2 \cdot 2 \cdot 2 \cdot 2}{2 \cdot 2} = \frac{\not{2} \cdot \not{2} \cdot 2 \cdot 2 \cdot 2 \cdot 2}{\not{2} \cdot \not{2}} = \frac{2 \cdot 2 \cdot 2 \cdot 2}{1} = 2^4 = 16$ $2^3 = 8 \neq 16$

$\frac{2^a}{2^b} = 2^{\frac{a}{b}}$ False $\frac{2^a}{2^b} = 2^{a-b} \longrightarrow (a-b)$ 2s times each other $a - b$ (the difference between a and b) (2^a: a 2s times each other; 2^b: b 2s times each other)

$\frac{1}{3^2} = \frac{1}{9}$ True

$\frac{2^2}{2^5} = 2^{-3}$ True $\frac{2 \cdot 2}{2 \cdot 2 \cdot 2 \cdot 2 \cdot 2} = \frac{\not{2} \cdot \not{2}}{\not{2} \cdot \not{2} \cdot 2 \cdot 2 \cdot 2} = \frac{1}{2 \cdot 2 \cdot 2} = \frac{1}{2^3} = \frac{1}{8}$ $\frac{2^2}{2^5} = 2^{2-5} = 2^{-3}$ $\frac{1}{2^3} = 2^{-3}$ $\frac{2^2}{2^5} = \frac{4}{32} = \frac{1}{8} = 2^{-3}$

$\frac{1}{2^a} = 2^{-a}$ True

$\left(2^3\right)^2 = 2^6$ True $\left(2^3\right)^2 = 2^3 \cdot 2^3 = (2 \cdot 2 \cdot 2)(2 \cdot 2 \cdot 2) = 2^6$ $\left(2^3\right)^2 = 8^2 = 64 = 2^6$

$\left(2^a\right)^b = 2^{a \cdot b}$ True $\left(2^a\right)^b = \underbrace{2^a \cdot 2^a \cdot \ldots \cdot 2^a}_{b\,(2^a)\text{s multiplied together}} = 2^{a \cdot b} \longrightarrow a \cdot b$ 2s times each other

$3^3 = 27$
$3^2 = 9$
$3^1 = 3$
$3^0 = 1$
$3^{-1} = \frac{1}{3}$
$3^{-2} = \frac{1}{3^2}$
$3^{-p} = \frac{1}{3^p}$

Facilitation Notes

This version of the problem string lists short notes for important teacher moves during the string. After you've done the string yourself and studied the relationships involved, you might make similar notes for the things you want a reminder of or deem important.

	No paper and pencil needed. *Today's string is about statements, true or false?*
$2^3 \cdot 2^2 = 2^6$	*True or False? Quick. Elicit both lots of 2s and 8·4.* *Make it about equivalencies, not things to do.* *Is one non-example enough to prove false?*
$2^a \cdot 2^b = 2^{a \cdot b}$	*How can we generalize? Press for justification. Model.*
$\frac{2^6}{2^2} = 2^3$	*True or False? Quick-ish. Elicit both lots of 2s and 64/4.* *Keep it about equivalencies, not things to do.*
$\frac{2^a}{2^b} = 2^{\frac{a}{b}}$	*How can we generalize? Press for justification. Model.*
$\frac{1}{3^2} = \frac{1}{9}$	*True or False? Very quick.* *This might be helpful later on.*
$\frac{2^2}{2^5} = 2^{-3}$	*True or False? Elicit lots of 2s, 4/32, and using the previous.* *Bring in patterns of decreasing exponents:* $3^3, 3^2, 3^1, 3^0, 3^{-1}, 3^{-2}$
$\frac{1}{2^a} = 2^{-a}$	*So, based on all of that, true or false?*
$\left(2^3\right)^2 = 2^6$	*True or False? Quick-ish. Elicit both lots of 2s and* $(8)^2 = 64$.
$\left(2^a\right)^b = 2^{a \cdot b}$	*How can we generalize? Press for justification. Model.*

PROBLEM STRINGS LESSON 4.1

Exponential Functions and Transformations

At a Glance

$y=2^x$

$y=12\cdot 2^x$

$y=-5^x-25$

$y=13^{-x}+200$

$y=0.9^x+75$

$y=0.5^{x+20}$

$y=2^{(-x-20)}$

Objectives

The goal of this problem string is to help students develop facility with transformations of exponential functions, while becoming proficient with the parent function. Using the power of technology, students can quickly test their assumptions by trying and adjusting. As they are pressed to defend their window choices, students build skill using transformations with exponential functions.

Placement

This string is similar to the series of problem strings in chapter 3 that use finding appropriate viewing windows as a vehicle to build and strengthen students' understanding and facility with parent functions and transformations. You could use this problem string as students are learning about exponential functions, especially the effect of vertical stretches.

This problem string could come before or during the work of the textbook Lesson 4.1 Exponential Functions.

Guiding the Problem String

Ask students to predict the graph based on their knowledge of transformations before you start graphing on the display calculator. You might graph one transformation at a time, noting the effect of each as you go, before leaving your starting window. Press students for justification for the viewing windows based on transformations.

The first problem sets the stage for the end behavior of exponential functions. The second problem is a vertical stretch, where it may surprise students that the "anchor point" of (0, 1) also stretches to (0, 12). The third and fourth problems change the base and have different transformations. The last two problems introduce a base between 0 and 1.

See problem string 3.4 for a sample dialogue of a problem string with a similar structure.

About the Mathematics

Often students have experience vertically stretching and shrinking functions that have an "anchor point" at the origin (0, 0), like $y=x$, $y=x^2$, $y=|x|$. When these functions are vertically stretched or compressed, the anchor point does not move because the product of the scale factor and 0 is 0. However, the "anchor point" of an exponential function is (0, 1), so when the function is vertically stretched or compressed, the product of the scale factor, a, and 1 is a, stretching the anchor point to $(0, a)$.

When $b>1$, $y=a\cdot b^x$ is increasing and because $b^{-x}=\left(\frac{1}{b}\right)^x$, $y=a\cdot b^{-x}$ is decreasing. When $0<b<1$, $y=a\cdot b^x$ is decreasing.

Important Questions

Use the following as you plan how to elicit and model student strategies.

- *What is the general behavior of* $y = 2^x$ *and more generally* $f(x) = b^x$*?*
- *What effect is* $f(x) + b$ *or* $f(x + b)$ *on the graph of* $f(x)$*? Why? How does that help you find an appropriate viewing window?*
- *What effect is* $-f(x)$ *on the graph of* $f(x)$*? Why? How does that help you find an appropriate viewing window?*
- *What effect is* $f(-x)$ *on the graph of* $f(x)$*? Why? How does that help you find an appropriate viewing window?*
- *What effect is* $af(x)$ *on the graph of* $f(x)$ *when* $a > 1$*? Why? What is happening with the "anchor point" of* $(0, 1)$*? Why is that different than other functions which have an anchor point at* $(0, 0)$*? How does that help you find an appropriate viewing window?*
- *What happens to* $f(x) = b^x$ *when* $0 < b < 1$*? Why? Increasing or decreasing?*

How would you summarize some of the things that came up in this string today?

- *Exponential functions grow or decay rapidly.*
- *When a function f(x) reflects over the x-axis, all of the y-values take on the opposite sign. The y-values that were once positive are now negative. The y-values that were once negative are now positive. This is represented by* $-f(x)$*.*
- *When a function f(x) is reflected over the y-axis, all of the y-values that used to correspond to x, now correspond to* $-x$*. This is represented by* $f(-x)$*.*
- *We can combine reflections and translations. It makes sense to reflect over the axis first then shift.*

Sample Final Display

Your display could look like this at the end of the problem string:

$y = 2^x$	increasing, (0, 1)
$y = 12 \cdot 2^x$	increasing (0, 12)
$y = -5^x - 25$	base of 5, reflection over x, look down
$y = 13^{-x} + 200$	base of 13, reflection over the y-axis, look up
$y = 0.9^x + 75$	decreasing (reflection over the x-axis), look up
$y = 0.5^{x+20}$	decreasing (reflection over y-axis), look left
$y = 2^{(-x-20)}$	reflection over the y-axis, look left, equivalent to previous, ½ = 2^{-1}

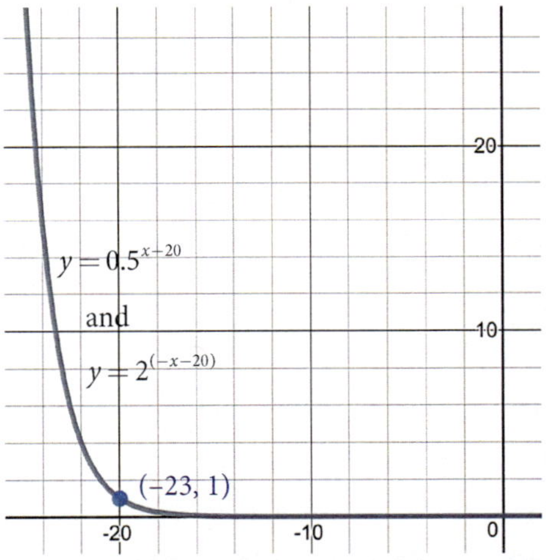

(continued)

Facilitation Notes

This version of the problem string lists short notes for important teacher moves during the string. After you've done the string yourself and studied the relationships involved, you might make similar notes for the things you want a reminder of or deem important.

$y=2^x$	*Quick. Establish end behavior and good viewing window.*
$y=12\cdot 2^x$	*Predict first. How does the 12 affect the previous function?* *Vertical stretch: y-values get bigger faster* *The anchor point stretches too!*
$y=-5^x-25$	*Predict for all the rest of the problems.* *What affect does the change of base from 2 to 5 have?* *The -5? The -25?* *Press for justification of windows using transformations.*
$y=13^{-x}+200$	*Repeat.*
$y=0.9^x+75$	*Linger. What is happening? Why decreasing?*
$y=0.5^{x+20}$	*Repeat.* *Connect decreasing to reflection over y-axis.*
$y=2^{(-x-20)}$	*Linger. What is happening? Why is it equivalent to the previous?*

PROBLEM STRINGS
LESSON

4.2 Exponential Decay

At a Glance	
$y = 30(0.8185)^x$ $y = 34(0.8185)^x$ $y = 25(0.8185)^x$ $y = 30(0.75)^x$ $y = 30(0.95)^x$ $(0, 32)\ (1, 24)\ (2, 18)\ (3, 14)\ (4, 10)\ (5, 8)$	**Objectives** The goal of this problem string is to solidify the learning about exponential decay functions that took place in the Radioactive Decay Investigation in which students simulate exponential decay data and write an exponential decay function to model the scenario. **Placement** This string could come right after or the day after students have used their collected simulation data to write an exponential function for simulated data of atoms that underwent radioactive decay. You could deliver this string any time after the Radioactive Decay Investigation in textbook Lesson 4.1 Exponential Functions. **Guiding the Problem String** The first problem is based on the data from the answers in the textbook teacher edition. You could change the numbers to match the data your class collected for (stage, number standing), where for each stage, each standing student rolls a die and those getting a one sit down. The purpose of the first problem is to provide an anchor to the investigation that students can refer to throughout the string. Use the display grapher to graph the functions. Take a little time on the second problem to predict and discuss similarities and differences. Push on answers for justification. The third problem should go quickly. The fourth and fifth problems give functions with different rates. Take time to investigate and make sense of the effect of higher and lower rates between 0 and 1. The last problem requires students to find a continuous exponential function that models the discrete data. Encourage students to think about the relationship between 32 and 24 before they just start dividing, because 24 is ¾ of 32. **About the Mathematics** The sequence of (stage, number standing) is a discrete geometric sequence that can be modeled by a continuous exponential function.

(continued)

Important Questions

Use the following as you plan how to elicit and model student strategies.

- *What is a in* $y = a \cdot b^x$ *? Where does it show up in the graph? What does it mean in the function? In the corresponding geometric sequence?*
- *What is b in* $y = a \cdot b^x$ *? Where does it show up in the graph? What does it mean in the function? In the corresponding geometric sequence?*
- *Given a geometric sequence, how can you find b?*

How would you summarize some of the things that came up in this string today?

- *In an exponential function* $y = a \cdot b^x$, *a is the y-intercept and also a vertical stretch of* $y = b^x$. *It corresponds to the* a_0 *term in the geometric sequence.*
- *In an exponential function* $y = a \cdot b^x$, *b is the rate that involves percent increase or percent decrease, and corresponds to the common multiplier in a geometric sequence.*
- *Rates, b, in* $y = a \cdot b^x$, *that correspond to the ratios (multipliers) in a geometric sequence, that are in between* 0 *and* 1 *create decreasing sequences.*
- *Higher rates, still between* 0 *and* 1, *make a decreasing sequence decrease slower than lower rates between* 0 *and* 1.

Sample Final Display

Your display could look like this at the end of the problem string:

$y = 30(0.8185)^x$ — y-intercept 30, rate 0.8185

$y = 34(0.8185)^x$ — higher y-intercept 34, same rate

$y = 25(0.8185)^x$ — lower y-intercept 25, same rate

$y = 30(0.75)^x$ — back to y-intercept 30, lower rate, decreases faster (keeps less, 75% as time increases)

$y = 30(0.95)^x$ — same y-intercept 30, higher rate, decreases slower (keeps more, 95% as time increases)

(0, 32) (1, 24) (2, 18) (3, 14) (4, 10) (5, 8) $\frac{24}{32} = \frac{3}{4} = 0.75, \frac{18}{24} = \frac{3}{4} = 0.75$

$y = 32(0.75)^x$ — y-intercept 32, rate 0.75

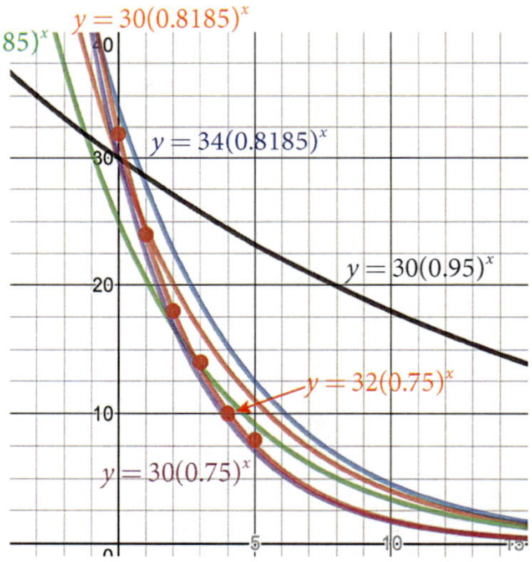

Facilitation Notes

This version of the problem string lists short notes for important teacher moves during the string. After you've done the string yourself and studied the relationships involved, you might make similar notes for the things you want a reminder of or deem important.

$y = 30(0.8185)^x$	*Remember data? (Connect to prior investigation.)* *What does this represent?* *Predict the graph. Display on grapher in good viewing window.*
$y = 34(0.8185)^x$	*What changed? What do we know about the class? The dice?* *Predict the graph. Compare.*
$y = 25(0.8185)^x$	*Repeat.*
$y = 30(0.75)^x$	*What changed? What do we know about the class? The dice?* *Predict the graph. Why does a lower rate behave that way?* *How many remain standing as time increases?*
$y = 30(0.95)^x$	*Repeat. Why does a higher rate behave that way?* *How many remain standing as time increases?*
$(0, 32)\,(1, 24)\,(2, 18)\,(3, 14)\,(4, 10)\,(5, 8)$	*Changing it up. Write a continuous function to model the discrete data.* *Predict the graph. Compare.*

PROBLEM STRINGS
LESSON
4.3

Exponent Relationships 2: Rational Exponents

At a Glance

$4^{\frac{3}{2}}$

$25^{\frac{3}{2}}$

$27^{\frac{4}{3}}$

$9^{\frac{3}{2}}$

$8^{\frac{2}{3}}$

$16^{\frac{3}{4}}$

Objectives

The goal of this problem string is for students to gain facility simplifying numbers raised to rational exponents, helping students realize the role in the notation of the numerator and denominator in the rational exponent.

Placement

This is the second in a series of four problems strings that work with exponent relationships. This string supports students in their work with rational exponents in expressions that simplify to whole numbers. You can use this problem string to help students gain facility with the notation and the meaning of the numerator and denominator, and the exponential and root relationships.

Use this problem string to support your work in textbook Lesson 4.3 Rational Exponents and Roots.

Guiding the Problem String

The problems in this string have whole number equivalents. The first problem can be simplified easily in either order, finding the root or raising to the numerator in the exponent first. The next three problems are easier to simplify by finding the root first. The first four problems simplify to a result larger than the original base because the exponent is larger than 1. The fifth problem can be simplified easily using either order. This and the final problem are the only problems in the string where the simplified result is less than the original base because the exponent is between 0 and 1. Encourage students throughout the problem string to verbalize their thinking. Limit your use of pronouns. Instead of "the square root of it is 3 so cube it and that's 27" say "the square root of 9 is 3 and 3 cubed is 27."

About the Mathematics

It is true that for $a > 0$, when $0 < \frac{m}{n} < 1$, then $a < a^{\frac{m}{n}} < a^m$. This is only meaningful for positive bases and some negative bases, when the nth root is real. An example is $(-8)^{\frac{2}{3}} = (-2)^2 = 4$ because the cube root of -8 is -2. However, an even root of -8, like $(-8)^{\frac{1}{2}} = \sqrt{-8} = i\sqrt{2}$ is complex and the discussion of "in between" is less meaningful. In this string, we limit the conversation to real numbers.

Important Questions

Use the following as you plan how to elicit and model student strategies.

- *What does this fractional exponent mean?*
- *How can we use the fractional exponent to estimate the result? What are some good benchmarks to compare to?*
- *Which did you do first, find the root or raise to the numerator in the rational exponent? Why?*
- *Do you think it will always be possible to take the root first and then raise to the numerator in the rational exponent? Only when the result is a whole number?*
- *Can you sometimes raise to the numerator in the rational exponent first and then find the root? When? Can you think of any other examples?*
- *What is true about the exponent when the power, $a^{\frac{m}{n}}$, is larger than the base, a? When the power is smaller than the base?*

How would you summarize some of the things that came up in this string today?

- *The denominator, n, in the rational exponent of* $a^{\frac{m}{n}}$ *represents the nth root of* a^m.
- *The numerator, m, in a rational exponent of* $a^{\frac{m}{n}}$ *represents the exponent of* $a^{\frac{1}{n}}$.
- *If the result is a whole number, you can take the root first and then raise to the numerator in the exponent. Sometimes you can raise to the numerator in the exponent first and then take the root.*
- *When the positive base, a, is raised to an exponent between* 0 *and* 1, *the result is between the base and the base raised to the numerator of the exponent. For* $a > 0$, when $0 < \frac{m}{n} < 1$, then $a < a^{\frac{m}{n}} < a^m$.

Sample Final Display

Your display could look like this at the end of the problem string:

$4^{\frac{3}{2}} = 8$	$4^1 < 4^{\frac{3}{2}} < 4^2$	$\left(4^3\right)^{\frac{1}{2}} = 64^{\frac{1}{2}} = 8$	$\left(4^{\frac{1}{2}}\right)^3 = 2^3 = 8$
$25^{\frac{3}{2}} = 125$	$25^1 < 25^{\frac{3}{2}} < 25^2$	$\left(25^{\frac{1}{2}}\right)^3 = 5^3 = 125$	
$27^{\frac{4}{3}} = 81$	$27^1 < 27^{\frac{4}{3}} < 27^2$	$\left(27^{\frac{1}{3}}\right)^4 = 3^4 = 81$	
$9^{\frac{3}{2}} = 27$	$9^1 < 9^{\frac{3}{2}} < 9^2$	$\left(9^{\frac{1}{2}}\right)^3 = 3^3 = 27$	
$8^{\frac{2}{3}} = 4$	$8^0 < 8^{\frac{2}{3}} < 8^1$	$\left(8^{\frac{1}{3}}\right)^2 = 2^2 = 4$	$\left(8^2\right)^{\frac{1}{3}} = 64^{\frac{1}{3}} = 4$
$16^{\frac{3}{4}} = 8$	$16^0 < 16^{\frac{3}{4}} < 16^1$	$\left(16^{\frac{1}{4}}\right)^3 = 2^3 = 8$	

Facilitation Notes

This version of the problem string lists short notes for important teacher moves during the string. After you've done the string yourself and studied the relationships involved, you might make similar notes for the things you want a reminder of or deem important.

$4^{\frac{3}{2}} = 8$	What is an estimate? Nice benchmarks? What is this equivalent to? Which did you do first, square root or cube? Why? Does it matter?
$25^{\frac{3}{2}} = 125$	Estimate? What is this equivalent to? Which first? Why?
$27^{\frac{4}{3}} = 81$	Repeat. Quick.
$9^{\frac{3}{2}} = 27$	Repeat. Repeat. Quick. I guess it's always easier to do the root first?
$8^{\frac{2}{3}} = 4$	Now what is your estimate? Why? Now which first? Sometimes you can do either first! When? Look back, when is the result smaller than the initial base?
$16^{\frac{3}{4}} = 8$	Now what is your estimate? Why? Did it follow the pattern? What's true of results when exponents are between 0 and 1? Exponents > 1?

PROBLEM STRINGS LESSON

4.4

Exponent Relationships 3: More Rational Exponents

At a Glance	
$3^{12} = (\_)^3$ * $3^{12} = (3^6)^{\_}$ * $9^{\frac{3}{2}} = (\_)^3$ * $9^{\frac{3}{2}} = (\_)^{\frac{1}{2}}$ * $4^x = 16$ $2^{2x} = 16$ $2^{2x} = 2$ $4^x = 2$ $9^x = 27$ $125^x = 25$ *optional problems	**Objectives** The goal of this is to build student facility with rational exponents by solving equations that include rational exponents. This string specifically targets the power of a power relationship. **Placement** This is the third in a series of four problems strings that work with exponent relationships. Use this string after students have learned about rational exponents. You can use this string after textbook Lesson 4.3 Rational Exponents and Roots. **Guiding the Problem String** The first four problems are optional. Use these to help your students if they need experience using the power of a power relationship to write equivalencies. These problems ask students to decompose the original problem which might seem backwards to students if they have only used exponent rules to simplify. The next four problems are in partners, where comparing the results is helpful to build toward the strategy needed for the last two problems, rewriting a number as a power so the equation consists of equivalent bases and therefore the exponents can be set equal to each other. The last two problems are clunkers, where students can use what they've been learning. Listen for students who begin to use relationships to rewrite the bases. Work throughout the string to keep students thinking and reasoning using relationships by emphasizing reasonableness and connections. **About the Mathematics** In an equation of two equivalent power expressions, if the bases are equivalent, the exponents are also equivalent. This can be used to solve equations where the powers can be written with equivalent bases. Fluency with this idea prepares students to reason with logarithms with the same base.

Sample Interactions

Use the following as you plan how to elicit and model student strategies. This is not meant as a script, but as a view into the relationships involved and the intent of the problem string.

Teacher: *Let's finish our work with rational exponents today with a problem string. Here's the first problem. How could we write 3 raised to the 12th as something raised to the third?* Brief think time. **Teacher:** *What are you thinking?* **Student:** *I don't remember the rule.* **Teacher:** *I bet you can just reason about it. What do you know about 3 to the 12th? And what do you know about something cubed?*	$3^{12} = (\_)^3$

Student: *Something cubed means something times itself three times.* **Teacher:** *So, there needs to be three of something? Of what?* **Student:** *Three to the fourth.*	$3^{12} = 3^4 \cdot 3^4 \cdot 3^4 = \left(3^4\right)^3$
Teacher: *How could we write 3 raised to the 12th as 3 to the 6th raised to some exponent? After you have an idea, turn and discuss it with your partner.* Brief partner discussion time. **Student:** *We think it's 3 to the 6th squared because six 3s times six 3s is twelve 3s.* **Teacher:** *All of those 3s multiplied together. Nice.*	$3^{12} = \left(3^6\right)^{-}$ $3^{12} = \left(3^6\right)^2 = 3^6 \cdot 3^6$
Teacher: *How could we write 9 raised to the three-halves as something cubed? We've just been working on rational exponents. I wonder if those concepts along with the first two problems of the string can help you here? After you have an idea, turn and discuss it with your partner.* Brief partner discussion time. **Student:** *If you have 9 to the one-half, that should work.* **Teacher:** *Why an exponent of one-half?* **Student:** *Because one-half times three is three-halves.* **Student:** *And also because that's what 9 to the three-halves means, take the square root and then cube that.* **Student:** *And I thought about the square root of 9, that's 3 and 3 cubed is 27. I checked that by putting it in my calculator and it checks out.*	$9^{\frac{3}{2}} = (\_)^3$ $9^{\frac{3}{2}} = \left(9^{\frac{1}{2}}\right)^3$
Teacher: *And how else could we write it? What if we wanted to know an equivalent way of writing something to the one-half?* **Student:** *That's just 9 to the 3rd, all to the one-half.*	$9^{\frac{3}{2}} = (\_)^{\frac{1}{2}}$ $9^{\frac{3}{2}} = \left(9^3\right)^{\frac{1}{2}}$
Teacher: *And for our next few problems, we're going to solve for x. So, solve for x if 4 raised to the 2x is 16. What is x? That's not hard, right? What is x?* **Student:** *It's just 2. Four squared is 16.* **Teacher:** *Right. I'll just note that.*	$4^x = 16$ $x = 2$ $4^2 = 16$
Teacher: *The second problem is 2 raised to the 2x equals 16. Solve for x.* Students think briefly.	$2^{2x} = 16$

(continued)

Teacher: *What are you thinking?* **Student:** *It's 2, because 2 to the what is 16? That's 4. So 2x = 4 so x is 2.* **Teacher:** *Since the 2s, the bases, are the same, the exponents have to be equivalent? That seems helpful.*	bases are the same, exponents are equivalent $2^{2x} = 16 = 2^{4}$ $2x = 4$ $x = 2$
Teacher: *Look at these two problems we just solved. We've got two different things equal to 16, so are they equivalent? Is 4 to the x equivalent to 2 raised to 2x? How could we reason about the equivalence?* *Now that you've thought about it for a bit, turn and work with a partner.*	$4^{x} = 2^{2x}$
Teacher: *What are you thinking?* **Student:** *We thought about how 2 raised to 2x is equivalent to 2 squared all raised to x. That's like 4 to x.* **Student:** *That makes sense. We went the other direction. Four is equivalent to 2 squared and that's like 2 raised to 2x.*	$2^{2x} = \left(2^{2}\right)^{x} = 4^{x}$ $4^{x} = \left(2^{2}\right)^{x} = 2^{2x}$
Teacher: *So, we kind of have a theme going of being able to rewrite things raised to exponents in equivalent ways. Nice. The next question is to solve for x when 2 raised to the 2x is equivalent to 2. Find x.* *Before I set you loose, does anyone have any initial thoughts?* **Student:** *I know x isn't very big. Maybe a fraction?* Students work briefly.	$2^{2x} = 2$
Teacher: *So, what's x? Why?* **Student:** *We found x is one-half because 2 without an exponent is raised to the 1, so 2x = 1, so x is one-half.*	$x = \frac{1}{2}$ $2^{2x} = 2^{1}$ $2x = 1$ $x = \frac{1}{2}$
Teacher: *The next problem is to find x if 4 to the x is 2. What's x? Any initial thoughts?* **Student:** *It's a fraction again because 4 is bigger than 2.* Students work briefly.	$4^{x} = 2$
Teacher: *Some of you are smiling. What's up?* **Student:** *Well, x is one-half again. I'm seeing connections.* **Teacher:** *Tell us about it.*	$x = \frac{1}{2}$

Student: *Well 4 is 2 squared so then it's 2 squared to the x and it's like the problem we did before.* **Student:** *And I knew that the square root of 4 is 2, so square root is the exponent ½.*	$4^x = (2^2)^x = 2^{2x} = 2$ $4^{\frac{1}{2}} = 2$
Teacher: *So, again there's this theme of writing things in equivalent ways to help think about finding x. Nice. The next problem is 9 to the x equals 27, find x. I wonder if there's some equivalencies that might help here?* *Let's reason about how big x is? What are some estimates?* **Student:** *Since 9 to the first is 9 and 9 squared is 81, x has to be in between 1 and 2.* **Teacher:** *Does anyone agree? Disagree? Okay, let's see what you find.* Students work while the teacher circulates and listens in.	$9^x = 27$
Teacher: *What were you guys working on? Get us started please.* **Student:** *I just kind of figured out that if you found the square root of 9, 3, then you cube 3 and that's 27.* **Teacher:** *So what is x?* **Student:** *It's one-half times 3, so three-halves.* **Teacher:** *So, kind of trying things, you were able to guess that? That's great! I wonder if there's a way to think about it that is a little more generalizable? Did anyone think to rewrite 9 and 27 as something equivalent?* **Student:** *We thought that 9 is 3 squared and 27 is 3 cubed.* **Student:** *Yeah, so if you write that as 3 squared all to the x, that's like 3 to the 2x, so 2x equals 3.*	$(9^{\frac{1}{2}})^3 = (3)^3 = 27$ $x = \frac{3}{2}$ $9^x = (3^2)^x = 3^{2x} = 27 = 3^3$ $2x = 3$ $x = 1.5 = \frac{3}{2}$
Teacher: *Great work. I wonder if that might influence how you solve for x when 125 to the x is equivalent to 25. Go!* Students work while the teacher circulates and listens in.	$125^x = 25$
Teacher: *Did anyone try rewriting with equivalencies?* **Student:** *Yes, we did. 125 is 5 cubed raised to the x and 25 is 5 squared. So 5 to the 3x is equal to 5 squared.* **Student:** *So, now 3x = 2 and x is ⅔.* **Teacher:** *Great. Did anyone look at it and guess some values for x?* **Student:** *I thought that was pretty cool how they were thinking about it so I tried and I know that 125 is 5 cubed, so I cube rooted 125 and then squared the 5. That's 25!* **Teacher:** *So, what's x?* **Student:** *Cube root and then square, that's 2 divided by 3, two-thirds.* **Teacher:** *Lots of nice reasoning using rational exponents today!*	$(5^3)^x = 5^2$ $5^{3x} = 5^2$ $3x = 2$ $x = \frac{2}{3}$ $(125^{\frac{1}{3}})^2 = (5)^2 = 25$ $x = \frac{2}{3}$

(continued)

Teacher: *How would you summarize some of the things that came up in this string today?*

Elicit the following:

- *Numbers raised to fractional exponents can be thought about in parts, first do one part and then the other.*
- *Thinking about equivalencies can be helpful, like* $4^x = (2^2)^x = 2^{2x}$.
- *Fractional exponents can represent roots, like* $x^{\frac{1}{2}} = \sqrt{x}$ *represents the square root of x and* $x^{\frac{1}{3}} = \sqrt[3]{x}$ *represents the cube root of x.*
- *When the bases are the same, the exponents are equivalent.*

Sample Final Display

Your display could look like this at the end of the problem string:

$3^{12} = (3^{4})^3$ $\quad$ $3^{12} = 3^4 \cdot 3^4 \cdot 3^4 = (3^4)^3$

$3^{12} = (3^{6})^{2}$ $\quad$ $3^{12} = (3^6)^2 = 3^6 \cdot 3^6$

$9^{\frac{3}{2}} = (9^{\frac{1}{2}})^3$ $\quad$ $9^{\frac{3}{2}} = (9^{\frac{1}{2}})^3$

$9^{\frac{3}{2}} = (9^{3})^{\frac{1}{2}}$ $\quad$ $9^{\frac{3}{2}} = (9^3)^{\frac{1}{2}}$

$4^x = 16$

$x = 2$ $\quad$ bases are the same, exponents are equivalent

$2^{2x} = 16$ $\quad$ $2^{2x} = 16 = 2^4$ $\quad$ $4^x = 2^{2x}$

$x = 2$ $\quad$ $2x = 4$ $\quad$ $2^{2x} = (2^2)^x = 4^x$ $\quad$ $4^x = (2^2)^x = 2^{2x}$

$x = 2$

$2^{2x} = 2$ $\quad$ $2^{2x} = 2^1$

$x = \frac{1}{2}$ $\quad$ $2x = 1$

$x = \frac{1}{2}$

$4^x = 2$ $\quad$ $4^x = (2^2)^x = 2^{2x} = 2$ $\quad$ $4^{\frac{1}{2}} = 2$

$x = \frac{1}{2}$

$9^x = 27$ $\quad$ $(9^{\frac{1}{2}})^3 = (3)^3 = 27$ $\quad$ $9^x = (3^2)^x = 3^{2x} = 27 = 3^3$

$x = \frac{3}{2}$ $\quad$ $2x = 3$

$x = 1.5 = \frac{3}{2}$

$125^x = 25$ $\quad$ $(5^3)^x = 5^2$ $\quad$ $(125^{\frac{1}{3}})^2 = (5)^2 = 25$

$x = \frac{2}{3}$ $\quad$ $5^{3x} = 5^2$

$3x = 2$

$x = \frac{2}{3}$

Facilitation Notes

This version of the problem string lists short notes for important teacher moves during the string. After you've done the string yourself and studied the relationships involved, you might make similar notes for the things you want a reminder of or deem important.

$3^{12} = (\_)^3$	The first 4 are optional—how fluent are students with power of power? How can we write 3 raised to the 12th as something raised to the third?
$3^{12} = (3^6)^{\_}$	Repeat. So if we have something raised to an exponent, we can rewrite it in an equivalent form?
$9^{\frac{3}{2}} = (\_)^3$	Repeat. What does an exponent of 3/2 mean?
$9^{\frac{3}{2}} = (\_)^{\frac{1}{2}}$	Repeat.
$4^x = 16$	Solve for x. Quick.
$2^{2x} = 16$	Solve for x. The previous and this both = 16. Are they equivalent? Write equivalencies using power of power.
$2^{2x} = 2$	Any intial thoughts? Between 0 and 1? Solve. Quick.
$4^x = 2$	Initial estimates? x is between 0 and 1? Turn and talk. Compare to previous. Rewriting using equivalencies can help!
$9^x = 27$	I wonder if equivalencies can help? What does 9 to the 3/2 mean?
$125^x = 25$	How can equivalencies help? What does 125 to the 2/3 mean?

PROBLEM STRINGS LESSON

4.4 Exponent Relationships 4

At a Glance

$$\frac{x^{15}y^{-2}}{x^{16}y^{-3}} = \frac{1}{xy}$$

$$\left(\frac{x^{100}z^{-48}}{y}\right)^{-1} = \frac{yz^{48}}{x^{100}}$$

$$\left(\frac{x^{3}y^{-14}z^{2}}{x}\right)^{-\frac{1}{2}} = \frac{xy^{7}}{z}$$

Objectives

The goal of this problem string is to have some fun with exponent relationships, strengthening students' ability to recognize and create equivalent expressions using many relationships simultaneously. By discussing and comparing different strategies, students become more adept at choosing efficient and clever approaches.

Placement

This is the fourth in a series of four problems strings that work with exponent relationships. This string serves to give students experience bringing many exponent relationships together to find equivalencies. Use this problem string after students have generalized the exponent relationships utilized in these problems.

You can use this problem string before or during textbook Lesson 4.4 Applications of Exponential and Power Equations.

Guiding the Problem String

This problem string is in the format of equations that are either true or false. As you present each problem, ask students to reason about relationships to decide if the statement is true or false. As students decide if the statement is true or false, press them to justify their reasoning. If the statement is false, ask for a true statement.

The first problem is false and requires students to make sense of negative exponents and division. The second problem is true, bringing in the relationship of a quotient raised to a negative exponent, where students need to acknowledge that y without a power is understood as y to the first power, y^1. The third problem is false, bringing in the relationship of a rational exponent.

For all of the problems, compare the idea of simplifying first and then dealing with the negative exponents or vice versa. Encourage students to understand both strategies and to be efficient and sophisticated in their choices rather than choosing the same strategy every time.

Be precise in your language. Instead of "since they're dividing, you can subtract," consider, "since the x's are a power divided by a power, you can subtract the exponents." Instead of "the negative exponent means they all flip," consider, "powers raised to negative exponents are equivalent to the reciprocal of the base raised to the opposite sign of the exponent." Instead of positional word like "over," consider, "numerator" and "denominator" to describe terms. Restate imprecise student comments when needed.

About the Mathematics

The strategies discussed in this problem string are not the absolute best moves nor the only way to approach the problem.

The equations in this string are statements of equivalency, true or false, but not an equation to solve for x. The purpose is for students to reason about equivalency using relationships.

Sample Interactions

Use the following as you plan how to elicit and model student strategies. This is not meant as a script, but as a view into the relationships involved and the intent of the problem string.

Teacher: *Let's kickoff today by warming up our brains with a problem string. You will not need paper and pencil for most of this string, maybe all of it. The format for this string is a bit different than many we've done. I'm going to give you a statement, an equation, and you decide if the statement is true, if the expressions are actually equivalent. For example, I could say that jelly beans are the same as chocolate chips. You would respond that my statement is not true. Okay, our first problem is this equation I'll write on the board. Is this a true statement? Are the expressions equivalent?* Brief think time.	$\frac{x^{15}y^{-2}}{x^{16}y^{-3}} = \frac{1}{xy}$
Teacher: *To start us off, true or false?* **Student:** *True.* **Student:** *False.* **Teacher:** *Excellent, we'll have a good debate. First, what are you thinking about the x's in both expressions?* **Student:** *That's easy. The x's divide out and you just have one x in the denominator so that part's good.* **Teacher:** *Everyone agree?* **Student:** *Yes, but I was trying to do the rule and subtract and I had a negative on the top and stuff. It would be a lot easier to just think about it that way.* **Teacher:** *Either way, we end up where we've dealt with the x's and we still need to deal with the y's.*	$\frac{x^{15}y^{-2}}{x^{16}y^{-3}} = \frac{\cancel{x^{15}}y^{-2}}{x\cdot\cancel{x^{15}}y^{-3}} = \frac{y^{-2}}{xy^{-3}}$ $\frac{x^{15}y^{-2}}{x^{16}y^{-3}} = \frac{x^{15-16}y^{-2}}{y^{-3}} = \frac{x^{-1}y^{-2}}{y^{-3}} = \frac{y^{-2}}{xy^{-3}}$
Student: *So, I did the subtraction thing again and got −2 minus −3 is, ... −2 plus 3, that's 1 so one y should be on top. Yes, so I think the original equation is false.* **Teacher:** *So, you decided to use the relationship of division of powers, subtracting the exponents. Did any of you think of it differently?*	$\frac{y^{-2}}{xy^{-3}} = \frac{y^{-2-(-3)}}{x} = \frac{y^{-2+3}}{x} = \frac{y^{1}}{x}$
Student: *I didn't want to mess with the negative exponents, so I moved the y's to make their exponents positive. So y squared on the bottom and y cubed on the top.* **Teacher:** *The y squared is in the denominator and the y cubed in the numerator?* **Student:** *Yes, so that's one y left in the numerator and I got y over x too. So it's false.* **Teacher:** *When the y's divide to 1, you end up with y divided by x. And you agree that the original statement is false.*	$\frac{y^{3}}{xy^{2}} = \frac{y}{x}$

(continued)

Teacher: *Let's look at what we've got up here. How would you describe the strategies? What were the first moves these students were thinking about? The first move often helps us see into their mind of how they decided to attack the problem, which relationships to use. Turn and talk to your partner about this.*

The teacher models the rest of their thinking and elicits a description of their strategy.

$$\frac{x^{15}y^{-2}}{x^{16}y^{-3}} = \frac{1}{xy}$$

False

$$\frac{x^{15}y^{-2}}{x^{16}y^{-3}} = \frac{\cancel{x^{15}}y^{-2}}{x\cdot\cancel{x^{15}}y^{-3}} = \frac{y^{-2}}{xy^{-3}}$$ divide the x's

$$\frac{y^{-2}}{xy^{-3}} = \frac{y^{-2-(-3)}}{x} = \frac{y^{-2+3}}{x} = \frac{y}{x}$$ subtract the exponents

$$\frac{y^{3}}{xy^{2}} = \frac{y}{x}$$ get the y's positive, then divide the y's

$$\frac{x^{15}y^{-2}}{x^{16}y^{-3}} = \frac{x^{15-16}y^{-2}}{y^{-3}} = \frac{x^{-1}y^{-2}}{y^{-3}} = \frac{y^{-2}}{xy^{-3}}$$ subtract the exponents

Teacher: *Which of the strategies do you wish your brain would be inclined to try the next time you hit a problem like this? Subtract exponents, a bunch of negatives, or get everything positive and divide to 1? Turn to your partner and discuss that please.*

Partners discuss while the teacher listens in.

Teacher: *The next problem is this equation I'll write on the board. Is it true or false?*

Think time.

$$\left(\frac{x^{100}z^{-48}}{y}\right)^{-1} = \frac{yz^{48}}{x^{100}}$$

Teacher: *What are you thinking? True? False? I am seeing that we disagree again. Did anyone work with the stuff on the inside of the parentheses first?*

Student: *I did. I decided I like that idea of getting everything positive, so I put the z's on the bottom.*

Teacher: *You know it's equivalent to put the z's in the denominator?*

Student: *Yeah, so then I multiplied the −1 through all of the exponents and got x to the −100th over y, z to the −48th. So, it's almost like the original, because the z's would go up and the x's would go down and you'd end up with z to the 48th over x to the 100th, y.*

$$\left(\frac{x^{100}z^{-48}}{y}\right)^{-1} = \left(\frac{x^{100}}{yz^{48}}\right)^{-1} = \frac{x^{-100}}{yz^{-48}} = \frac{z^{48}}{x^{100}y}$$

Teacher: *Like this? Any comments?*

Student: *But what about the y?*

Student: *What do you mean?*

Student: *When you did the exponent of −1, you didn't do it to the y. Isn't that y to the exponent of 1, so then it should be y to the −1. Which means the y is in the numerator. Right?*

Student: *Oh, right. Yes, I agree. So, actually I guess it's true.*

$$\left(\frac{x^{100}z^{-48}}{y^{1}}\right)^{-1} = \left(\frac{x^{100}}{y^{1}z^{48}}\right)^{-1} = \frac{x^{-100}}{y^{-1}z^{-48}} = \frac{yz^{48}}{x^{100}\cancel{y}} = \frac{yz^{48}}{x^{100}}$$

should be

Teacher: *These guys are thinking it's true. Did anyone work with the exponent of −1 first? I'm curious how that would work.*

Student: *Yes, that's what we were thinking. That exponent of −1 just made everything flip.*

Teacher: *What do you mean?*

Student: *That negative one exponent means that if you multiplied it through, the sign on every exponent changes, so you can think of them all moving from the numerator to the denominator and vice versa. So, you get y in the numerator and x to the 100th and z to the −48 in the denominator.*

$$\left(\frac{x^{100}z^{-48}}{y}\right)^{-1} = \frac{y}{x^{100}z^{-48}}$$

Student: *Oh, that's interesting. I'm thinking about that.*

Teacher: *It sounds like this is a new idea for some of you. Turn to a partner and discuss this for a bit.*

What are you thinking?

Student: *It is interesting that the negative exponent can affect everything like that. It makes sense. And it seems like it will be useful. From there, you just throw the z to the numerator. So, it is true.*

$$\left(\frac{x^{100}z^{-48}}{y}\right)^{-1} = \frac{y}{x^{100}z^{-48}} = \frac{yz^{48}}{x^{100}}$$

Teacher: *How would you describe each of these strategies?*

The teacher elicits descriptions and notes them on the display.

$$\left(\frac{x^{100}z^{-48}}{y^{1}}\right)^{-1} = \left(\frac{x^{100}}{y^{1}z^{48}}\right)^{-1} = \frac{x^{-100}}{y^{-1}z^{-48}} = \frac{yz^{48}}{x^{100}\cancel{y}} = \frac{yz^{48}}{x^{100}}$$

should be

multiply by the −1 exponent

$$\left(\frac{x^{100}z^{-48}}{y}\right)^{-1} = \frac{y}{x^{100}z^{-48}} = \frac{yz^{48}}{x^{100}}$$

reciprocate the whole expression because the −1 exponent

Teacher: *And the last problem of today's string. Is this equation true or false? And how do you know?*

If you want to use a pencil to keep track of your thinking, that's fine, but strive to use relationships and not get caught up in procedure.

$$\left(\frac{x^{3}y^{-14}z^{2}}{x}\right)^{-\frac{1}{2}} = \frac{xy^{7}}{z}$$

Teacher: *I noticed that you two started a strategy and then abandoned it. Tell us more about that please.*

Student: *Yes, we started by multiplying the −½ exponent to all of the exponents inside the parentheses, but it got tricky fast. The x cubed became x to the −3⁄2 and the x in the denominator would be x to the −½. Ick.*

Teacher: *So, you could do that, but you decided to try something else. Nice problem solving.*

$$\left(\frac{x^{3}y^{-14}z^{2}}{x}\right)^{-\frac{1}{2}} = \frac{x^{-\frac{3}{2}}\ldots\ldots}{x^{-\frac{1}{2}}}$$

seems like too much, try something else...

(continued)

Student: *We decided to clean things up in the parentheses first. Divide out the extra x's and put the y in the denominator. Then multiply the −1 exponent with all of the exponents. Then because they were all raised to negative exponents, we put them all where they weren't, the y in the numerator and the x and z in the denominator. So, it's false. The x should not be in the numerator.* **Teacher:** *How could we describe this approach?* **Student:** *Clean up inside the parentheses, get all the exponents positive, then multiply by the negative exponent.*	$\left(\frac{x^3y^{-14}z^2}{x}\right)^{-\frac{1}{2}}=\left(\frac{x^2z^2}{y^{14}}\right)^{-\frac{1}{2}}=\frac{x^{-1}z^{-1}}{y^{-7}}=\frac{y^7}{xz}$ inside () first, all exponents +, then × by the exponent
Teacher: *Did anyone use the negative exponent to reciprocate everything?* **Student:** *Yes, that's what we did first.* The teacher models the rest of their thinking and elicits a description of their strategy.	$\left(\frac{x^2y^{-14}z^2}{1}\right)^{\frac{1}{2}}=\left(\frac{1}{x^2y^{-14}z^2}\right)^{\frac{1}{2}}=\frac{y^7}{xz}$ reciprocate the whole expression because the −1 exponent
Teacher: *I heard you guys talking about how the y helped you. Say more about that please.* **Student:** *We noticed that the y to the −14 raised to the −½ ends up being y^7. That helped us think about doing the same thing to the rest of it.* The teacher models the rest of their thinking and elicits a description of their strategy.	$\left(\frac{x^2y^{-14}z^2}{1}\right)^{-\frac{1}{2}}=\frac{x^{-1}y^7z^{-1}}{1}=\frac{y^7}{xz}$ inside () first, then × by the exponent

Teacher: *How would you summarize some of the things that came up in this string today?*

Elicit the following:

- *Look at the whole equation first before deciding on a first move.*
- *It can be helpful to consider the effect of a negative exponent, reciprocating the entire rational expression.*
- *You can try simplifying inside the parentheses first.*
- *You don't have to simplify inside the parentheses first.*

Sample Final Display

Your display could look like this at the end of the problem string:

$$\frac{x^{15}y^{-2}}{x^{16}y^{-3}} = \frac{1}{xy}$$

False

$$\frac{x^{15}y^{-2}}{x^{16}y^{-3}} = \frac{\cancel{x^{15}}y^{-2}}{x \cdot \cancel{x^{15}}y^{-3}} = \frac{y^{-2}}{xy^{-3}}$$

divide the x's

subtract the exponents

$$\frac{y^{-2}}{xy^{-3}} = \frac{y^{-2-(-3)}}{x} = \frac{y^{-2+3}}{x} = \frac{y^{1}}{x}$$

get the y's positive, then divide the y's

$$\frac{y^{3}}{xy^{2}} = \frac{y}{x}$$

$$\frac{x^{15}y^{-2}}{x^{16}y^{-3}} = \frac{x^{15-16}y^{-2}}{y^{-3}} = \frac{x^{-1}y^{-2}}{y^{-3}} = \frac{y^{-2}}{xy^{-3}}$$

subtract the exponents

$$\left(\frac{x^{100}z^{-48}}{y}\right)^{-1} = \frac{yz^{48}}{x^{100}}$$

True

$$\left(\frac{x^{100}z^{-48}}{y^{1}}\right)^{-1} = \left(\frac{x^{100}}{y^{1}z^{48}}\right)^{-1} = \frac{x^{-100}}{y^{-1}z^{-48}} = \frac{yz^{48}}{x^{100}\cancel{y}} = \frac{yz^{48}}{x^{100}}$$

should be

multiply by the −1 exponent

$$\left(\frac{x^{100}z^{-48}}{y}\right)^{-1} = \frac{y}{x^{100}z^{-48}} = \frac{yz^{48}}{x^{100}}$$

reciprocate the whole expression because the −1 exponent

$$\left(\frac{x^{3}y^{-14}z^{2}}{x}\right)^{-\frac{1}{2}} = \frac{xy^{7}}{z}$$

False

$$\left(\frac{x^{3}y^{-14}z^{2}}{x}\right)^{-\frac{1}{2}} = \frac{x^{-\frac{3}{2}}\ldots\ldots}{x^{-\frac{1}{2}}}$$

seems like too much, try something else...

$$\left(\frac{x^{3}y^{-14}z^{2}}{x}\right)^{-\frac{1}{2}} = \left(\frac{x^{2}z^{2}}{y^{14}}\right)^{-\frac{1}{2}} = \frac{x^{-1}z^{-1}}{y^{-7}} = \frac{y^{7}}{xz}$$

inside () first, all exponents +, then × by the exponent

$$\left(\frac{x^{2}y^{-14}z^{2}}{1}\right)^{\frac{1}{2}} = \left(\frac{1}{x^{2}y^{-14}z^{2}}\right)^{\frac{1}{2}} = \frac{y^{7}}{xz}$$

reciprocate the whole expression because the −1 exponent

$$\left(\frac{x^{2}y^{-14}z^{2}}{1}\right)^{-\frac{1}{2}} = \frac{x^{-1}y^{7}z^{-1}}{1} = \frac{y^{7}}{xz}$$

inside () first, then × by the exponent

(continued)

Facilitation Notes

This version of the problem string lists short notes for important teacher moves during the string. After you've done the string yourself and studied the relationships involved, you might make similar notes for the things you want a reminder of or deem important.

	No paper and pencil needed. Today's string is about statements—true or false?
$\frac{x^{15}y^{-2}}{x^{16}y^{-3}} = \frac{1}{xy}$	Is this equation true or false? What are you thinking about the xs? The ys? How could we describe your first moves, strategy?
${\frac{x^{100}z^{-48}}{y}}^{-1} = \frac{yz^{48}}{x^{100}}$	True or false? Did anyone work on the inside of the () first? The exponent of −1 first? Turn and talk. How could we describe your strategies?
$\left(\frac{x^{3}y^{-14}z^{2}}{x}\right)^{-\frac{1}{2}} = \frac{xy^{7}}{z}$	True or false? Did anyone start something and abandon? Why? Did anyone clean up inside the () first? Use the neg exponent to reciprocate? Multiply by the exponent of −1/2? How could we describe your strategies?

PROBLEM STRINGS LESSON

4.5 Inverse Functions

At a Glance

$f(4)=\_\_$

$f^{-1}(2)=\_\_$

$f(\frac{1}{2})=\_\_$

$f(\_\_)=0$

$f(\_\_)=1$

$f(\frac{1}{4})=\_\_$

$f^{-1}(3)=\_\_$

$f^{-1}(x)=\_\_$

Objectives

The purpose of this problem string is to reinforce inverse functions and function notation by students working with the graph of a function and its inverse. Since the inverse is the exponential function $y=2^x$, the string also serves as an informal introduction or reinforcement of logarithms, in this case $y=\log_2 x$.

Placement

This string could be used to introduce or quickly review function notation. It also previews the notion of logarithms because the unnamed given function is $y=\log_2 x$, but students reason about it using the graph.

You could use this problem string during or after textbook Lesson 4.5 Building Inverses of Functions.

Guiding the Problem String

The problems in this string alternate between using $f(x)$ and $f^{-1}(x)$ notation and between asking for x or y-values. Help keep students grounded by pushing for justification about which function they are using and which value they are given.

Limit your use of pronouns. Instead of "So the inverse of it is this over there," say, "So the inverse of f at 2 is 4 and that is here (pointing) on the graph."

The first two problems should go quickly but use them to set the stage. Physically point to the ordered pairs as you plot the points. Make sure you label everything.

The next five problems follow the same format except the partner problem is not listed, so be aware that when the problem asks for $f(\frac{1}{2})=\_\_$, after students find and justify that answer, you will follow with "So, now what do we know about the inverse of f?"

The last problem asks students to generalize about using the points they have found for $f^{-1}(x)$. Depending on your goal for this problem string, you could do this quickly, spend some time finding the function if needed, or use it as a jumping off point to introduce logarithms.

About the Mathematics

The inverse function notation, where $f^{-1}(x)$ is the inverse of $f(x)$, is social knowledge. We use it by convention. Therefore, do not make students guess about it, just tell them. The sense of an inverse function and how to use it to denote ordered pairs and the relationship between and function and its inverse is logical mathematical knowledge and must be constructed through experience.

(continued)

Sample Interactions

Use the following as you plan how to elicit and model student strategies. This is not meant as a script, but as a view into the relationships involved and the intent of the problem string.

Enter the function $y = \log_2 x$ in the display grapher, without the equation showing. Alternatively project just the grid using the display grapher and sketch the function on the display as shown.

Teacher: *Let's get warmed up with a problem string. The problems in today's string are based on a function I'm going to call f(x) and it looks like this red graph here. How would you describe this function? Turn and briefly describe the function f to your partner.*

Students turn and briefly talk.

Teacher: *What did your partner say?*

Student: *It's red. It's increasing. It's in quadrants one and four. It looks like it might be approaching the y-axis and not crossing over it. It kind of looks like the square root function, but we can't see the starting point."*

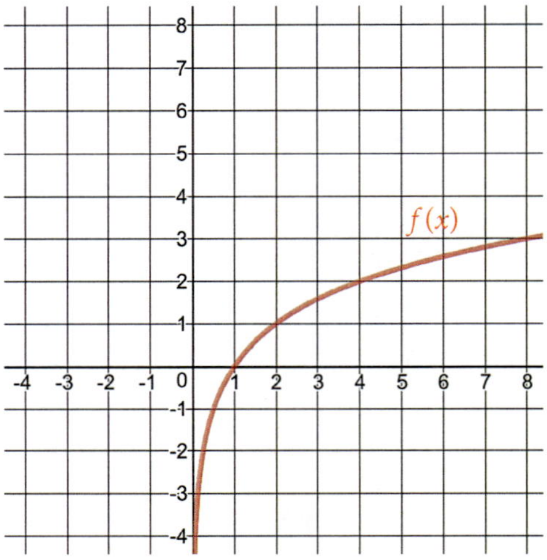

Teacher: *The first problem of today's string is to find f(4). What is the function value, the y-value, at the x-value of 4?*

$f(4) = \_\_$

Student: *Two, because if you go to where x is 4, y is 2.*

Teacher: *I'll write that ordered pair here and plot it on the graph.*

$f(4) = \underline{\ 2\ }$ (4, 2)

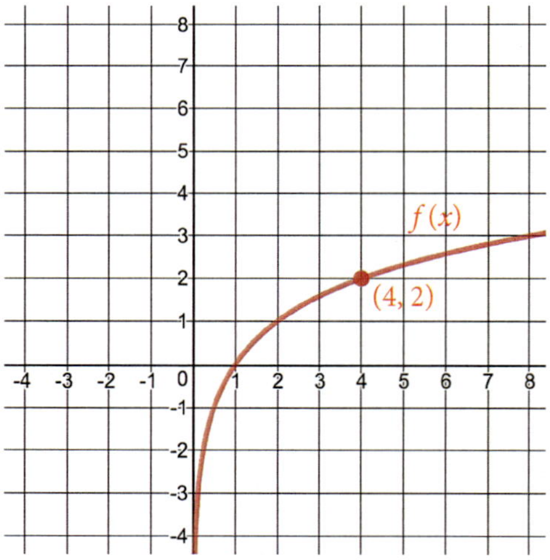

Teacher: *The next problem is what is f inverse of 2,* $f^{-1}(2) = \_\_$ *? What does this mean?*

Student: *What does that little −1 mean?*

Teacher: *This notation means the function that is the inverse of f. So what does it mean to ask about the inverse of f at 2?*

Student: *The inverse function is where the x- and the y-coordinates are switched, right?*

Student: *So, if the point (4, 2) is on f, then (2, 4) is on the inverse of f.*

The teacher records the point (2, 4) and plots it on the graph.

$f(4) = \underline{2}$ (4, 2)

$f^{-1}(2) = \underline{4}$ (2, 4)

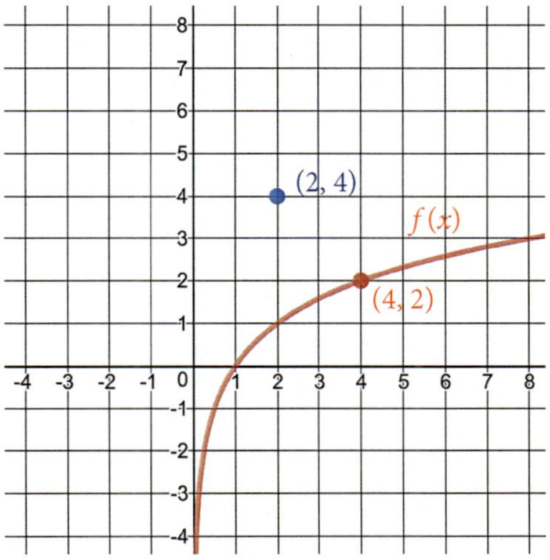

Teacher: *The next problem is to find the function f at one-half,* $f(\frac{1}{2}) = \_\_$.

The teacher elicits responses about the function at 0.5, writes the ordered pair, and plots and labels the point.

Teacher: *What does that mean about the inverse of f, or you can say f inverse?*

The teacher writes the point in inverse notation, writes the ordered pair, and plots and labels the point.

$f(4) = \underline{2}$ (4, 2)

$f^{-1}(2) = \underline{4}$ (2, 4)

$f(\frac{1}{2}) = \underline{-1}$ $(\frac{1}{2}, -1)$ $f^{-1}(-1) = \frac{1}{2}$ $(-1, \frac{1}{2})$

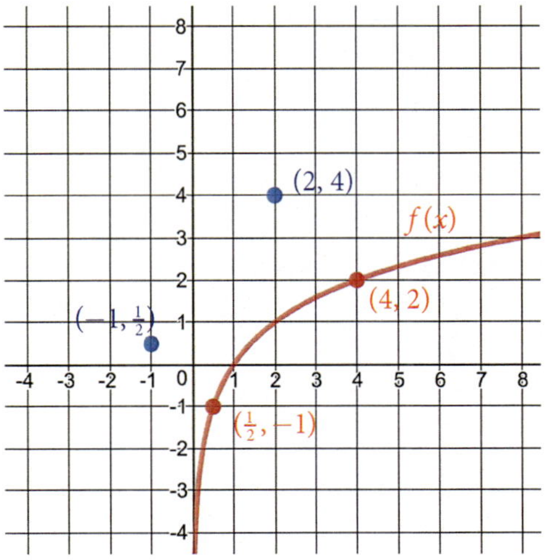

(continued)

Teacher: *The next problem is a bit different. This time we know the function value is 0, but we don't know where, f (____) = 0. What does this mean?*

Student: *The y-value is 0, so it looks like the x-value must be 1.*

Teacher: *So the red function at the x-value of 1 is 0? Great.*

The teacher writes the ordered pair and plots and labels the point.

Teacher: *What does that mean about f inverse?*

The teacher writes the point in inverse notation, writes the ordered pair, and plots and labels the point.

$f(4) = \underline{2}$ (4, 2)

$f^{-1}(2) = \underline{4}$ (2, 4)

$f(\frac{1}{2}) = \underline{-1}$ $(\frac{1}{2}, -1)$ $f^{-1}(-1) = \frac{1}{2}$ $(-1, \frac{1}{2})$

$f(\underline{1}) = 0$ (1, 0) $f^{-1}(0) = 1$ (0, 1)

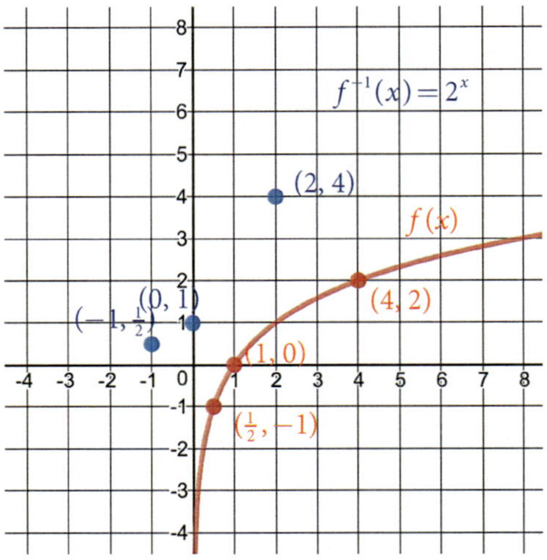

The teacher repeats with the next problem, f (__) = 1, writing, plotting, and labeling points.

The teacher then follows quickly with the next problem which establishes a bit more second quadrant behavior for the inverse, $f(\frac{1}{4}) = \_\_$, writing, plotting, and labeling points.

Teacher: *Hey, some of these points on the inverse look familiar. Hmmm ...*

$f(4) = \underline{2}$ (4, 2)

$f^{-1}(2) = \underline{4}$ (2, 4)

$f(\frac{1}{2}) = \underline{-1}$ $(\frac{1}{2}, -1)$ $f^{-1}(-1) = \frac{1}{2}$ $(-1, \frac{1}{2})$

$f(\underline{1}) = 0$ (1, 0) $f^{-1}(0) = 1$ (0, 1)

$f(\underline{2}) = 1$ (2, 1) $f^{-1}(1) = 2$ (1, 2)

$f(\frac{1}{4}) = \underline{-2}$ $(\frac{1}{4}, -2)$ $f^{-1}(-2) = \frac{1}{4}$ $(-2, \frac{1}{4})$

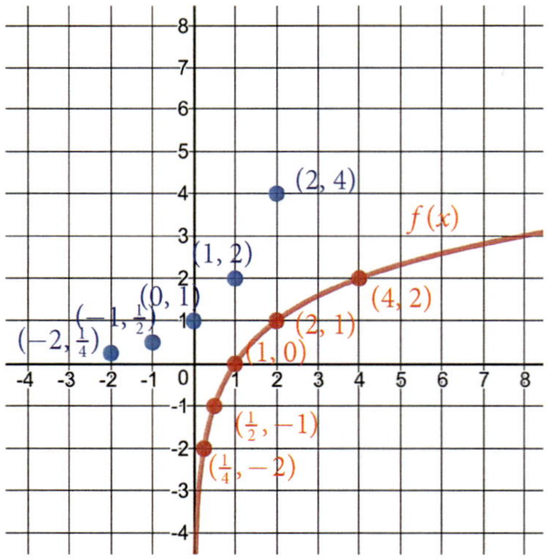

Teacher: *Let's change things up in a different way. This problem is to find f inverse at 3,* $f^{-1}(3) = \_\_$*. What does that mean?*

Student: *That's easy. It looks like at 3, the function is about one and a half.*

Teacher: *For which function, f or the inverse of f?*

Student: *What do you mean?*

Student: *I think that we aren't supposed to look for the x-value of 3 on the red function.*

Teacher: *What do we know?*

Student: *We know that the inverse of f has a point with x-value of 3.*

Teacher: *Does that tell us anything about f?*

Student: *I think that means that the y-value of f, the red function, is 3. So we need the x-value of that point. Will you point to the place where the red function has a y of 3?*

Teacher: *This point right here?*

Student: *Yes, that is the point (8, 3). Which means that the blue point is (3, 8).*

$f(4) = \underline{2}$ (4, 2)

$f^{-1}(2) = \underline{4}$ (2, 4)

$f(\frac{1}{2}) = \underline{-1}$ $(\frac{1}{2}, -1)$ $f^{-1}(-1) = \frac{1}{2}$ $(-1, \frac{1}{2})$

$f(\underline{1}) = 0$ (1, 0) $f^{-1}(0) = 1$ (0, 1)

$f(\underline{2}) = 1$ (2, 1) $f^{-1}(1) = 2$ (1, 2)

$f(\frac{1}{4}) = \underline{-2}$ $(\frac{1}{4}, -2)$ $f^{-1}(-2) = \frac{1}{4}$ $(-2, \frac{1}{4})$

$f^{-1}(3) = \underline{8}$ (8, 3) (3, 8)

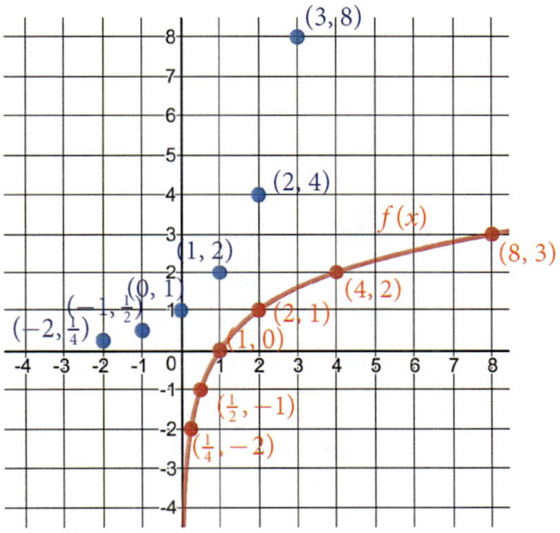

Teacher: *For the last problem today, let's get a little general. I wonder what you can tell me about the inverse of f at any x value? What is* $f^{-1}(x) = \_\_$*?*

Student: *What do you mean?*

Teacher: *If we could fill in points at every x-value, what would the inverse of f look like?*

Student: *You mean, like connect the points?*

$f^{-1}(x) = \_\_$

(continued)

Student: *Do you mean that if we found each inverse of every point that's on the red f right now, then we'd have the inverse of f?*

Teacher: *Yes, but I'm also wondering if the values that you see here for f inverse look like any functions we know? Turn to your partner and see what you can find looking at the relationships between the blue points.*

Students turn and talk to a partner while the teacher listens for students who are noticing the (0, 1) and (0, 2) points.

Teacher: *What do you think?*

Student: *It looks like an exponential function.*

Student: *We think it's $y = 2^x$. The y-values are doubling each time.*

Teacher: *Does everyone agree? Does anyone have any more evidence?*

Student: *It goes through the (0, 1) and then doubles. And if you go backwards, those values are halving.*

Teacher: *Let's type it in and see. Yep, make sense?*

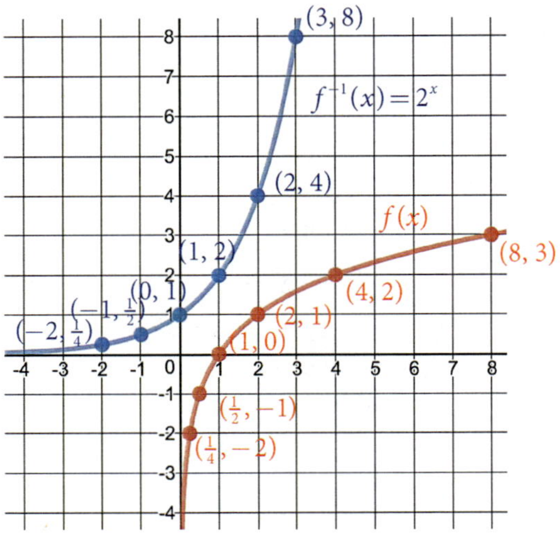

Teacher: *So, if the blue function is the exponential function $y = 2^x$, what kind of relationships are happening in our original function f? They are inverse functions right?*

Student: *It's almost like it's 2 to the ys?*

Student: *Yeah, like the x values are 2 raised to the y-values.*

Teacher: *Could I write that thinking this way: $x = 2^y$?*

Teacher: *How would you summarize some of the things that came up in this string today?*

Elicit the following:

- *The notation for the inverse function of f(x) is $f^{-1}(x)$.*
- *If (a, b) is a point on f, then (b, a) is a point on $f^{-1}(x)$ and vice versa.*
- *The inverse looks like a reflection of the function over the line y = x.*

Sample Final Display

Your display could look like this at the end of the problem string:

$f(4) = \underline{2}$	$(4, 2)$		
$f^{-1}(2) = \underline{4}$	$(2, 4)$		
$f(\frac{1}{2}) = \underline{-1}$	$(\frac{1}{2}, -1)$	$f^{-1}(-1) = \frac{1}{2}$	$(-1, \frac{1}{2})$
$f(\underline{1}) = 0$	$(1, 0)$	$f^{-1}(0) = 1$	$(0, 1)$
$f(\underline{2}) = 1$	$(2, 1)$	$f^{-1}(1) = 2$	$(1, 2)$
$f(\frac{1}{4}) = \underline{-2}$	$(\frac{1}{4}, -2)$	$f^{-1}(-2) = \frac{1}{4}$	$(-2, \frac{1}{4})$
$f^{-1}(3) = \underline{8}$	$(8, 3)$	$(3, 8)$	
$f^{-1}(x) = \underline{2^x}$			

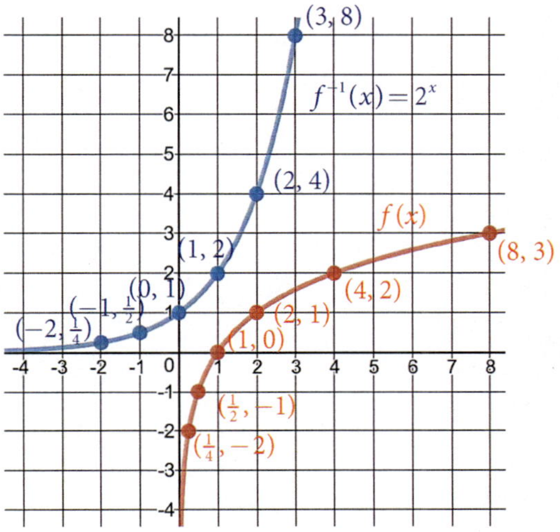

Facilitation Notes

This version of the problem string lists short notes for important teacher moves during the string. After you've done the string yourself and studied the relationships involved, you might make similar notes for the things you want a reminder of or deem important.

$f(4) = \_\_\_$	*Given that this graph is f(x) ...* *Write ordered pair, plot and label point. Use one color for all things f(x).*
$f^{-1}(2) = \_\_\_$	*This notation means the inverse of f(x).* *Write ordered pair, plot and label point. Use a different color for all things f^{-1}.*
$f(\frac{1}{2}) = \_\_\_$	*Repeat. What does this mean about f^{-1}?*
$f(\_\_\_) = 0$	*Let's change things up a bit. Now what?* *What does this mean about f^{-1}?*
$f(\_\_\_) = 1$	*Repeat. What does this mean about f^{-1}?*
$f(\frac{1}{4}) = \_\_\_$	*Quick. Use to establish some of the second quadrant behavior of f inverse.* *Some of these points look familiar...*
$f^{-1}(3) = \_\_\_$	*Let's change things up in a different way. Now what?* *What does this mean about f?*
$f^{-1}(x) = \_\_\_$	*Let's get a little general – what is the inverse function?* *How do you know?* *If that's true, that it's 2^x, what relationships are happening in it's inverse, f?*

PROBLEM STRINGS
LESSON
4.6 Introducing Logarithms

At a Glance

$L\,2, 8 = 3$

$L\,5, 25 = 2$

$L\,3, 81 = 4$

$L\,2, 16 =$ ___

$L\,9, 81 =$ ___

$L\,5, 125 =$ ___

$L\,11, 11 =$ ___

$\log_8 64 =$ ___

$\log_{42} 42 =$ ___

$\log_{10} 1{,}000 =$ ___

$\log_{16} 4 =$ ___

$\log_a c = b \Leftrightarrow$

Objectives

The goal of this string is to give students an experience thinking about logarithmic relationships, the relationships between numbers being raised to exponents and their powers, and ease students into logarithm notation.

Placement

This problem string can serve as an introduction to logarithms and logarithmic notation. Students should have prior experience with exponential relationships.

You can use this problem string to introduce textbook Lesson 4.6 Logarithms.

Guiding the Problem String

The first problems introduce a fictional notation system that encourages students to find the relationship between a number and a power of that number. As students are considering that relationship for different numbers, the traditional logarithm notation is introduced. Help student make the connection. Encourage them to be thinking about numbers and powers of those numbers and the exponents involved. The last five problems are given in conventional logarithm notation. Help students connect that to their previous work. The last problem asks students to generalize the relationship between the logarithm notation and the exponential relationships.

About the Mathematics

We have used a non-traditional notation system to help introduce the conventional logarithm notation. We do not intend that students do anything with the *L* notation after this string, but it might be helpful to pull back on if students get confused in the future. We choose *L* for the obvious connection to the word logarithm, but we chose the positioning because it can help students think about the exponential relationships involved, rather than rote memorization of the new and rather oblique logarithm notation.

In the relationships of $a^b = c$, a is the base, b is the exponent, and c is the power. Sometimes teachers say "*a* raised to the power of *b*," but we will use "*a* raised the exponent of *b*" to differentiate it from using "power" to describe *c*. For example, "Two raised to the fourth is 16, so 16 is a power of 2."

Sample Interactions

Use the following as you plan how to elicit and model student strategies. This is not meant as a script, but as a view into the relationships involved and the intent of the problem string.

Teacher: *The problem string today starts a little differently than usual. I'm going to use a capital L to represent a mystery function. You will look at three problems and figure out the relationships that are involved. So, for these problems, what is L doing?* *After you have studied and have some ideas, turn and talk with your partner.* Students think, then turn and talk. The teacher circulates, listening in and taking note of students who are thinking about exponents.	$L\,2, 8 = 3$ $L\,5, 25 = 2$ $L\,3, 81 = 4$

Teacher: *I heard you two wondering about some things. Tell us about that please.*

Student: *We think that L has something to do with raising to exponents.*

Student: *Yeah, 2 to the 3rd is 8, 5 squared is 25.*

Student: *And 3 to the 4th is 81.*

Teacher: *So there's something going on with exponents? I'm going to record those statements. Who agrees? What does L do?*

Student: *You're supposed to figure out the exponent. What's the exponent that relates the two numbers?*

$L\,2, 8 = 3$	$2^3 = 8$	How are 2 and 8 related? 8 is a power of 2.
$L\,5, 25 = 2$	$5^2 = 25$	What does L do? Find the exponent that relates the two numbers.
$L\,3, 81 = 4$	$3^4 = 81$	

Teacher: *Okay, so we have this weird function, that for now we're calling L, that relates numbers to a power of the number. Neat. The next problem is to use L to figure out that relationship between 2 and 16,* $L\,2, 16 =$ ___ Brief think time.	$L\,2, 16 =$ ___
Teacher: *First, what question is this asking?* **Student:** *Two to what is 16.* **Teacher:** *What do you think?* **Student:** *It's 4.* **Teacher:** *Why?* **Student:** *Because 2 times 2 times 2 times 2 is 16.* **Teacher:** *How would you write that with exponents?* **Student:** *Two raised to the 4th is 16.*	2 to what exponent is 16? 4 $2^4 = 16$

(continued)

The teacher repeats with the next three problems, asking students to interpret the L function, find the missing exponents and write the exponential relationships.	$L\,9, 81 = \underline{\ 2\ }$ $9^2 = 81$ $L\,5, 125 = \underline{\ 3\ }$ $5^3 = 125$ $L\,11, 11 = \underline{\ 1\ }$ $11^1 = 11$

Teacher: *So, interestingly in history, people wondered about these relationships. If you had a number and a power of that number, what would the exponent be? And they call that function a logarithm. And they write it a little differently than we have been. Let's go back through these last few problems and rewrite them the way that mathematicians have decided to use notation.*

So, starting with the fourth problem, $L\,2, 16 = 4$, *mathematicians have decided to write that as* $\log_2 16 = 4$. *Use this relationship to write the next three problems as logarithms.*

Students work while the teacher circulates, correcting the notation as necessary. Notation is social—just tell students. As students finish, the teacher records them on the board.

$L\,2, 16 = \underline{\ 4\ }$	$2^4 = 16$	2 to what power is 16? 4	$\log_2 16 = 4$
$L\,9, 81 = \underline{\ 2\ }$	$9^2 = 81$	$\log_9 81 = 2$	
$L\,5, 125 = \underline{\ 3\ }$	$5^3 = 125$	$\log_5 125 = 3$	
$L\,11, 11 = \underline{\ 1\ }$	$11^1 = 11$	$\log_{11} 11 = 1$	

Teacher: *Great! So the next problem is in that logarithm notation. You say it like this, "the logarithm base 8 of 64." What is the logarithm base 8 of 64?* *What does this mean?* **Student:** *So, since eight squared is 64 then the logarithm base 8 of 64 is 2. The exponent is 2.*	$\log_8 64 = \underline{\quad}$ $8^? = 64, ? = 2$ "the logarithm base 8 of 64" $\log_8 64 = \underline{\ 2\ }$ $8^? = 64, ? = 2$
The teacher repeats with the next three problems, asking students to interpret the logarithm notation, write the exponential relationships, and find the missing exponents. The teacher draws back on the L notation as needed, striving to help students make connections between the logarithm notation and exponential relationships.	$\log_{42} 42 = \underline{\ 1\ }$ $42^? = 42, ? = 1$ $\log_{10} 1{,}000 = \underline{\ 3\ }$ $10^? = 1{,}000, ? = 3$ $\log_{16} 4 = \underline{\ \frac{1}{2}\ }$ $16^? = 4, ? = \frac{1}{2}$
Teacher: *For the last problem today, let's get a little general. If we have this strange logarithm notation, where the logarithm base a of c is b, what does that imply? This symbol means that this statement implies something and vice versa.* **Student:** *It's like you're asking about the relationship between a and a power of a, c. So a to the exponent of b is c.*	$\log_a c = b \Leftrightarrow$ $\log_a c = b \Leftrightarrow a^b = c$

Teacher: *How would you summarize some of the things that came up in this string today?*

Elicit the following:

- *There is a notation to represent a number and a power of that number. It's called a logarithm.*
- *This notation means* $\log_a c = b \Leftrightarrow a^b = c$.
- *You say "the logarithm base a of c is b."*
- *a is the base, b is the exponent, and c is the power of a.*

Sample Final Display

Your display could look like this at the end of the problem string:

$L\ 2, 8 = 3$	$2^3 = 8$	How are 2 and 8 related? 8 is a power of 2.	
$L\ 5, 25 = 2$	$5^2 = 25$	What does L do? Find the exponent that relates the two numbers.	
$L\ 3, 81 = 4$	$3^4 = 81$		
$L\ 2, 16 = \underline{4}$	$2^4 = 16$	2 to what power is 16? 4	$\log_2 16 = 4$
$L\ 9, 81 = \underline{2}$	$9^2 = 81$	$\log_9 81 = 2$	
$L\ 5, 125 = \underline{3}$	$5^3 = 125$	$\log_5 125 = 3$	
$L\ 11, 11 = \underline{1}$	$11^1 = 11$	$\log_{11} 11 = 1$	
$\log_8 64 = \underline{2}$	$8^? = 64, ? = 2$	"the logarithm base 8 of 64"	
$\log_{42} 42 = \underline{1}$	$42^? = 42, ? = 1$		
$\log_{10} 1{,}000 = \underline{3}$	$10^? = 1{,}000, ? = 3$		
$\log_{16} 4 = \underline{\frac{1}{2}}$	$16^? = 4, ? = \frac{1}{2}$		
$\log_a c = b$	$a^b = c$		

(continued)

Facilitation Notes

This version of the problem string lists short notes for important teacher moves during the string. After you've done the string yourself and studied the relationships involved, you might make similar notes for the things you want a reminder of or deem important.

$L\,2,8=3$ $L\,5,25=2$ $L\,3,81=4$	Present first 3 problems all at once. How are 2 and 16 related? 16 is a power of 2. What does L do?
$L\,2,16=\underline{\;4\;}$	2 to what exponent is 16? 4.
$L\,9,81=\underline{\;2\;}$ $L\,5,125=\underline{\;3\;}$ $L\,11,11=\underline{\;1\;}$	Present next 3 problems all at once. Write the exponential relationship. Then bring students back together and introduce logarithm notation. So, mathematicians call this relationship a "logarithm". Here is the way they decided to write it. It represents the same relationships we have been working with. Have students write problems 4-7 in log notation. This is social.
$\log_8 64=\underline{\;2\;}$	This problem is in log notation. What does it mean? We say "logarithm base 8 of 64." Write the exponential relationship and find the exponent, not necessarily in that order.
$\log_{42} 42=\underline{\;1\;}$ $\log_{10} 1{,}000=\underline{\;3\;}$ $\log_{16} 4=\underline{\;\frac{1}{2}\;}$	Repeat with these 3 problems. Interpret log notation, write the exponential relationship, find the missing exponents, not necessarily in that order. Encourage students to think in terms of relationships.
$\log_a c=b \qquad a^b=c$	Let's get a little general. If we know this part, what does that imply?

PROBLEM STRINGS
LESSON

4.8 Sequences or Series?

At a Glance	
$a_0 = 4, a_n = a_{n-1} + 2$ $a_0 = 4, a_n = a_{n-1} \cdot 2$ $a_0 = 16, a_n = a_{n-1} - 2$ $a_0 = 16, a_n = a_{n-1} \cdot 0.5$	**Objectives** The goal of this problem string is to quickly help students solidify the difference between a sequence and a partial sum of a series by engaging students in finding the first four terms of the sequence and finding the fourth partial sum of the series.
	Placement This problem string can be used to introduce the idea of partial sums of series or to solidify the difference between a sequence and series. You can use this problem string after textbook Lesson 4.8 Partial Sums of Geometric Series.
	Guiding the Problem String This problem string should go quickly. The problems alternate between arithmetic and geometric sequences with similar definitions. Use these problems to help students differentiate between a sequence, a list of terms, and a series, the sum of a sequence. In the string, students are finding the fourth partial sum of each sequence. Encourage students to find the sum using additive relationships. Bring out that the sums of the first and last terms are equivalent to the sum of the middle terms of arithmetic sequences.
	About the Mathematics A sequence is a list of terms. A series is the indicated sum of terms of a sequence.

Important Questions

Use the following as you plan how to elicit and model student strategies.

- *What is an arithmetic sequence?*
- *What is a geometric sequence?*
- *What is the difference between a sequence and a series?*

How would you summarize some of the things that came up in this string today?

- *A sequence is a list of terms. You can represent it by listing all of the terms or describing them using sequence notation.*
- *A series is the sum of a sequence. It's not a list. It's a sum.*
- *We can find partial sums of a series by adding up the specified number of terms of the sequence.*
- *We can be in clever how we add up the terms of the sequence by looking for friendly combinations.*

(continued)

Sample Final Display

Your display could look like this at the end of the problem string:

$a_0 = 4, a_n = a_{n-1} + 2$	4, 6, 8, 10	$S_4 = 4+6+8+10 = 28$ 10 + 10 + 8	$S_4 = 4+6+8+10 = 28$ 14 + 14
$a_0 = 4, a_n = a_{n-1} \cdot 2$	8, 16, 32, 64	$S_4 = 8+16+32+64 = 120$ 40 + 80	
$a_0 = 16, a_n = a_{n-1} - 2$	16, 14, 12, 10	$S_4 = 16+14+12+10 = 52$ 30 + 22	$S_4 = 16+14+12+10 = 52$ 26 + 26
$a_0 = 16, a_n = a_{n-1} \cdot 0.5$	16, 8, 4, 2	$S_4 = 16+8+4+2 = 30$ 20 + 10	
	sequence—list of terms	series—sum of terms of a sequence	

partial sum—sum of a number of terms of a sequence

Facilitation Notes

This version of the problem string lists short notes for important teacher moves during the string. After you've done the string yourself and studied the relationships involved, you might make similar notes for the things you want a reminder of or deem important.

$a_0 = 4, a_n = a_{n-1} + 2$	*Here is the first problem. What do we call this? What kind of sequence is it?* *Find the first 4 terms of this sequence.* *Find the S_4, the sum of the first four terms, called the fourth partial sum of the series.* *Be clever! How can you group the numbers to find the sum easily?*
$a_0 = 4, a_n = a_{n-1} \cdot 2$	*What kind of sequence is this?* *Find the first 4 terms of this sequence.* *Find the S_4. Be clever!*
$a_0 = 16, a_n = a_{n-1} - 2$	*Repeat.* *Quick.*
$a_0 = 16, a_n = a_{n-1} \cdot 0.5$	*Repeat.* *What is the difference between a sequence and a series?*

PROBLEM STRINGS LESSON

5.0 Solving Quadratic Equations 1

At a Glance

$$x^2+4x-5=0$$

$$x^2-4x-5=0$$

$$-x^2+4x+5=0$$

$$-x^2-4x+5=0$$

Objectives

The goal of this Graphing Quadratic Equations series of problem strings is to help students gain facility with deciding on an efficient strategy to solve quadratic equations. This string helps students construct three strategies: factor using the zero product property, factor out a -1 and then factor using the zero product property, and factor using factor pairs.

Placement

This is the first in a series of four problem strings. Use this string if your students need work on factoring quadratic trinomials, need to construct the strategy of factoring using factor pairs, or if they have not considered trying to factor an equivalent form of a quadratic trinomial. This string could be delivered before lessons on solving quadratic equations using the quadratic formula. Students need experience with the zero product property.

You could deliver this string before textbook Chapter 5 or any time before 5.2 The Quadratic Formula. If you deliver it during 5.2 Completing the Square, be aware that a strategy students may use is to complete the square.

Guiding the Problem String

This string should go quickly, especially the first two problems. Elicit connections between the algebraic and graphic solutions. Use the optional second problem if students are not factoring using factor pairs. Spend more time on the third problem, wondering about the equivalent solutions from the first problem. If students do not see the equivalence, write and ask if this is true: $-(x^2-4x-5)=-x^2+4x+5$. By the end of the string, help students verbalize the strategy of solving quadratic equations by factoring using the zero product property, factoring using factor pairs, or factoring out a -1 and then factoring using the zero product property.

About the Mathematics

The absolute values of the coefficients stay the same from problem to problem, so students can concentrate on what happens as the signs change. The first two equations are easily factored, and because the last two problems are not easily factored, students might try to factor unconventionally or use the equivalence from the graph solutions to factor out a -1 and factor the resulting expression.

When factoring the quadratic x^2+bx+c using the zero product property, you consider the factors of c that sum to b. In the factor using factor pairs strategy, you solve $x^2+bx=c$ and therefore $x(x+b)=c$, by considering the factors of c that work.

Sample Interactions

Use the following as you plan how to elicit and model student strategies. This is not meant as a script, but as a view into the relationships involved and the intent of the problem string.

(continued)

The teacher has the demonstration grapher on, potentially suggesting that students might also graph during the string. **Teacher:** *Let's start with solving this equation:* $x^2+4x-5=0$ *. For this equation, what is x?* Students work, the teacher circulates, looking for students who are factoring and graphing, as well as taking note of any other strategies.	$x^2+4x-5=0$
Teacher: *What is x for this equation?* **Student:** *I got* −5 *and* 1. **Teacher:** *You got x* = −5, 1. *How?* **Student:** *I factored it into* $(x+5)(x-1)$. **Teacher:** *How does that help? What does factoring do for you?* **Student:** *It's equal to* 0 *so one of the factors have to be* 0. **Teacher:** *If two factors multiply to* 0, *then at least one of the factors must equal* 0, *so if* $x+5=0$, *then* $x=-5$. *And if* $x-1=0$, *then* $x=1$.	$x^2+4x-5=0$ $(x+5)(x-1)=0$ $x+5=0 \quad x-1=0$ $x=-5 \quad x=1$
Teacher: *I noticed that some of you factored a different way. Tell us about that.* **Student:** *I put the* 5 *on the other side and then factored out the x.* **Teacher:** *Why would you do that?* **Student:** *Now I know that there are two things times each other that are* 5. *I just need to try two factors of* 5. *Since* 1 *works,* 1 *times* 5 *is* 5, *then* −5 *will also work.*	$x^2+4x-5=0$ $x^2+4x=5$ $x(x+4)=5$ $1(1+4)=5,\quad -5(-5+4)=5$ $x=1,-5$
Teacher: *Let's slow that down a little. Where are the two factors?* **Student:** *The x is a factor and the x + 4 is a factor. So, I think of all of the factors of* 5. **Student:** *Ah, and* 1 *times* 5, 1 + 4, *is* 5 *so you know that* 1 *works. But how did you know that* −5 *was the other one?* **Teacher:** *Well, you could try them all until you tried* −5 *and found that it worked. But that's not what you said, was it?* **Student:** *No, once I know* 1 *works, then I think back to the beginning equation and the* −5 *there. The two factors have to multiply to be that* −5. *So,* 1 *times* −5 *is* −5. **Teacher:** *Okay, let's focus on the part where you factored out the x. Who understood that part and could say it again for us, please?* **Student:** *Since x times x* + 4 *has to be* 5, *you can just try factors of* 5 *into the x until it works. You find an x so that it times itself and* 4 *makes* 5.	$1 \times -5 = -5$ $x^2+4x-5=0$ $x^2+4x=5$ $x(x+4)=5$ $1(1+4)=5,\quad -5(-5+4)=5$ $x=1,-5$

Teacher: *I noticed a few of you had graphing calculators out. Why?*

Student: *I graphed it and found the x-intercepts, the same* -5, 1.

Teacher: *So you graphed the* $y =$ *the expression. Why the x-intercepts?*

Student: *Because that's where it's 0.*

Teacher: *So, you can factor two ways or graph to find when the y-values are 0. Nice.*

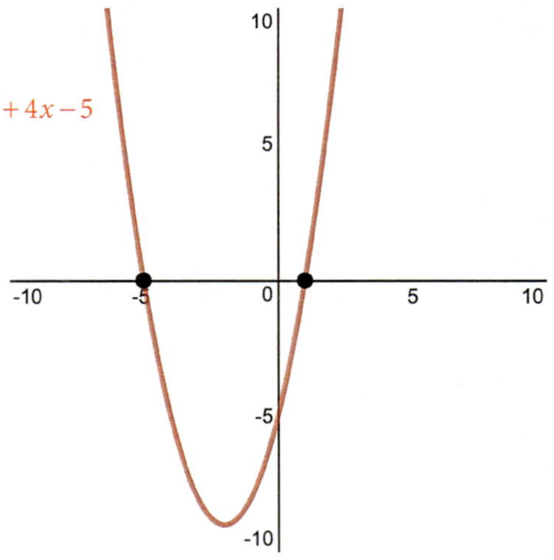

Teacher: *Next problem:* $x^2 - 4x - 5 = 0$. *What is x?*

The teacher again chooses a student who factored and used the zero product property, used factor pairs, and a student who graphed to solve.

$x^2 - 4x - 5 = 0$

$(x-5)(x+1) = 0$

$x - 5 = 0 \quad x + 1 = 0$

$x = 5 \quad x = -1$

$x^2 - 4x - 5 = 0$

$x^2 - 4x = 5$

$x(x-4) = 5$

$x = 5, -1$

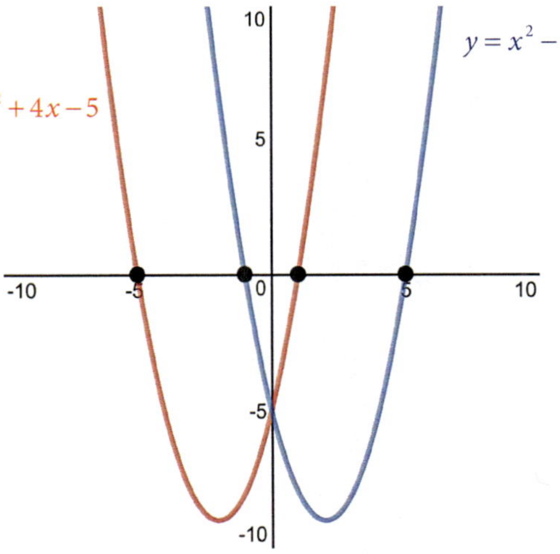

(continued)

Teacher: *How about this one?* $-x^2+4x+5=0$ *What is x?* **Teacher:** *I noticed that many of you were graphing this time. Interesting results. What did you notice?* **Student:** *It has the same x-intercepts as the last equation,* -5, 1. *The graph is just upside down.* **Student:** *Yeah, and then I tried to factor it and I got it to work, though it's not quite what I'm used to:* $(-x+5)(x+1)$. **Teacher:** *Do your factors result in the same solutions?* **Student:** *Yes, I still get* -5, 1. **Teacher:** *How can these two equations be different and the two graphs be different but they have the same solution? Tell us about that.* **Student:** *The graphs both hit the x-axis at the same points, but one has a positive* x^2 *so opens up and the other a negative* x^2 *so opens down.*	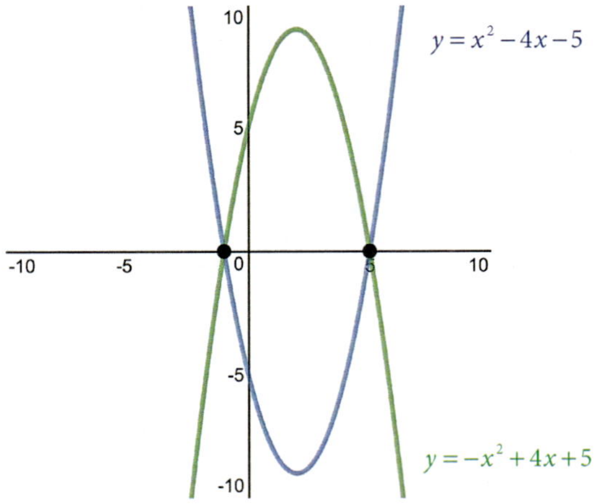
Teacher: *That describes the graphs, what about the equations? How are they related?* **Student:** *They are! You can factor out the negative* 1 *and what's left is the same, so then just factor that same result to find where either one equals* 0. **Teacher:** *Does anyone agree? Disagree? Does it make sense to factor out the* -1, *and then factor? Is that helpful?*	$-x^2+4x+5=0$ $(-x+5)(x+1)=0$ $-x+5=0 \quad x+1=0$ $x=5 \quad x=-1$ $-x^2+4x+5=-1(x^2-4x-5)$
Teacher: *What about that other factoring strategy? I noticed that you didn't do very much work. Tell us about that.* **Student:** *I noticed that you could switch everything to the other side and it was the very same as the second problem, so the answers are the same.* **Teacher:** *So, you could create an equivalent equation that has the same solution. Nice!*	$x^2-4x-5=0$ $x^2-4x=5$ $x(x-4)=5$ $x=5,-1$ $-x^2+4x+5=0$ $5=x^2-4x$ $5=x(x-4)$ been here before...

Teacher: *Let's try another one. Solve* $-x^2-4x+5=0$.

Student: *It works! It's easier to take out the* -1 *and then factor that expression.*

Student: *And the graph is just reflected, so it has the same x-intercepts too.*

Teacher: *What solutions did you find? What is x?*

Student: *I found* -5, 1 *because I factored out the* -1 *and then factored* $x^2+4x-5=0$ *and we've already solved that. That was the first problem.*

Student: *I factored this problem and got the same solutions. It's* $(-x+1)(x+5)$.

Student: *Right, so the solutions would be the same.*

Student: *And I saw that if I moved everything over to the opposite side, it's the same as the first problem.*

$-x^2-4x+5=0$

$x=-5,1$

$-x^2-4x+5=0$

$(-x+1)(x+5)=0$

$-x+1=0 \quad x+5=0$

or

$-x^2-4x+5=-1(x^2+4x-5)$

$=-1(x+5)(x-1)$

$x+5=0 \quad x-1=0$

$x=-5,1$

$-x^2-4x+5=0$

$5=x^2+4x$

$5=x(x+4)$

see first problem...

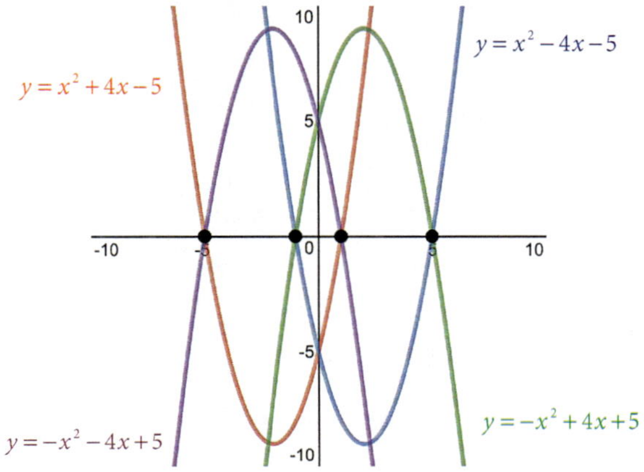

Teacher: *How would you summarize some of the things that came up in this string today about the equations, the graphs, the x-intercepts?*

Elicit the following:

- *The numbers in the equations are all the same, except the signs.*
- *The x-intercepts are the same if the solutions are the same, even if the parabolas are not the same orientation.*

Sample Final Display

Your display could look like this at the end of the problem string:

(continued)

$x^2+4x-5=0$

$x=-5,1$

$x^2+4x-5=0$
$(x+5)(x-1)=0$
$x+5=0 \quad x-1=0$
$x=-5 \quad x=1$

$1 \times -5 = -5$
$x^2+4x-5=0$
$x^2+4x=5$
$x(x+4)=5$
$1(1+4)=5, \quad -5(-5+4)=5$
$x=1,-5$

$x^2-4x-5=0$

$x=-1,5$

$x^2-4x-5=0$
$(x-5)(x+1)=0$
$x-5=0 \quad x+1=0$
$x=5 \quad x=-1$

$x^2-4x-5=0$
$x^2-4x=5$
$x(x-4)=5$
$x=5,-1$

$-x^2+4x+5=0$

$x=-1,5$

$-x^2+4x+5=0$
$(-x+5)(x+1)=0$
$-x+5=0 \quad x+1=0$
$x=5 \quad x=-1$

$-x^2+4x+5=-1(x^2-4x-5)$

$-x^2+4x+5=0$
$5=x^2-4x$
$5=x(x-4)$
been here before...

$-x^2-4x+5=0$

$x=-5,1$

$-x^2-4x+5=0$
$(-x+1)(x+5)=0$
$-x+1=0 \quad x+5=0$
or
$-x^2-4x+5=-1(x^2+4x-5)$
$=-1(x+5)(x-1)$
$x+5=0 \quad x-1=0$
$x=-5,1$

$-x^2-4x+5=0$
$5=x^2+4x$
$5=x(x+4)$
see first problem...

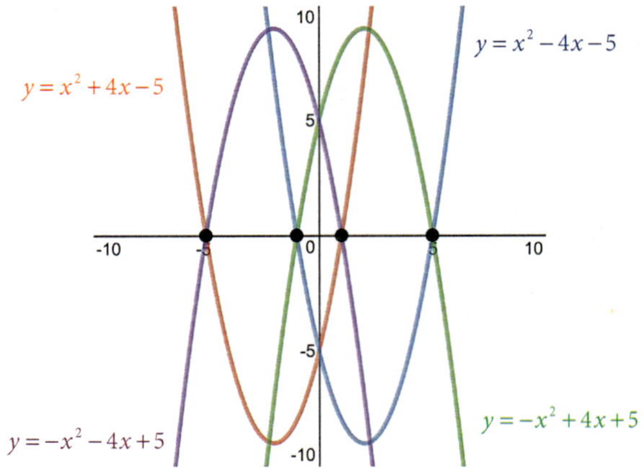

Facilitation Notes

This version of the problem string lists short notes for important teacher moves during the string. After you've done the string yourself and studied the relationships involved, you might make similar notes for the things you want a reminder of or deem important.

$x^2+4x-5=0$	*Have demo grapher on and ready. Quick. Elicit and model factoring, graphing. Connnect factoring and graphing.*
$x^2-4x-5=0$	*Quick. Elicit and model factoring, graphing. Connect to prior problem, notice "reflection" of zeros.*
$-x^2+4x+5=0$	*How would you factor this one? Wonder about equivalence with prior problem.*
$-x^2-4x+5=0$	*Model factoring and graphing. Connect to first problem.*

PROBLEM STRINGS LESSON

5.1 Graphing Quadratic Functions 1

At a Glance

$y = x^2$

$y = 3x$

$y = x^2 + 3x$

$y = x(x+3)$*

$y = x^2 - 3x$

*optional problem

Objectives

The goal of this Graphing Quadratic Functions series of problem strings is to help students develop a network of understandings about quadratic functions, connecting and using multiple representations. This string begins to develop the graphing strategies of adding ordinates and factoring to find zeros.

Placement

This is the first in a series of 2 strings to develop the graphing strategies of adding ordinates and factoring to find zeros. This string works with the form $y = x^2 + bx$. The next string will continue the work and allow students the opportunity to apply the strategies to the form $y = x^2 + bx + c$.

You could deliver this string after students have experience with the parent function $y = x^2$ and transformations and as students are graphing quadratic functions.

This string would be a nice pre-chapter problem string for textbook Chapter 5: Quadratic Functions and Relations.

Guiding the Problem String

This problem string is an opportunity to assess students' prior understanding. Listen, watch, and ask probing questions. As you deliver the third problem, wonder aloud about the combination of the first two problems: What does it even mean to add a parabola and a line? The fourth problem is provided if no students suggest using the factored form to graph the third problem. For the last problem, explore with students the adding ordinates strategy both by adding $-3x$ to x^2 and by subtracting $3x$ from x^2.

About the Mathematics

The strategy we are calling adding ordinates comes from the (abscissa, ordinate) language of ordered pairs. It simply means combining the y-values from the parts of a combination function. In this string, we look at combining a linear function with a quadratic function. For $y = x^2 + 3x$, you look to add all of the y-values of $y = x^2$ to the y-values of $y = 3x$ at each respective x-value. For example, (1, 1) for $y = x^2$ and (1, 3) for $y = 3x$ becomes (1, 4) for $y = x^2 + 3x$.

The ideas therein, that the quadratic function dominates the whole function, thus making the combination function also a quadratic function, portends that general meaning of polynomials, that is that the term of highest degree dominates the polynomial.

But this form, $y = x^2 + bx$, also begs to be factored into $y = x(x+b)$, where the roots are $x = 0, -b$. The vertex can then be found by considering symmetry between the roots or as always by $\frac{-b}{2a}$ for the x-coordinate, though we acknowledge the use of $\frac{-b}{2a}$ if students use it, we don't model it for the rest of the class yet as typically it is not widely developed at the curricular point where we would deliver this string. However, if it has been developed, then it might be appropriate to model for the whole class.

(continued)

Sample Interactions

Use the following as you plan how to elicit and model student strategies. This is not meant as a script, but as a view into the relationships involved and the intent of the problem string.

Teacher: *You have seen this function before, $y = x^2$. Sketch a quick graph on your paper and label some important points.* Students work and the teacher circulates, looking for students who have graphed the points shown. **Teacher:** *Please tell us which points you chose.* **Student:** (0, 0), (1, 1), (2, 4) **Student:** *And the points on the other side,* (−1, 1), (−2, 4).	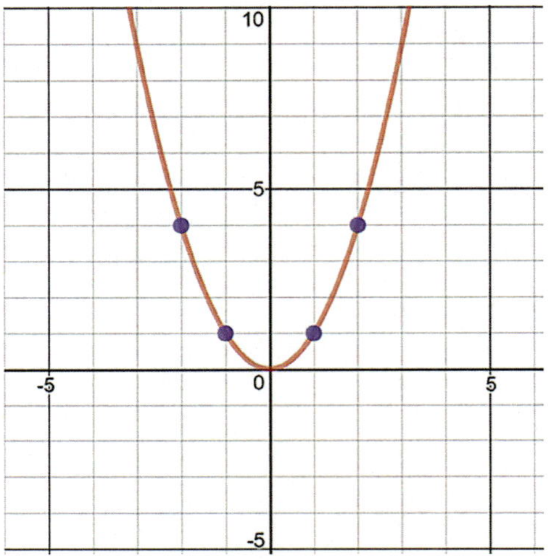
Teacher: *Okay, great. What about this function, $y = 3x$? What does it look like? Label a couple of important points.*	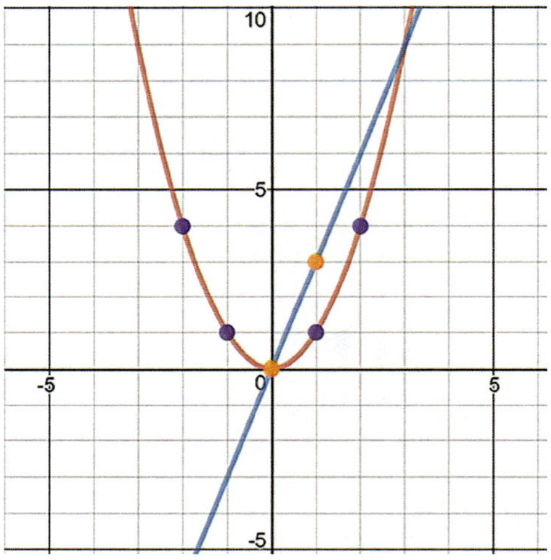

Teacher: *I wonder what the function would look like that is a combination of the two functions we have already? That is our next problem:* $y = x^2 + 3x$ *. Predict first. What do you think the graph might look like? Turn and talk about your prediction.*

Students partner talk, while the teacher listens in.

Teacher: *What does* $y = x^2 + 3x$ *look like? Sketch a graph.*

Students work and the teacher circulates, looking for students who are adding ordinates and asking them to share. The teacher uses braces to highlight the *y*-values of each component function, $y = x^2$ and $y = 3x$ and how they add together to be the *y*-value of the combination function, $y = x^2 + 3x$.

(If no students add ordinates, the teacher could ask:

How do the points at $x = 1$ on all three functions relate? What about at other *x*-values? Can we think about the function $y = x^2 + 3x$ as the addition of all of the $y =$ values of $y = x^2$ and $y = 3x$?)

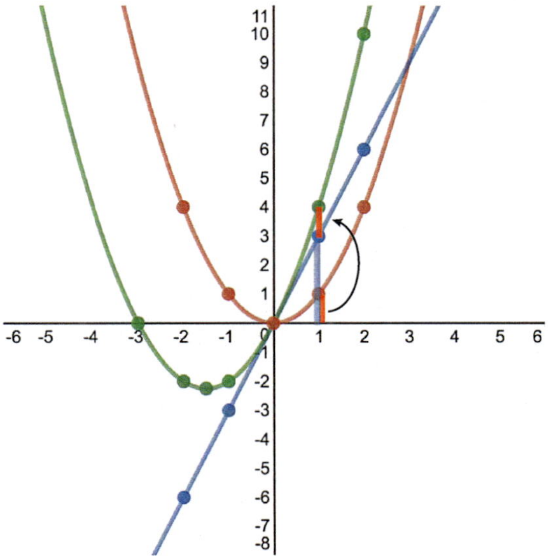

The teacher continues to have students share adding ordinates, highlighting the values for 3–4 key points.

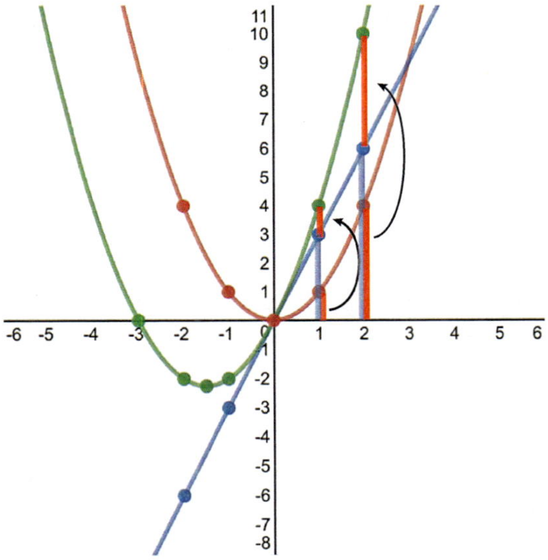

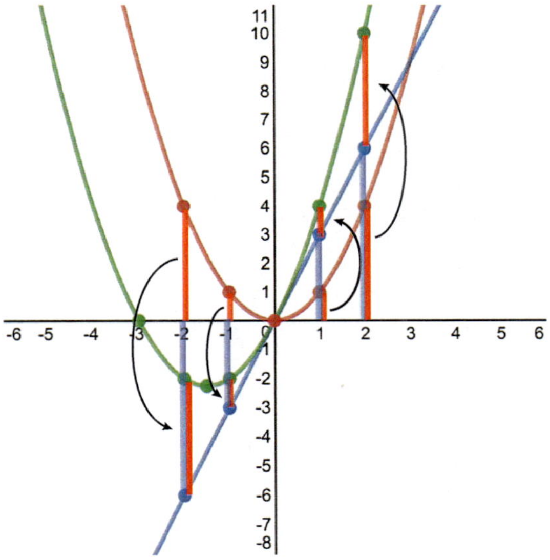

(continued)

Teacher: *So, what's happening? Why does adding a parabola to a line result in a parabola? Think about that. When your partner is ready, turn and talk to your partner about your thinking.*

Students discuss in pairs while the teacher listens in.

Teacher: *Let's focus on the first quadrant? Why does the combination function increase? In other words, why is the green function positive and increasing in the first quadrant?*

Why does the combination function increase again over in the second and third quadrants? Shouldn't it be decreasing because the line is decreasing?

Which function is growing or decreasing faster in the third and fourth quadrant?

Based on all of that, what could we say about a parabola plus a line?

What's happening at $x = 0$ and $x = -3$? Why?

The teacher records student generalizations.

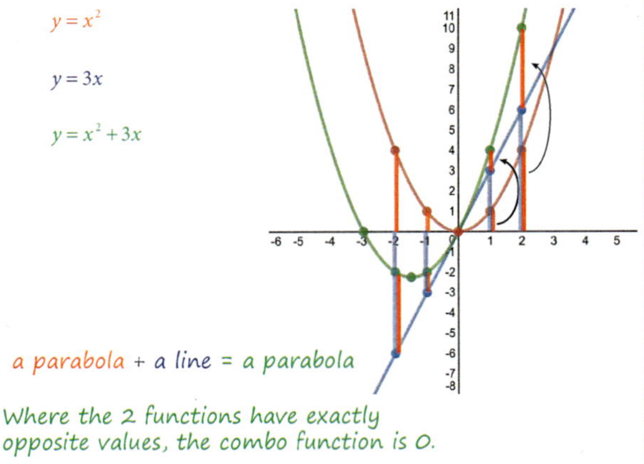

No students had used a "factoring to find zeros" strategy, so the teacher uses the optional next problem, $y = x(x+3)$.

Teacher: *The next problem today is $y = x(x+3)$. Tell me about the graph. I wonder which strategy you might use.*

Students work, while the teacher circulates to finds students to share as the teacher models solving for the zeros $x = 0, -3$ and then using symmetry to find the vertex. Notice the purple x's to mark the x-intercepts and the vertex.

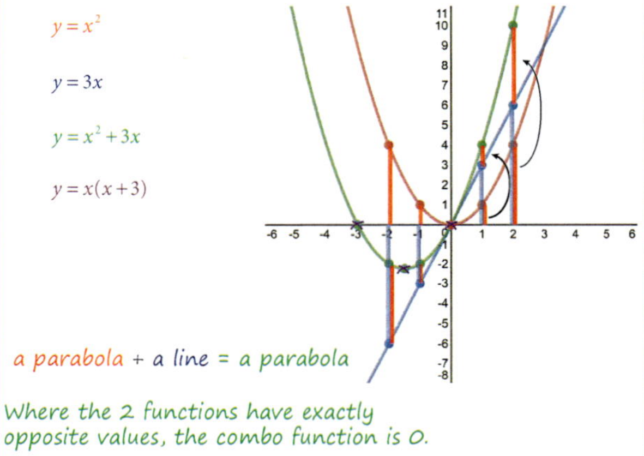

Teacher: *The last problem today is* $y = x^2 - 3x$ *. Tell me about the graph. I wonder which strategy you might use.*

Students work, then the teacher models both an adding ordinates strategy and a factoring strategy. When discussing the adding ordinates strategy, the teacher elicits from students both the ideas of subtracting $3x$ from x^2 and adding $-3x$ to x^2.

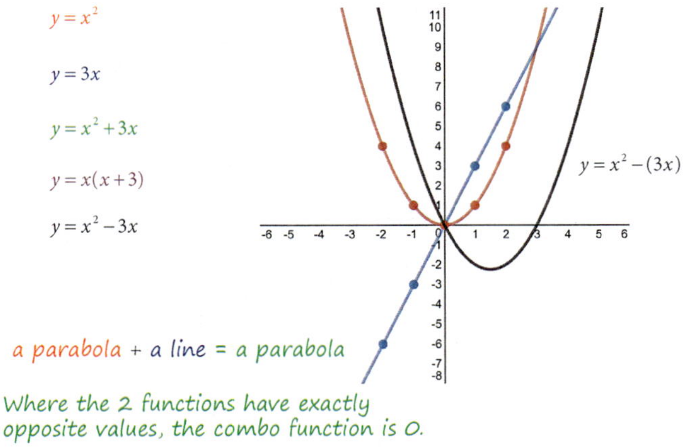

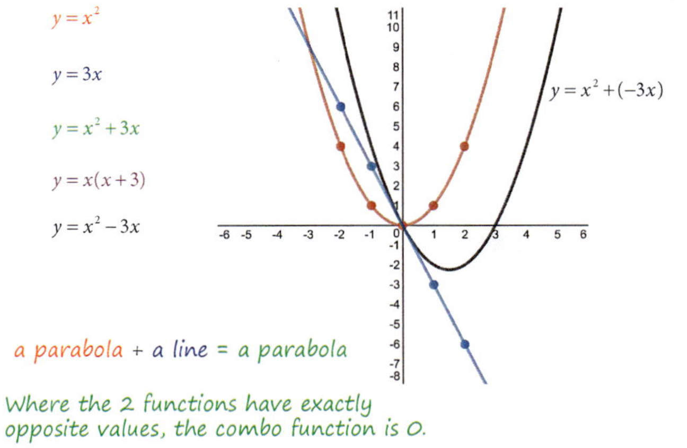

Teacher: *What about the factoring strategy? Let's put that up here too.*

What are the x-intercepts?

How did you find the vertex?

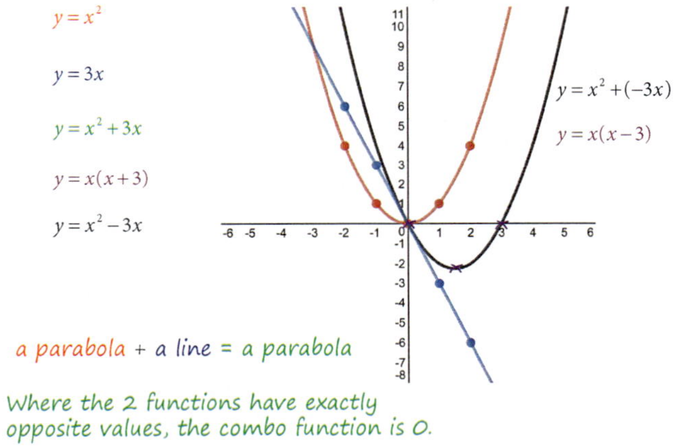

Teacher: *How would you summarize some of the things that came up in this string today?*

Elicit the following:

- *A quadratic function combined with a linear function behaves like a quadratic function and that is why the combination function is a quadratic function.*
- *The long run behavior of the linear part pulls on the long run behavior of the quadratic part but does not change the overall long run behavior of the combination function.*
- *You can find the graph of a combination function by adding the y-values of the parts.*
- *If the expression factors nicely, you can use the factors to find the x-intercepts and halfway between x-intercepts is the vertex.*

Sample Final Display

Your display could look like this at the end of the problem string:

(continued)

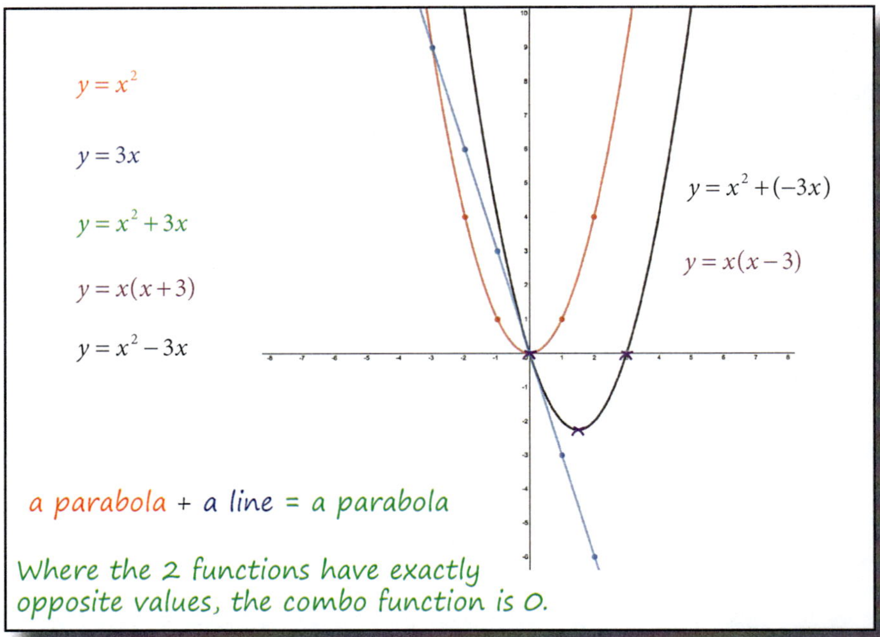

Facilitation Notes

This version of the problem string lists short notes for important teacher moves during the string. After you've done the string yourself and studied the relationships involved, you might make similar notes for the things you want a reminder of or deem important.

$y = x^2$	Seen before. Quickly graph, note important points. Sketch graph on board.
$y = 3x$	Seen before too! Quickly graph, important points. Add graph to previous on board.
$y = x^2 + 3x$	I wonder what it would look like to combine? Sketch. Could you use the important points to help? How? How does end behavior of parts affect end behavior of combo? If student uses factored form, share, add zeros to graph.
$(y = x(x+3))$	Optional Sketch a graph. Share. How is this connected to previous? Why? Which strategy do you prefer, factor-zeros or adding functions?
$y = x^2 - 3x$	I wonder how you'd sketch this? Share. Add to graph. Share factor-zeros strategy. Wondere about -3x versus +(-3x). How is this problem connected to previous? Why? How does end behavior of parts affect end behavior of combo?

PROBLEM STRINGS LESSON

5.1 Graphing Quadratic Functions 2

At a Glance

$y = x^2$

$y = 6x$*

$y = x^2 + 6x$

$y = x^2 - 6x$

$y = x^2 + 6x + 7$

$y = x^2 - 6x + 7$

$y = x^2 + 4x + 5$

*optional problem

Objectives

The goal of the Graphing Quadratic Functions series of problem strings is to help students develop a network of understandings about quadratic functions, connecting and using multiple representations. This string continues to develop the graphing strategies of adding ordinates and factoring to find zeros.

Placement

This is the second in a series of two strings to develop the graphing strategies of adding ordinates (combining a quadratic and linear function) and factoring to find zeros. The previous string worked with the form $y = x^2 + bx$. This string builds on that work by adding the transformation of shifting $y = x^2 + bx$ up or down c units to get the graph of $y = x^2 + bx + c$.

You could deliver this string after the previous string, Graphing Quadratic Functions 1. You could also choose to skip the previous string. If you do, you will probably need to take more time in the first few problems of this string. Students need prior experience with the quadratic parent function and transformations.

This string would be a nice pre-chapter problem string for Chapter 5: Quadratic Functions and Relations or as a follow up to 5.1 Equivalent Quadratic Forms.

Guiding the Problem String

This problem string is an opportunity to assess students' prior understanding. Listen, watch, and ask probing questions. The second problem is optional. You can probably skip it if your students worked the previous string. If this is the first string for these strategies, as you deliver the third problem wonder aloud about the combination of the first two problems: What does it even mean to add a parabola and a line? The fourth problem is not factorable, but it is a shift up three units from a previous problem. You might wonder aloud how $y = x^2 + 6x$ and $y = x^2 + 6x + 7$ are related. The last problem is an opportunity for students to try their hand at a completely new function and for you to assess students' depth of understanding.

Focus the conversation on what is helpful to kids—and the idea that we are playing with lots of ways to envision and solve, but some might be more helpful to some students than others, and that's the focus of our work today. Was a strategy helpful? Why? Why not?

About the Mathematics

The strategy we are calling "adding ordinates" comes from the (abscissa, ordinate) language of ordered pairs. It simply means adding the y-values from the parts of a combination function. In this string, we look at combining a linear function with a quadratic function. The ideas therein, that the quadratic function dominates the whole function thus making the combination function also a quadratic function, portends that general meaning of polynomials, that is that the term of highest degree dominates the polynomial.

This form, $y = x^2 + bx$, also begs to be factored into $y = x(x + b)$, where the roots are $x = 0, -b$. The vertex can then be found by considering symmetry between the roots. The x-value of the vertex must be halfway between the roots. The y-value can be found by evaluating the function at the x-value.

When presented with $y = x^2 + bx + c$, we can recognize it as a vertical shift, $y = c$ added to $y = x^2 + bx$. This relationship allows us to use the "factor-zeros, then shift" strategy.

(continued)

Important Questions

Use the following as you plan how to elicit and model student strategies. See Solving Quadratic Equations 1 for ideas on leading the first part of this problem string.

- *Sketch a graph of* $y = x^2$. *What are some important points on the quadratic parent function?*
- *Sketch a graph of* $y = x^2 + 6x$. *I wonder what it even means to add a quadratic function to a linear function. Hmmm ...*
- *What about those of you who factored,* $y = x^2 + 6x = x(x+6)$ *? Discuss your thinking.*
- *I wonder how everything we just discussed for* $y = x^2 + 6x$ *might impact how you will sketch a graph of* $y = x^2 - 6x$ *?*
- *Did anyone think about subtracting* $y = 6x$ *from* $y = x^2$*? Let's model that thinking on the board.*
- *Did anyone think about adding* $y = -6x$ *to* $y = x^2$*? So, you're thinking about* $y = x^2 + (-6x)$ *Let's model that thinking on the board.*
- *Did anyone factor* $y = x^2 - 6x = x(x-6)$*? Let's model that thinking on the board.*
- *Some of you are noticing relationships between the graphs of* $y = x^2 + 6x$ *and* $y = x^2 - 6x$. *Translation? Reflection over the y-axis?*
- *What about this function,* $y = x^2 + 6x + 7$*? How would you graph it? What have we done today that might influence you?*
- *How is* $y = x^2 + 6x + 7$ *related to* $y = x^2 + 6x$*? I saw some of you not do very much work at all, you just used the graph of* $y = x^2 + 6x$. *Please explain how and why.*
- *So, are you saying that to graph something that's not factorable over the reals,* $y = x^2 + 6x + 7$, *you can consider the first part,* $y = x^2 + 6x = x(x+6)$ *and use the zeros of x = 0, –2 to graph that part,* $y = x^2 + 6x$ *and then just shift that all up 7? Turn to your partner and talk about that idea.*
- *I wonder if that will affect how you graph the next problem,* $y = x^2 - 6x + 7$*? You can graph it any way you want. When you're done, step back and really think about the relationships and decide what you think is a really slick strategy that is efficient and sophisticated. Look for connections between all of the graphs so far.*

Elicit the following:

- *Shift the graph of* $y = x^2 - 6x$ *up 7, translate the graph of* $y = x^2 + 6x + 7$ *to the right, or reflect the graph of* $y = x^2 + 6x + 7$ *over the y-axis. Which of these can you do because you already have parts up here? Which would you choose from scratch if you didn't have these helper graphs?*
- *Here's a problem to try from scratch. Graph* $y = x^2 + 4x + 5$.

How would you summarize some of the things that came up in this string today?

- *A quadratic function combined with a linear function behaves like a quadratic function that is why the combination function is a quadratic function.*
- *The long run behavior of the linear part pulls on the long run behavior of the quadratic part but does not change the overall long run behavior of the combination function.*
- *You can find the graph of a combination function by adding the y-values of the parts.*
- *If it factors nicely, you can use the factors to find the x-intercepts and then halfway in between those is the vertex.*
- *If a quadratic does not factor, you can factor the first two terms to find interim zeros, plot that interim function and then shift everything up or down c.*

Sample Final Display

Your display could look like this at the end of the problem string:

$y = x^2$

$y = x^2 + 6x$
$= x(x+6)$

$y = x^2 - 6x$
$= x(x-6)$

$y = x^2 + 6x + 7$
$= (x^2 + 6x) + 7$

$y = x^2 - 6x + 7$
$= (x^2 - 6x) + 7$

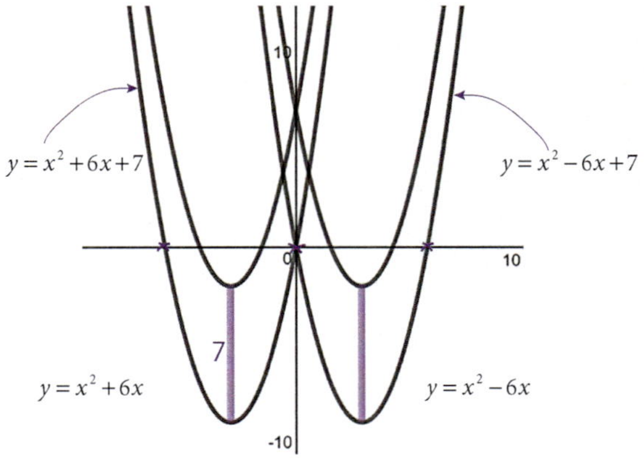

$y = x^2$

$y = x^2 + 6x$
$= x(x+6)$

$y = x^2 - 6x$
$= x(x-6)$

$y = x^2 + 6x + 7$
$= (x^2 + 6x) + 7$

$y = x^2 - 6x + 7$
$= (x^2 - 6x) + 7$

$y = x^2 + 4x + 5$
$= (x^2 + 4x) + 5$
$= x(x+4) + 5$

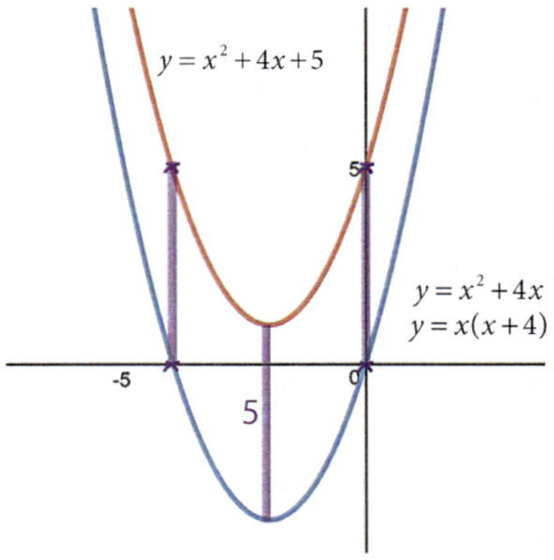

Facilitation Notes

This version of the problem string lists short notes for important teacher moves during the string. After you've done the string yourself and studied the relationships involved, you might make similar notes for the things you want a reminder of or deem important.

$y = x^2$	Seen before. Quickly graph, note important points. Sketch graph on board.
$y = 6x$	Optional. Seen before too! Quickly graph, important points. Add graph to previous on board.
$y = x^2 + 6x$	I wonder what it would look like to combine? Sketch. Could you use the important points to help? How? How does end behavior of parts affect end behavior of combo? Find factored form, share, add zeros to graph.
$y = x^2 - 6x$	Sketch a graph. Share. How is this connected to previous? Why? Shift? Reflection? Which strategy do you prefer, factor-zeros or adding functions?
$y = x^2 + 6x + 7$	I wonder how you'd sketch this? How is this problem connected to previous? Share factor-zeros then shift strategy. Add to graph.
$y = x^2 - 6x + 7$	Repeat. Share factor-zeros then shift strategy. Add to graph. Shift? Reflection?
$y = x^2 + 4x + 5$	(New display.) How about a new, unrelated function? Share factor-zeros then shift strategy.

PROBLEM STRINGS
LESSON

5.2 Complete the Square

At a Glance

	x	-2
x		
-2		

	___	___
___	x^2	$3x$
___	$3x$	9

	___	___
___	x^2	$5x$
___	$5x$	

	___	___
___	x^2	$-5x$
___	$-5x$	

$x^2 + 20x +$ ____

$x^2 + 5x +$ ____

$x^2 - 5x +$ ____

$x^2 + bx +$ ____

Objectives

The goal of this problem string is to develop students' spatial sense of completing the square.

Placement

This string could come before students learn the technique of completing the square to build the concept spatially before students attack it algebraically. Alternatively, this string could be used to back up the algebraic work with a spatial interpretation.

You could deliver this string any time before students begin section 5.2 Completing the Square in the textbook or as a back up to the Investigation in 5.2, to make sure that students have solidified the concepts.

Guiding the Problem String

The first four problems are delivered as (special rectangular) square diagrams with missing dimensions. As students fill in the dimensions and areas, record the equivalent algebraic expressions. Help students make connections between the square diagrams where the corner square is missing, the algebraic notation, and the phrase "complete the square". Then deliver the last four problems as algebraic expressions and model on the square diagrams so that when the blank is filled in, the resulting quadratic expression represents the products in a square diagram.

About the Mathematics

All of these rectangular diagrams are square because the dimensions end up the square of a linear binomial, $(x+d)(x+d)=(x+d)^2$. When drawing, make sure that the x-lengths are equal and the d lengths are equal so that the resulting x^2 and d^2 areas are square.

	x	d
x	x^2	dx
d	dx	d^2

When viewed spatially like this, completing the square becomes less of a memorized procedure and more of an act of taking an incomplete square and completing it. To do that, you must split the bx term in $y = ax^2 + bx + c$ to fill two equivalent spaces in the square diagram. Often students who learn to complete the square from only an algebraic perspective have no idea why they divide b by two or why you then square that, other than it works.

This string does not attempt to do anything about keeping an expression or an equation equivalent by then subtracting whatever was added to complete the square. That bit will need to be done elsewhere, like it is in the Investigation in 5.2.

Sample Interactions

Use the following as you plan how to elicit and model student strategies. This is not meant as a script, but as a view into the relationships involved and the intent of the problem string.

Teacher: *This first problem is a rectangular diagram. If the dimensions of a rectangle are something like what you see here, what would go in the middle? What is x times x? And x times –2? How do those relate to this rectangular diagram? Fill in the middle please.*

Students work briefly.

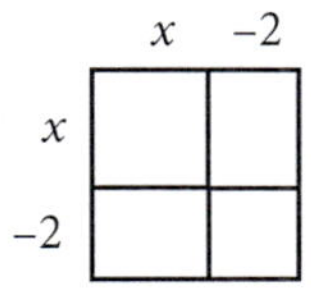

Teacher: *Let's use multiplication to fill in the inside together quickly.*

Teacher: *How would we write the algebraic equivalent of the distributive property?*

Teacher: *Did you notice that this is a square? How could you know? So we can write the algebra as a square too.*

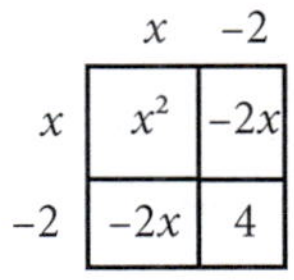

$x^2 - 4x + 4 = (x-2)(x-2) = (x-2)^2$

Teacher: *The second problem has the middle filled in. You supply the outside values. Also write the corresponding algebra representation.*

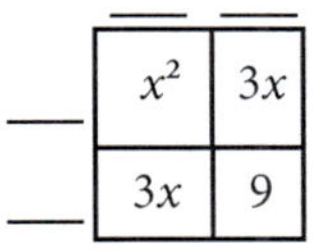

Teacher: *How would you describe this rectangle? Is it also a square? How do you know?*

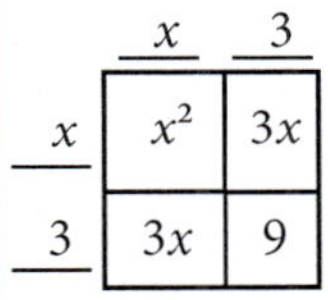

$x^2 + 6x + 9 = (x+3)(x+3) = (x+3)^2$

Teacher: *The third problem has a missing inside piece. A puzzle! What could it be? Fill in the missing piece and the outside values and the corresponding algebra representation.*

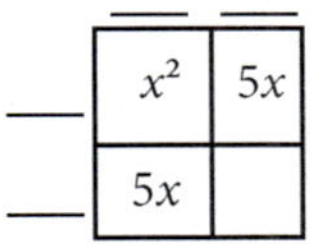

Teacher: *How did you know it was 25? How does that 25 relate to the 5x terms? What does it mean that both of the inside rectangles contained 5x? Does that imply that the outside rectangle is a square? Will it always? How do the 5x terms relate to the 10x term in the algebraic* $x^2 + 10x + 25$*?*

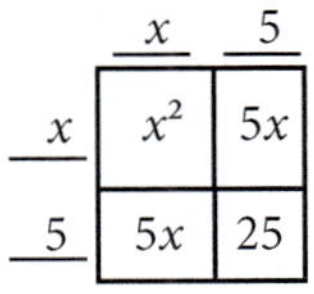

$x^2 + 10x + 25 = (x+5)^2$

Teacher: *If the only things that change in the next problem are that the inside rectangles contain –5x, how does that change everything? What changes and what stays the same?*

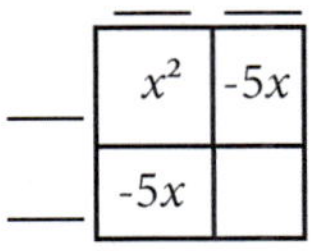

(continued)

Teacher: *Is it still a square? What is making it a square? How do the $-5x$ terms relate to the $-10x$ term in the algebraic $x^2 - 10x + 25$?*

	x	-5
x	x^2	$-5x$
-5	$-5x$	25

$x^2 - 10x + 25 = (x-5)^2$

Teacher: *Now the problem looks like the algebraic representation. Your job is to draw the corresponding rectangular diagram and fill in the blank in order to create a square.*

$x^2 + 20x +$ ______

Teacher: *What did you add to the expression so that the rectangular diagram became a square? How did you know? What do the 10x terms have to do with the 20x term? Will they always be half of that term? Why? How does the quadratic expression relate to a binomial squared? How does $(x + 10)^2$ relate to the square diagram?*

$x^2 + 20x +$ <u>100</u>

	x	10
x	x^2	$10x$
10	$10x$	100

$x^2 + 20x + 100 = (x+10)^2$

Teacher: *This problem is like the last one. Please draw the corresponding rectangular diagram, fill in the blank in order to create a square, and write the corresponding algebraic expression.*

$x^2 + 5x +$ ______

Teacher: *How did you find 6.25 squared? Did you use fractions? Decimals? How does this square and it's corresponding algebraic expression compare to the previous problem? What is a difference between an x-term that is even, like 20x, and an x-term that is odd, like 5x, when you are completing a square?*

$x^2 + 5x +$ <u>6.25</u>

	x	2.5
x	x^2	$2.5x$
2.5	$2.5x$	6.25

(doubling/halving)

$\frac{5}{2} \cdot \frac{5}{2} = \frac{25}{4}$ or $2.5 \cdot 2.5 = 5 \cdot 1.25 = 10 \cdot 0.625 = 6.25$

$x^2 + 5x + \frac{25}{4} = (x + \frac{5}{2})^2$

$x^2 + 5x + 6.25 = (x + 2.5)^2$

Teacher: *The next problem should go quickly. What will change? What won't? Why?*

$x^2 - 5x +$ ______

Teacher: *What changed? What didn't? Why?*

$x^2 - 5x +$ ______

	x	-2.5
x	x^2	$-2.5x$
-2.5	$-2.5x$	6.25

$x^2 - 5x + \frac{25}{4} = (x - \frac{5}{2})^2$

$x^2 - 5x + 6.25 = (x - 2.5)^2$

Teacher: *For the last problem today let's get a little general. Instead of having a diagram with numbers, this one has letters that stand for numbers. Let's suppose all you know is that you want to complete a square and you've got x^2 and some number, b, times x. Where do these parts go in the rectangular diagram? What else do you need to complete a square?*	$x^2 + bx +$ _____

Teacher: *How would you summarize some of the things that came up in this string today?*

Elicit the following:

- *In a square, the dimensions are the same.*
- *If you have the b-term, you split it in half in the square diagram, because it's a square.*
- *When you add the corner piece in the rectangle diagram to make a square, it also makes a square in the algebra, a binomial squared.*

(continued)

Sample Final Display

Your display could look like this at the end of the problem string:

	x	-2
x	x^2	$-2x$
-2	$-2x$	4

$$x^2-4x+4=(x-2)(x-2)=(x-2)^2$$

	x	3
x	x^2	$3x$
3	$3x$	9

$$x^2+6x+9=(x+3)(x+3)=(x+3)^2$$

	x	5
x	x^2	$5x$
5	$5x$	25

$$x^2+10x+25=(x+5)^2$$

	x	-5
x	x^2	$-5x$
-5	$-5x$	25

$$x^2-10x+25=(x-5)^2$$

$x^2 + 20x +$ 100

	x	10
x	x^2	$10x$
10	$10x$	100

$$x^2+20x+100=(x+10)^2$$

$x^2 + 5x +$ 6.25

	x	2.5
x	x^2	$2.5x$
2.5	$2.5x$	6.25

$$x^2+5x+\tfrac{25}{4}=(x+\tfrac{5}{2})^2$$

$$x^2+5x+6.25=(x+2.5)^2$$

$x^2 - 5x +$ 6.25

	x	-2.5
x	x^2	$-2.5x$
-2.5	$-2.5x$	6.25

$$x^2-5x+\tfrac{25}{4}=(x-\tfrac{5}{2})^2$$

$$x^2-5x+6.25=(x-2.5)^2$$

$x^2 + bx + \dfrac{b^2}{4}$

	x	$\tfrac{b}{2}$
x	x^2	$\tfrac{b}{2}x$
$\tfrac{b}{2}$	$\tfrac{b}{2}x$	$\tfrac{b^2}{4}$

$$x^2+bx+\frac{b^2}{4}=(x+\tfrac{b}{2})(x+\tfrac{b}{2})=(x+\tfrac{b}{2})^2$$

Facilitation Notes

This version of the problem string lists short notes for important teacher moves during the string. After you've done the string yourself and studied the relationships involved, you might make similar notes for the things you want a reminder of or deem important.

	x	-2
x		
-2		

Quick. Fill in the interior.
Write the algebraic representation.
Connect to a binomial squared.

	___	___
___	x^2	$3x$
___	$3x$	9

Quick. Fill in the exterior.
Write the algebraic representation.
How would you describe this rectangle? A square?
How do you know?

	___	___
___	x^2	$5x$
___	$5x$	

Fill in the missing interior and exterior.
Write the algebraic representation.
How did you find 25? How does 25 relate to the 5x terms?
Do the congruent 5x pieces imply a square? Will it always?
How do the 5x and 10x terms relate in the algebraic expression?

	___	___
___	x^2	$-5x$
___	$-5x$	

Quick. Fill in the missing interior and exterior.
Write the algebraic representation.
What changed and what stayed the same from the previous problem?
How do the −5x terms relate to the −10x term in the algebraic expression?

$x^2 + 20x +$ ____

Now the problem looks like the algebraic representation.
Draw the corresponding rectangular diagram and fill in to create a square.
What did you add? Will it always be half of the coefficient of the x-term?

$x^2 + 5x +$ ____

Repeat. How did you find 2.5·2.5? Decimals (double/half?), Fractions?
What do you think about the odd x coefficient versus the previous even?

$x^2 - 5x +$ ____

Quick. What will change? What won't? Why?

$x^2 + bx +$ ____

Let's get a little general. What if I want to take part of a square and make it into a square?

PROBLEM STRINGS
LESSON

5.2 Solving Quadratic Equations 2

At a Glance

$x^2 = 49$

$(x+2)^2 = 49$

$(x+3)^2 + 13 = 49$

$(x-2)^2 - 36 = 0$*

$(x+1)^2 - 2 = 0$

$(x-4)^2 - 8 = 0$

*optional problem

Objectives

The goal of this problem string is to develop students' facility with solving equations of the type they will create by completing the square.

Placement

This string could come before students learn the technique of completing the square to motivate a reason for learning to complete the square. In other words, if students have some experience solving equations of this type, they might be more inclined to complete the square, knowing the result is solvable.

You could deliver this string any time before students begin section 5.2 Completing the Square in the textbook.

Guiding the Problem String

The first problem sets up the need for two solutions for some quadratic equations, where $x = \pm 7$ This should be quick, but wonder aloud about –7 if students do not bring it up. The next two problems may take some work to get students to consider both solutions. Elicit three strategies: Solve by inspection, factor using zero product property, and using factor pairs. The last two problems bring in solutions that are not perfect squares and make the conversation interesting about which strategy to choose. If students struggle with either factor strategy, encourage students to consider which strategy might be more efficient, or even possible.

When students end up with equations like $x + 2 = -7$, if students are struggling with the integers, consider modeling the relationships on an open number line.

About the Mathematics

When quadratic equations are in the form, $a(x-d)^2 + m = n$, it can be convenient to solve the equation using properties of equality and inverse operations to get to the point where you can take the square root of both sides of the equation. This can be called "solving by inspection." Students often refer to it as taking the square root of both sides.

Alternatively, you can multiply everything out, gather the terms on one side, and factor the resulting quadratic trinomial into the product of two linear binomials, if it's factorable. This is what we refer to as "factor using the zero product property".

Alternatively, you can gather the x^2 and x terms onto one side and the constant term to the other side of the equation and then factor into this form $x(x-d) = p$. Now the task is to find factors of p that work. This is what we refer to as "using factor pairs".

Later we will compare these strategies with the quadratic formula and discuss which form of the equation lends itself to each strategy.

Sample Interactions

Use the following as you plan how to elicit and model student strategies. This is not meant as a script, but as a view into the relationships involved and the intent of the problem string.

Teacher: *Let's start with a quick question. If a number times itself, called squared, is 49, what is the number?* **Student:** *That's easy, 7?* **Teacher:** *How do you know?* **Student:** *Because 7 times 7 is 49.* **Teacher:** *Is there any other number multiplied by itself that is 49? What about negative 7? We can write that with a square root symbol and call it taking the square root of both sides of the equation. Technically, when you take the square root of x^2, we write the absolute value of x. This is a way of noting that x could be either the square root of 49 or the opposite of the square root of 49, 7 or −7. So, just like you can add, subtract, multiply, or divide both sides of an equation by the same thing and maintain equivalence, you can also take the square root of both sides of an equation.*	$x^2 = 49$ $\sqrt{x^2} = \sqrt{49}$ $\lvert x \rvert = \sqrt{49}$ $x = \pm\sqrt{49} = \pm 7$
Teacher: *I noticed that a few of you did something else. Tell us about that.* **Student:** *I made it $x^2 - 49$ and factored that to x + 7 times x − 7. So, x is 7 and −7.*	$x^2 - 49 = 0$ $(x-7)(x+7) = 0$ $x = -7, 7$
Teacher: *Two fine ways to find x. I wonder which you like more? Here is the next problem, first add 2 to x, then square the result and that is equal to 49. Solve for x.* Students work and the teacher circulates, looking for students who only finds one solution and those who find 2, and for students who use the factor using the zero product property strategy, and who use the factor pair strategy. **Teacher:** *What did you get?* **Student:** *I found 5.* **Teacher:** *Did anyone get anything different?* **Student:** *I got 5 but also −9.*	$(x+2)^2 = 49$ $x = -9, 5$
Teacher: *Tell us how you got the 5.* **Student:** *I just thought about what squared is 49. That's 7. So 5 and 2 is 7.* **Teacher:** *And how did the rest of you get −9? Someone who took the square root of both sides?* **Student:** *I took the square root of both sides, so x + 2 is both positive and negative 7.* **Teacher:** *I'll write both of those separately.* **Student:** *So, what plus 2 is −7, that is −9.* **Teacher:** *And I can model that on an open number line.* *So, you used inverse operations, like undoing the square by taking the square root to solve for x. Nice.*	$\sqrt{(x+2)^2} = \sqrt{49}$ $x + 2 = \pm\sqrt{49}$ $x + 2 = 7 \quad x + 2 = -7$ $x = 5 \quad x = -9$ 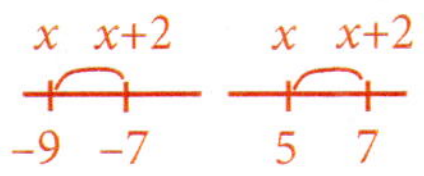

(continued)

Teacher: *Did anyone multiply first? Square the binomial?* **Student:** *I did and then I put everything on the same side and factored. Since it's equal to 0, you can have the factors equal 0.* **Teacher:** *So you factored and used the zero product property. Great.*	$(x+2)^2=49$ $x^2+4x+4=49$ $x^2+4x-45=0$ $(x+9)(x-5)=0$ $x=-9,5$
Student: *I multiplied it out the same way but when I got to the second line, I took the 4 to the other side, so I ended up with $x^2+4x=45$. Then I factored out the x. I thought about what number times that number and 4 is 45. First I found 5, then also −9.* **Teacher:** *You did not use the zero product property, you factored in a different way and used factor pairs of 45 to help you. Three great strategies. I wonder which one you want to use for this next problem?*	$x^2+4x=45$ $x(x+4)=45$ $x=-9,5$

The teacher repeats similarly with the third and optional fourth problem, modeling the three strategies.

$(x+3)^2+13=49$

$x=-9,3$

$(x+3)^2=36$

$\sqrt{(x+3)^2}=\sqrt{36}$

$x+3=\pm\sqrt{36}$

$x+3=-6 \quad x+3=6$

$x=-9 \quad x=3$

$(x+3)^2-36=0$

$x^2+6x+9-36=0$

$x^2+6x-27=0$

$(x+9)(x-3)=0$

$x=-9,3$

$x^2+6x=27$

$x(x+6)=27$

$x=3,-9$

Teacher: *Which of these strategies is feeling efficient for these problems?*

Student: *Both factoring strategies are not bad, but you have to multiply the binomial squared first.*

Student: *Yeah, none of these equations are already multiplied out.*

Student: *I have a hard time with the first strategy, taking the square root of both sides because I keep forgetting that there can be two solutions.*

Student: *I don't like to factor, it feels like too much guessing to me. So, I like the first one better.*

Teacher: *What shall we call the first strategy?*

Student: *Taking the square root of both sides and solving?*

Teacher: *I wonder which strategy you'll like for the next problem.* Students work and the teacher circulates, looking for students who were factoring but now turn to taking the square root.	$(x+1)^2-2=0$

Teacher: *I saw you factor for the other problems, but for this one, you didn't. What's up?*

Student: *I couldn't get factoring to work.*

Teacher: *Why not?*

Student: *Well, now that I have an answer by taking the square root, I can tell you that it's because it wasn't going to factor. The answers are not integers, they're square roots.*

Teacher: *Let's get your solution up here and talk about it.*

The student tells the teacher about solving by taking the square root of both sides as the teacher models. Then the teacher asks about what happened when students attempted to factor.

Student: *The only factors of* 1 *are* 1 *and* −1. *Neither works!*

Student: *Also, in the original problem, there is that* 2. *That is not a perfect square. In the problems before, the numbers were perfect squares.*

$(x+1)^2-2=0$

$x=-1\pm\sqrt{2}$

$(x+1)^2-2=0$

$\sqrt{(x+1)^2}=\sqrt{2}$

$x+1=\pm\sqrt{2}$

$x=-1+\sqrt{2}\quad x=-1-\sqrt{2}$

~~$(x+1)^2-2=0$~~

~~$x^2+2x+1-2=0$~~

~~$x^2+2x-1=0$~~

~~$(x+\_\_)(x-\_\_)=0$~~

?????

~~$x^2+2x=1$~~

~~$x(x+2)=1$~~

~~$x=$~~

????

Student: *So, it looks like sometimes the square root strategy might be the best one. And maybe that has to do with the constant being a perfect square or not?*

Teacher: *Interesting. ... Let's try the last problem.*

The teacher repeats similarly with the last problem, modeling the square root (inverse operations, properties of equality) strategy.

$(x-4)^2-8=0$

$x=4\pm2\sqrt{2}$

$(x-4)^2-8=0$

$\sqrt{(x-4)^2}=\sqrt{8}$

$x-4=\pm2\sqrt{2}$

$x=4+2\sqrt{2}\quad x=4-2\sqrt{2}$

Teacher: *How would you summarize some of the things that came up in this string today?*

Elicit the following:

- *If you are not going to solve by graphing with technology, there are three strategies so far for solving quadratic equations.*
- *If the equation is already in the form to take the square root of both sides, that might be efficient.*
- *You can solve by factoring two ways, zero product property and finding factor pairs.*
- *If it won't factor, try the square root strategy.*

(continued)

Sample Final Display

Your display could look like this at the end of the problem string:

$x^2 = 49$

$x = 7, -7$

$\sqrt{x^2} = \sqrt{49}$

$|x| = \sqrt{49}$

$x = \pm\sqrt{49} = \pm 7$

$x^2 - 49 = 0$

$(x-7)(x+7) = 0$

$x = -7, 7$

$(x+2)^2 = 49$

$x = -9, 5$

$\sqrt{(x+2)^2} = \sqrt{49}$

$x+2 = \pm\sqrt{49}$

$x+2 = 7 \quad x+2 = -7$

$x = 5 \quad x = -9$

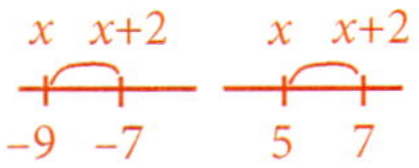

$(x+2)^2 = 49$

$x^2 + 4x + 4 = 49$

$x^2 + 4x - 45 = 0$

$(x+9)(x-5) = 0$

$x = -9, 5$

$x^2 + 4x = 45$

$x(x+4) = 45$

$x = -9, 5$

$(x+3)^2 + 13 = 49$

$x = -9, 3$

$(x+3)^2 = 36$

$\sqrt{(x+3)^2} = \sqrt{36}$

$x+3 = \pm\sqrt{36}$

$x+3 = -6 \quad x+3 = 6$

$x = -9 \quad x = 3$

$(x+3)^2 - 36 = 0$

$x^2 + 6x + 9 - 36 = 0$

$x^2 + 6x - 27 = 0$

$(x+9)(x-3) = 0$

$x = -9, 3$

$x^2 + 6x = 27$

$x(x+6) = 27$

$x = 3, -9$

$(x+1)^2 - 2 = 0$

$x = -1 \pm \sqrt{2}$

$(x+1)^2 - 2 = 0$

$\sqrt{(x+1)^2} = \sqrt{2}$

$x+1 = \pm\sqrt{2}$

$x = -1 + \sqrt{2} \quad x = -1 - \sqrt{2}$

~~$(x+1)^2 - 2 = 0$~~

~~$x^2 + 2x + 1 - 2 = 0$~~

~~$x^2 + 2x - 1 = 0$~~

~~$(x + \_\_)(x - \_\_) = 0$~~

?????

~~$x^2 + 2x = 1$~~

~~$x(x+2) = 1$~~

~~$x =$~~

????

$(x-4)^2 - 8 = 0$

$x = 4 \pm 2\sqrt{2}$

$(x-4)^2 - 8 = 0$

$\sqrt{(x-4)^2} = \sqrt{8}$

$x - 4 = \pm 2\sqrt{2}$

$x = 4 + 2\sqrt{2} \quad x = 4 - 2\sqrt{2}$

Facilitation Notes

This version of the problem string lists short notes for important teacher moves during the string. After you've done the string yourself and studied the relationships involved, you might make similar notes for the things you want a reminder of or deem important.

$x^2 = 49$	*If a number times itself, squared, is 49, what is the number? What about −7? Talk about absolute value with variables. Did anyone factor? Could you?*
$(x+2)^2 = 49$	*What if you add 2 first, then square and it's 49, what's x? Elicit taking the square root, factoring with zero product property, and factoring with factor pairs.*
$(x+3)^2 + 13 = 49$	*Quicker. Model three strategies. Which strategy is feeling efficient for these problems?*
$(x-2)^2 - 36 = 0$	*Optional. If time.*
$(x+1)^2 - 2 = 0$	*Hmmm.... what to do when factoring doesn't work? Why doesn't factoring work?*
$(x-4)^2 - 8 = 0$	*After solved and modeled, what are the 3 strategies used today? When might you use each one?*

PROBLEM STRINGS LESSON 5.3

The Quadratic Formula

At a Glance

$$\frac{-4 \pm 8}{2}$$

$$\frac{-4}{2} \pm \frac{8}{2}*$$

$$\frac{-4 \pm 8}{8}$$

$$\frac{-4 \pm \sqrt{8}}{2}$$

$$\frac{-6 \pm \sqrt{27}}{3}$$

$$\frac{-6 \pm \sqrt{27}}{9}$$

Optional string with complex numbers:

$$\sqrt{4}, \sqrt{-4}$$

$$\left(\sqrt{8}\right), \sqrt{-8}, \sqrt{-18}$$

$$\frac{2 \pm \sqrt{-8}}{-2}$$

$$\frac{2 \pm \sqrt{-8}}{-8}$$

$$\frac{-3 \pm \sqrt{-18}}{6}$$

$$\frac{-3 \pm \sqrt{-18}}{3}$$

$$\frac{-3 \pm \sqrt{-18}}{2}$$

*optional problem

Objectives

The goal of this problem string is to develop efficiency using the quadratic formula by helping students develop two strategies for simplifying the solutions.

Placement

This string could come before students begin solving problems using the quadratic formula. This string could also be used after students have used the quadratic formula to address common errors and develop efficiency. The optional complex number string could be helpful when students are solving quadratic equations with complex solutions.

You could deliver this string before students begin working problems in section 5.3 The Quadratic Formula in the textbook or to support the work in 5.3, to help students avoid common arithmetic errors and to develop efficiency. You could deliver the optional complex number string when students are solving quadratic equations in 5.4 Complex Numbers.

Guiding the Problem String

If no students solve the first problem by splitting the fraction, insert the optional second problem, then notice the equivalent answers and help students connect. The next problem has a denominator that will not divide evenly, so the solutions will be fractions. The next three problems alternate the same way, while also introducing radicals. Throughout the problem string, remember that one goal is to give students quick feedback on their arithmetic. While you're discussing strategy as a whole class, students are getting valuable experience and feedback. Circulate and help students make sense of common errors. If students are making the same mistakes, consider inserting some "sister" problems for more practice or doing another "sister" string the next day.

About the Mathematics

Students may be more familiar with combining fractions with the same denominator, than splitting up a fraction, but this can be a helpful strategy when simplifying the solutions to the quadratic formula.

Students might be use to this $\frac{2}{7}+\frac{4}{7}=\frac{6}{7}$ but this $\frac{6}{7}=\frac{1}{7}+\frac{5}{7}$ is also helpful.

Sample Interactions

Use the following as you plan for ideas on how to elicit and model student strategies. This is not meant as a script, but as a view into the relationships involved and the intent of the problem string.

Teacher: *When you are solving quadratic equations using the quadratic formula, you'll end up with expressions like this first problem. Let's work together today to simplify finding the solutions to quadratic equations. I'm not going to give you the equation, just the quadratic formula with the correct values already substituted in and $b^2 - 4ac$ has already been figured. We are sort of taking a peek in the middle of someone solving. I wonder if we can find an equivalent and simpler way to express this fraction. So, seek to simplify this first problem and then we'll share strategies.* Students work briefly while the teacher circulates, looking for students who figure the numerator first and then divide and for students who split the fraction first and noting any other strategies.	$\frac{-4 \pm 8}{2}$
Teacher: *Let's start with those of you who figured out the numerator first. You guys were taking the grouping symbol of that fraction bar seriously. Tell us about that.* **Student:** *I did the −4 plus 8 first to get 4 and divided that by 2 so 2. Then I did the −4 minus 8 to get −12 and divided that by 2 so −6.* **Teacher:** *Did everyone follow that? Any questions? How would you restate this strategy generally?* **Student:** *Do the numerator first, then divide.*	$= \frac{-4+8}{2} = \frac{4}{2} = 2$ $= \frac{-4-8}{2} = \frac{-12}{2} = -6$
Teacher: *Not everyone did that. What were you others doing?* **Student:** *I put the −4 over the 2 and then plus and minus the 8 over the 2.* **Teacher:** *Ah, you split up the fraction first and found −4 divided by 2 and then added and subtracted 8/2.* **Student:** *Yeah, so that's −2 plus and −4 which is 2 and −6.* **Teacher:** *How would you generalize or describe this strategy?* **Student:** *Split up the fraction first, then simplify.* **Teacher:** *Does that make sense to everyone? We got the same answer. Can you just split up a fraction like that? How do you know? What might allow us to do that? Think about that and we'll come back to it.* *Both strategies seem to work just fine for this problem. Do you have a favorite now that you've considered both? I wonder if each strategy will come in handy for different problems. ...*	$\frac{-4}{2} \pm \frac{8}{2} = -2 \pm 4 = -6, 2$
Teacher: *The second problem is the same idea—barging in as someone is solving using the quadratic formula. Pick it up for them please.*	$\frac{-4 \pm 8}{8}$

(continued)

Lesson 5.3 • The Quadratic Formula (continued)

Teacher: *How would you solve that problem if you were simplifying the numerator first?* *How did you simplify the fractions?*	$\frac{-4+8}{8}=\frac{4}{8}=\frac{1}{2}$ $\frac{-4-8}{8}=\frac{-12}{8}=-\frac{3}{2}$
Teacher: *How would you solve that problem by splitting up the fraction first?* *Why is it okay to split up the fraction that way?* *Which strategy do you like best for this problem?*	$\frac{-4}{8}\pm\frac{8}{8}=-\frac{1}{2}\pm1=-\frac{3}{2},\frac{1}{2}$
Teacher: *For the third problem we jump in before the student has figured all of the square root. Finish simplifying please.*	$\frac{-4\pm\sqrt{8}}{2}$
Teacher: *How would you solve that problem by simplifying the numerator first?* *How did you reason about simplifying the square root of 8?* *Why did you factor out the 2? How did you know to do that?*	$\frac{-4\pm2\sqrt{2}}{2}=\frac{2(-2\pm1\sqrt{2})}{2}=-2\pm\sqrt{2}$
Teacher: *How would you solve that problem by splitting up the fraction first?* *Which strategy do you like best for this problem?* *So far, we keep getting exactly the same answer no matter which strategy. Maybe it doesn't matter. ...*	$-\frac{4}{2}\pm\frac{\sqrt{8}}{2}=-\frac{4}{2}\pm\frac{2\sqrt{2}}{2}=-2\pm\sqrt{2}$
Teacher: *Here is the third problem. Go!*	$\frac{-6\pm\sqrt{27}}{3}$
Teacher: *How would you solve that problem by simplifying the numerator first?* *How would you solve that problem by splitting up the fraction first?* *Which strategy do you like best for this problem?*	$\frac{-6\pm3\sqrt{3}}{3}=\frac{3(-2\pm1\sqrt{3})}{3}=-2\pm\sqrt{3}$ $-\frac{6}{3}\pm\frac{\sqrt{27}}{3}=-2\pm\frac{3\sqrt{3}}{3}=-2\pm\sqrt{3}$
Teacher: *And last but not least, what do you think about this problem? Any ideas which strategy you might want to try?*	$\frac{-6\pm\sqrt{27}}{9}$

Teacher: *How would you solve that problem by simplifying the numerator first?* *How would you solve that problem by splitting up the fraction first?* *Which strategy did you find more helpful for this problem? Why?*	$\frac{-6\pm3\sqrt{3}}{9}=\frac{3(-2\pm1\sqrt{3})}{9}=\frac{-2\pm\sqrt{3}}{3}$ $\frac{-6\pm\sqrt{27}}{9}=-\frac{6}{9}\pm\frac{3\sqrt{3}}{9}=-\frac{2}{3}\pm\frac{\sqrt{3}}{3}$

Teacher: *How would you summarize some of the things that came up in this string today?*

Elicit the following:

- *You can split up a fraction where you split the numerator by adding or subtracting but the denominator stays the same.*
- *You can simplify the quadratic formula by figuring the numerator first and then dividing.*
- *Alternatively, you can simplify the quadratic formula by splitting the fraction and simplifying each result.*
- *If the denominator isn't going to divide evenly, you might as well leave the fraction together because you'll end up putting it back together if you don't.*
- *Look for common factors in the two numerator terms that will divide with factors of the denominator.*

Sample Final Display

Your display could look like this at the end of the problem string:

$\frac{-4\pm8}{2}=-6,2$	$\frac{-4\pm8}{2}$: $=\frac{-4+8}{2}=\frac{4}{2}=2$ $=\frac{-4-8}{2}=\frac{-12}{2}=-6$	$\frac{-4}{2}\pm\frac{8}{2}=-2\pm4=2,-6$
$\frac{-4\pm8}{8}=-\frac{3}{2},\frac{1}{2}$	$\frac{-4+8}{8}=\frac{4}{8}=\frac{1}{2}$ $\frac{-4-8}{8}=\frac{-12}{8}=-\frac{3}{2}$	$\frac{-4}{8}\pm\frac{8}{8}=-\frac{1}{2}\pm1=-\frac{3}{2},\frac{1}{2}$
$\frac{-4\pm\sqrt{8}}{2}=-2\pm\sqrt{2}$	$\frac{-4\pm2\sqrt{2}}{2}=\frac{2(-2\pm1\sqrt{2})}{2}=-2\pm\sqrt{2}$	$-\frac{4}{2}\pm\frac{\sqrt{8}}{2}=-\frac{4}{2}\pm\frac{2\sqrt{2}}{2}=-2\pm\sqrt{2}$
$\frac{-6\pm\sqrt{27}}{3}=-2\pm\sqrt{3}$	$\frac{-6\pm3\sqrt{3}}{3}=\frac{3(-2\pm1\sqrt{3})}{3}=-2\pm\sqrt{3}$	$-\frac{6}{3}\pm\frac{\sqrt{27}}{3}=-2\pm\frac{3\sqrt{3}}{3}=-2\pm\sqrt{3}$
$\frac{-6\pm\sqrt{27}}{9}=\frac{-2\pm\sqrt{3}}{3}=-\frac{2}{3}\pm\frac{\sqrt{3}}{3}$	$\frac{-6\pm3\sqrt{3}}{9}=\frac{3(-2\pm1\sqrt{3})}{9}=\frac{-2\pm\sqrt{3}}{3}$	$\frac{-6\pm\sqrt{27}}{9}=-\frac{6}{9}\pm\frac{3\sqrt{3}}{9}=-\frac{2}{3}\pm\frac{\sqrt{3}}{3}$

(continued)

Facilitation Notes

This version of the problem string lists short notes for important teacher moves in the string. After you've done the string yourself and studied the relationships involved, you might make similar notes for the things you want a reminder of or deem important.

$\frac{-4 \pm 8}{2}$	If someone was simiplifying after putting values in to the quadratic formula, finish simplifying for them. How did you simplify? (Simplify numerator first, split up fraction first)
$\frac{-4 \pm 8}{8}$	Simplify. How did you simplify? (Simplify numerator first, split up fraction first) Does it make sense to split up the fraction that way? Why? Does that always work?
$\frac{-4 \pm \sqrt{8}}{2}$	Simplify. How did you simplify? (Simplify numerator first, split up fraction first) How did you deal with the square root of 8?
$\frac{-6 \pm \sqrt{27}}{3}$	Anyone simplify the numerator first? Split up the fraction first? Which strategy do you like best for this problem?
$\frac{-6 \pm \sqrt{27}}{9}$	Anyone simplify the numerator first? Split up the fraction first? Which strategy do you like best for this problem?

PROBLEM STRINGS LESSON

5.4 Solving Quadratic Equations 3

At a Glance

$$x^2+4x+5=0$$

$$x^2-4x+5=0$$

$$-x^2-4x-5=0$$

$$-x^2+4x-5=0$$

Objectives

The goal of this series of problem strings is to help students gain facility with deciding on an efficient strategy to solve quadratic equations. This string helps students consider what happens in the quadratic formulas that results in complex solutions and helps students make connections between the equations, graphs, and solutions. While students are making generalizations, they are also getting practice solving quadratic equations with quick feedback.

Placement

This is the third in a series of four problem strings. This string could be delivered as students have solved quadratic equations by completing the square and are working to solve quadratic equations with complex solutions using the quadratic formula.

You could deliver this string as students are working on 5.4 Complex Numbers or 5.5 Solving Quadratic Equations in the textbook.

Guiding the Problem String

These problems all need about the same amount of time. During the first problem, spend time reminding students of completing the square and the quadratic formula as needed. As students solve the next problems elicit connections between what is happening under the radical in the quadratic formula to result in different kinds of solutions. In the second problem, why are the solutions complex numbers and what do graphs of those quadratics look like? By the end of the string, help students verbalize that a negative number in the radical of a solution to a quadratic equation means complex roots and graphs of parabolas that do not intersect the x-axis.

About the Mathematics

The absolute values of the coefficients stay the same from problem to problem, so students can concentrate on what happens as the signs change. When the discriminant is non-negative, $b^2-4ac\geq 0$, the solution is a real number. When the discriminant is negative, $b^2-4ac<0$, the solution is a complex number.

Functions have zeros, graphs have x-intercepts, equations have solutions and roots.

Because of the coefficients, students may choose to complete the square or use the quadratic formula, probably based on which strategy comes more easily to them. When $a\neq 1$, students may choose the quadratic formula. Later strings will use different coefficients to draw attention to choosing efficient strategies. The one strategy that is not efficient for these problems is graphing, though graphing does confirm that the solutions are complex numbers.

(continued)

Sample Interactions

Use the following as you plan how to elicit and model student strategies. This is not meant as a script, but as a view into the relationships involved and the intent of the problem string.

<table>
<tr>
<td>
Teacher: Let's start with solving this equation: $x^2+4x+5=0$. For this equation, what is x?
Students work and the teacher circulates, looking for students who are trying to factor, completing the square, using the quadratic formula, or graphing and taking note of any other strategies.
Teacher: What is x?
Student: $x=-2\pm i$.
</td>
<td>
$x^2+4x+5=0$
$x=-2\pm i$
</td>
</tr>
<tr>
<td>
Teacher: So, I noticed that a few of you tried to factor. Would you please describe how you knew the quadratic wouldn't factor?
Student: I couldn't find a way to make it work with 5 and 4. Five only factors into 1 and 5, but since both the 5 and 1 are positive, you can't add to get a 4.
Teacher: Great. And some of you completed the square. Tell me about that?
Student: Sure, I put x^2 in the square and split up the 4x into 2x and 2x. That means to make a square I still need the corner to be 4. That's an x + 2 by x + 2. So, since I added 4 to complete the square, then I subtracted 4, and I ended up with $(x+2)^2+1=0$ Then $x+2=\pm\sqrt{-1}$ and $x=-2\pm i$.
</td>
<td>
$x^2+4x+5=0$
$x=-2\pm i$

$(x^2+4x+\underline{4})+5-\underline{4}=0$
$(x+2)^2+1=0$
$(x+2)^2=-1$
$x+2=\pm\sqrt{-1}$
$x=-2\pm i$

<table>
<tr><td></td><td>x</td><td>2</td></tr>
<tr><td>x</td><td>x^2</td><td>2x</td></tr>
<tr><td>2</td><td>2x</td><td>4</td></tr>
</table>
</td>
</tr>
<tr>
<td>
Teacher: Nice. And how about using the quadratic formula?
A student walks through substituting the values into the quadratic formula and simplifying while the teacher quickly records.
Teacher: So, both strategies found the same solutions. Which strategy do you wish your brain would go to for this problem next time?
</td>
<td>
$x=\frac{-4\pm\sqrt{4^2-4(1)(5)}}{2(1)}=\frac{-4\pm\sqrt{16-20}}{2}$
$=\frac{-4\pm\sqrt{-4}}{2}=\frac{-4\pm 2i}{2}=-2\pm i$
</td>
</tr>
<tr>
<td colspan="2">
Student: I like completing the square because once I've completed the square, I can just solve and I don't have to remember that long formula.
Student: I like the quadratic formula because I can just plug stuff in and solve. Completing the square is still weird to me.
Student: It seems like either strategy is just as easy if you know how to do both.
Teacher: I wonder what a problem would be like so that one strategy would be an obvious better choice? That will come up in our work so you might be thinking about it. Today, concentrate on understanding what is happening in both strategies and the connections to the graph.
</td>
</tr>
</table>

Teacher: *What does the graph of the quadratic function on the left of the equation look like? Does it support the same solutions?*

The teacher displays a graph of $y = x^2 + 4x + 5$.

Student: *The parabola doesn't intersect the x-axis!*

Student: *So, yes, the complex solutions make sense. It's interesting that the vertex looks like it's back at x = −2.*

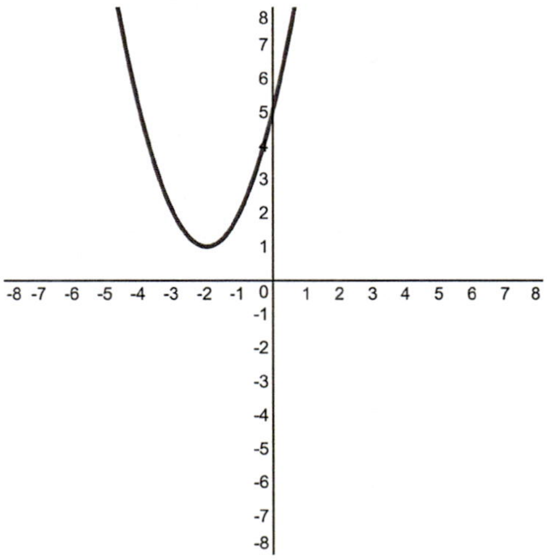

Teacher: *That is interesting. Okay, next problem in the string. What is x for this equation* $x^2 - 4x + 5 = 0$*?*

Student: *I got* $x = 2 \pm i$.

Students work and the teacher circulates, looking for students who are completing the square, using the quadratic formula, or graphing and taking note of any other strategies.

Teacher: *This problem is a lot like the last one. I noticed that some of you just tweaked your work from the last problem. Will you tell us how the completing the square compared to the last problem?*

Student: *This time, instead of splitting up 4x, I split up −4x, so the square is an x − 2 by x − 2. But −2 squared is still 4, so everything is the same except the x − 2, and so the answer is just* $x = 2 \pm i$*, instead of −2.*

Teacher: *So, one sign change between the two equations changes the answer by a sign change. Interesting. Does that play out with the quadratic formula?*

$x^2 + 4x + 5 = 0$

$x = -2 \pm i$

$(x^2 + 4x + 4) + 5 - 4 = 0$

$(x + 2)^2 + 1 = 0$

$(x + 2)^2 = -1$

$x + 2 = \pm\sqrt{-1}$

$x = -2 \pm i$

$x^2 - 4x + 5 = 0$

$x = 2 \pm i$

$(x^2 - 4x + 4) + 5 - 4 = 0$

$(x - 2)^2 + 1 = 0$

$(x - 2)^2 = -1$

$x - 2 = \pm\sqrt{-1}$

$x = 2 \pm i$

Teacher: *Instead of walking all the way through the quadratic formula, I'd like you to compare this work to the problem before. What changes, what stays the same, and how does that affect the solutions?*

Student: *It's just the b that changes, from positive 4 to −4.*

Student: *Everything else stays the same.*

Student: *The b only shows up in 2 places in the quadratic formula and since negative 4 squared is the same as 4 squared, the negative really only affects the very first part of the formula, now the first 4 ends up being positive.*

Teacher: *And that agrees with the students who completed the square. I wonder what the graph looks like.*

$x^2 + 4x + 5 = 0$

$x = -2 \pm i$

$$x = \frac{-4 \pm \sqrt{4^2 - 4(1)(5)}}{2(1)} = \frac{-4 \pm \sqrt{16 - 20}}{2}$$

$$= \frac{-4 \pm \sqrt{-4}}{2} = \frac{-4 \pm 2i}{2} = -2 \pm i$$

$x^2 - 4x + 5 = 0$

$x = 2 \pm i$

still +16

$$x = \frac{-(-4) \pm \sqrt{(-4)^2 - 4(1)(5)}}{2(1)} = \frac{4 \pm \sqrt{16 - 20}}{2}$$

$$= \frac{4 \pm \sqrt{-4}}{2} = \frac{4 \pm 2i}{2} = 2 \pm i$$

(continued)

The teacher adds to the display, a graph of $x^2 - 4x + 5 = 0$.

Student: *It looks like the first one, just translated.*

Student: *Or reflected over the y-axis.*

Student: *And the vertex is at x equals positive 2.*

Teacher: *I wonder if there is a connection? For now, let's focus on the x-intercepts, when the function values are 0. That's what we were solving for, right?*

Student: *The solutions are complex—the graph does not have x-intercepts.*

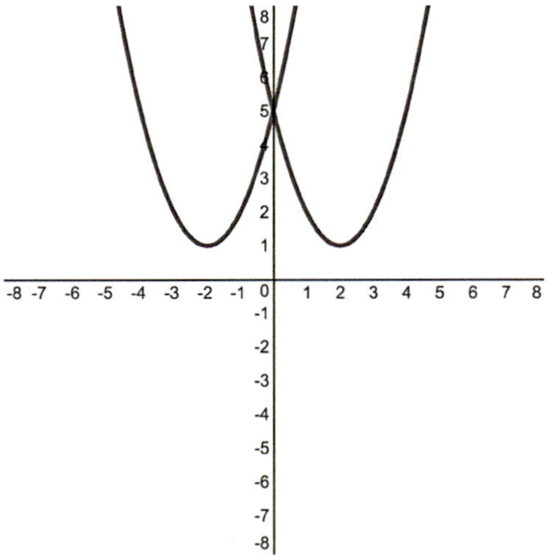

Teacher: *Okay, next problem: $-x^2 - 4x - 5 = 0$. What is x?*

Students work and the teacher circulates.

Student: *It's the same as the first problem, $x = -2 \pm i$.*

The teacher is prepared to take time here to clarify the completing the square (factoring the −1, etc.).

Teacher: *Tell me how you approached this problem?*

Student: *Well, I was wondering if this has the same solutions as the first problem because the two expressions are the same, except all opposite signs. And I found that the two equations do have the same solution.*

Teacher: *Does that mean that the parabolas are the same? What's the same? The parabolas, the solutions, the expressions?*

Student: *The parabolas are reflections.*

Student: *The expressions are not equivalent but their solutions are the same. The y-values are the same, 0, at the same x-values. The expressions are kind of opposite.*

Teacher: *If you multiplied one expression by −1, would that give you the other one? Sure enough.*

After modeling both completing the square and using the quadratic formula, the teacher asks about strategy choice.

Teacher: *Which of these strategies do you wish your brain would be inclined to think of for problems like this?*

Student: *For this problem, I like the quadratic formula, because with completing the square, you have to mess with the −1, factor it out, remember to add the 4, not subtract it.*

Student: *One thing that's nice about completing the square is that you end up with a form of the expression that you could graph. Once I know what the vertex is and whether the parabola opens up or down, I know if the solutions are real or complex.*

$$-(x^2+4x)-5=0$$
$$-(x^2+4x+4)-5+4=0$$
$$-(x+2)^2-1=0$$
$$(x+2)^2=-1$$
$$x+2=\pm\sqrt{-1}$$
$$x=-2\pm i$$

$$x=\frac{-(-4)\pm\sqrt{(-4)^2-4(-1)(-5)}}{2(-1)}=\frac{4\pm\sqrt{16-20}}{-2}$$
$$=\frac{4\pm\sqrt{-4}}{-2}=\frac{4\pm 2i}{-2}=-2\pm i$$

$$-x^2-4x-5=-1(x^2+4x+5)$$

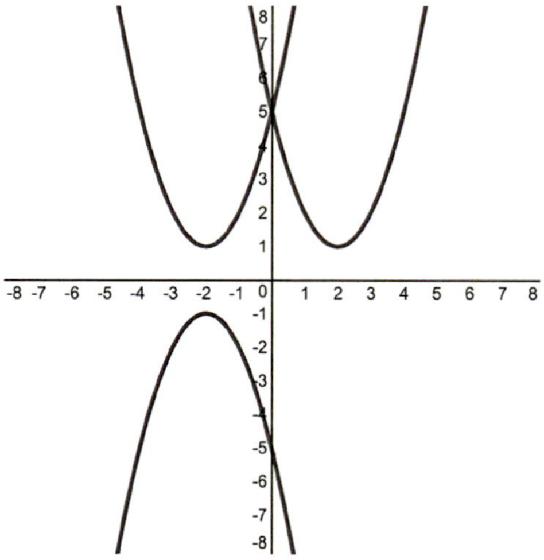

Teacher: *The last problem today is* $-x^2+4x-5=0$*? What is x?*

Student: *It's* $x=2\pm i$.

The teacher leads a discussion about the connections between all four problems, the solutions, and the graphs.

$-x^2+4x-5=0$

$x=2\pm i$

$-(x^2-4x)-5=0$

$-(x^2-4x+4)-5+4=0$

$-(x-2)^2-1=0$

$(x-2)^2=-1$

$x-2=\pm\sqrt{-1}$

$x=2\pm i$

$$x=\frac{-4\pm\sqrt{4^2-4(-1)(-5)}}{2(-1)}=\frac{-4\pm\sqrt{16-20}}{-2}$$

$$=\frac{-4\pm\sqrt{-4}}{-2}=\frac{-4\pm 2i}{-2}=2\pm i$$

$-x^2+4x-5=-(x^2-4x+5)$

every time it's 16-20

$(-4)^2=4^2$, $\sqrt{-4}=2i$

All of the solutions are complex,

none of the graphs have x-intercepts

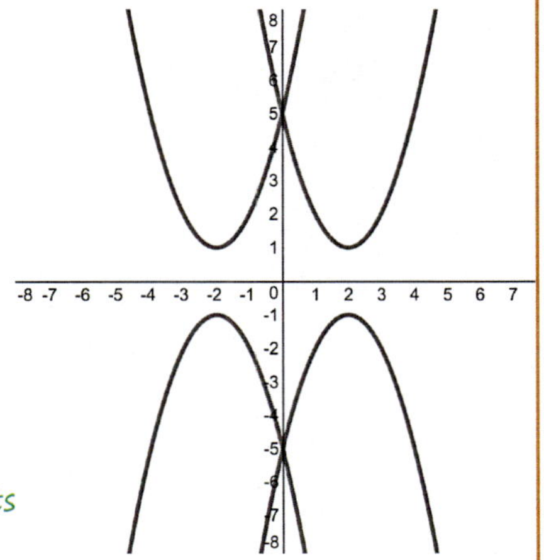

Teacher: *How would you summarize some of the things that came up in this string today?*

Elicit the following:

- *If there's a negative under the radical, the solutions are complex and the graph has no x-intercepts.*
- *Every time it ended up with 16 – 20, so it is always the square root of –4 and that's 2i.*
- *Even though the graphs don't have x-intercepts, there are still patterns between them.*
- *The quadratic formula only gives the solutions. Completing the square also helps you graph because you get the vertex form.*

Sample Final Display

Your display could look like this at the end of the problem string:

(continued)

$x^2+4x+5=0$

$x=-2\pm i$

$(x^2+4x+4)+5-4=0$
$(x+2)^2+1=0$
$(x+2)^2=-1$
$x+2=\pm\sqrt{-1}$
$x=-2\pm i$

$$x=\frac{-4\pm\sqrt{4^2-4(1)(5)}}{2(1)}=\frac{-4\pm\sqrt{16-20}}{2}=\frac{-4\pm\sqrt{-4}}{2}=\frac{-4\pm 2i}{2}=-2\pm i$$

	x	2
x	x^2	2x
2	2x	4x

$x^2-4x+5=0$

$x=2\pm i$

$(x^2-4x+4)+5-4=0$
$(x-2)^2+1=0$
$(x-2)^2=-1$
$x-2=\pm\sqrt{-1}$
$x=2\pm i$

same

still +16

$$x=\frac{-(-4)\pm\sqrt{(-4)^2-4(1)(5)}}{2(1)}=\frac{4\pm\sqrt{16-20}}{2}=\frac{4\pm\sqrt{-4}}{2}=\frac{4\pm 2i}{2}=2\pm i$$

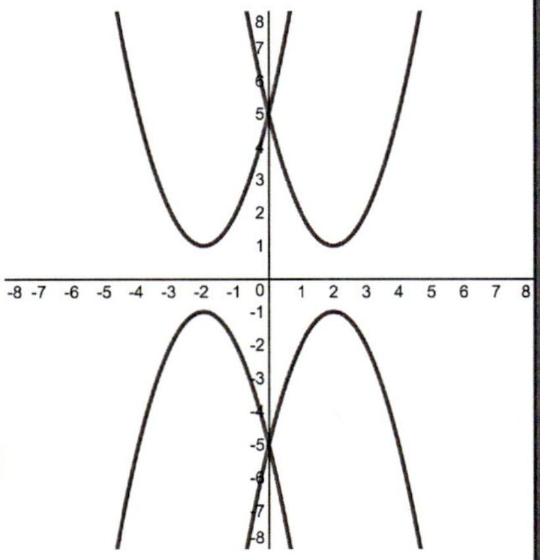

$-x^2-4x-5=0$

$x=-2\pm i$

$-(x^2+4x)-5=0$
$-(x^2+4x+4)-5+4=0$
$-(x+2)^2-1=0$
$(x+2)^2=-1$
$x+2=\pm\sqrt{-1}$
$x=-2\pm i$

same

$$x=\frac{-(-4)\pm\sqrt{(-4)^2-4(-1)(-5)}}{2(-1)}=\frac{4\pm\sqrt{16-20}}{-2}=\frac{4\pm\sqrt{-4}}{-2}=\frac{4\pm 2i}{-2}=-2\pm i$$

same except (−)

$-x^2-4x-5=-1(x^2+4x+5)$

$-x^2+4x-5=0$

$x=2\pm i$

$-(x^2-4x)-5=0$
$-(x^2-4x+4)-5+4=0$
$-(x-2)^2-1=0$
$(x-2)^2=-1$
$x-2=\pm\sqrt{-1}$
$x=2\pm i$

$$x=\frac{-4\pm\sqrt{4^2-4(-1)(-5)}}{2(-1)}=\frac{-4\pm\sqrt{16-20}}{-2}=\frac{-4\pm\sqrt{-4}}{-2}=\frac{-4\pm 2i}{-2}=2\pm i$$

$-x^2+4x-5=-(x^2-4x+5)$

every time it's 16-20
$(-4)^2=4^2$, $\sqrt{-4}=2i$
All of the solutions are complex,
none of the graphs have x-intercepts

Facilitation Notes

This version of the problem string lists short notes for important teacher moves during the string. After you've done the string yourself and studied the relationships involved, you might make similar notes for the things you want a reminder of or deem important.

$x^2+4x+5=0$	Elicit and model factoring, complete the square, quad formula. Quickly graph with display calc.
$x^2-4x+5=0$	Elicit and model complete the square, quad formula. Add graph to previous. Connect to previous problem.
$-x^2-4x-5=0$	Repeat. Quad form preferred because of $-x^2$? How is this the same, different from previous two? Predict graph, then add.
$-x^2+4x-5=0$	Repeat. How is this the same, different from previous two? Predict graph, then add.

PROBLEM STRINGS LESSON

5.5 Solving Quadratic Equations 4

At a Glance

$x^2+3x+2=0$

$3x^2+5x+20=0$

$x^2-4x+1=0$

$-4.9x^2+76.2x+9.67=0$

$x^2+8x=9$

Objectives

The goal of this series of problem strings is to help students develop efficient strategies for solving quadratic equations. This particular string is intended to help students learn to choose wisely between the many strategies.

Placement

This is the fourth in a series of four strings to develop and choose efficiency strategies for solving quadratic equations. This string should follow the other three strings in the series. Students should have had experience solving quadratic equations by graphing, using factoring and the zero product property, using factor pairs, completing the square, and using the quadratic formula.

This string would be a nice as a follow up to textbook section 5.5 Solving Quadratic Equations or as a review at the end of Chapter 5: Quadratic Functions and Relations.

Guiding the Problem String

This problem string is an opportunity to assess students' grasp of different strategies, flexibility using different strategies, and connections between multiple representations of quadratic functions. Listen, watch, and ask probing questions to determine where more work is needed.

The first problem, $x^2+3x+2=0$, is easily solved using any method and so provides a nice jumping off point to compare many strategies. The second problem, $3x^2+5x+20=0$ does not appear to be easy to try to factor and in fact, has complex roots so solving demands either the quadratic formula or completing the square. However, based on the *a* and *b* values of 3 and 5 respectively, completing the square would be a bit unwieldy. The third problem, $x^2-4x+1=0$, does not factor, but a graph would intersect the *x*-axis (using the $y=(x(x-4))+1$ strategy (see 5.1 Graphing Quadratic Functions 2), so either completing the square or the quadratic formula could be argued to work nicely. The fourth problem with its delightful real world projectile motion coefficients is a great candidate to solve approximately using technology to look for *x*-intercepts on the graph. The last problem is already formatted to use factor pairs to solve $x(x+8)=9$, but could also quickly be ready to solve by factoring and using the zero-product property. It could, of course, be solved with heavier handed strategies like completing the square or the quadratic formula, but the discussion should focus on efficiency—choosing a strategy that works well for each problem.

About the Mathematics

Some quadratic equations are easily solved with many different strategies. Some are efficiently solved with only one or two strategies. All quadratic equations can be solved exactly using the quadratic formula. All quadratic equations with real solutions can be solved approximately by graphing with technology. For some quadratic equations, it is easy to complete the square, with the added benefit of having the vertex form of the function $f(x)$ as well as the solutions to $f(x) = 0$. For many quadratic equations that are found in textbooks, it is efficient to solve by factoring using the zero-product property or by using factor pairs.

(continued)

Important Questions

Use the following as you plan how to elicit and model student strategies.

Consider starting this string by having students help you list ways they can solve quadratic equations. Keep this list up during the string.

- *What are all of the strategies that could work for this problem? Which do you choose and why?*
- *What is it about the form of the equation that nudges you toward a particular strategy?*
- *What is it about the numbers that nudges you toward a particular strategy?*
- *After you've solved the equation, look back at the numbers and structure and consider what strategy you wish your brain would be more inclined toward the next time you run into such a problem.*
- *Which strategy(ies) would solve any quadratic equation?*
- *Which strategy(ies) only solve quadratic equations with real solutions?*
- *Which strategy(ies) only solve quadratic equations with integer solutions?*
- *Which strategy(ies) work well for equations with unwieldy numbers?*
- *What do you look for when choosing a good/efficient strategy?*

How would you summarize some of the things that came up in this string today?

- *You can solve anything with the quadratic formula and completing the square.*
- *You might not want to solve equations with lots of decimals with the quadratic formula and completing the square. Maybe graph those.*
- *You might not want to solve equations with an $a \neq 1$ and an odd b value by completing the square. Try something else.*
- *Try to factor, either by using the zero-product property or factor pairs.*
- *If factoring doesn't work, quickly see if you can get a sense of the types of answers by estimating the graph. If it clearly won't intersect the x-axis, expect complex solutions.*
- *If you're going to need to use a graph of the function, you might as well either graph it using a calculator or complete the square so you have the vertex.*

Sample Anchor Chart

Your anchor chart could look like this at the end of the problem string:

Choosing a Smart Strategy

If it's easy use factor pairs or factor.	$x^2+3x+2=0$ $x(x+8)=9$
Ugly coefficients? Graph, especially if you're going to get approximations anyway.	$-4.9x^2+76.2x+9.67=0$
If a = 1 and b is nice, complete the square.	$x^2-4x+1=0$
You can always use the quadratic formula, but it might not be worth it. Try other things first.	$3x^2+5x+20=0$

Sample Final Display

Your display could look like this at the end of the problem string:

$x^2+3x+2=0$

$x=-2,-1$

$(x+1)(x+2)=0$

$x+1=0 \quad x+2=0$

$x=-1,-2$

$x(x+3)=-2$

$x=-1,-2$

$3x^2+5x+20=0$

$x=\dfrac{-5\pm i\sqrt{215}}{6}$

$$x=\frac{-5\pm\sqrt{(5)^2-4(3)(20)}}{2(3)}=\frac{-5\pm\sqrt{25-240}}{6}$$

$$=\frac{-5\pm\sqrt{-215}}{6}=\frac{-5\pm i\sqrt{215}}{6}$$

$x^2-4x+1=0$

$x=2\pm\sqrt{3}$

$$x=\frac{-(-4)\pm\sqrt{(-4)^2-4(1)(1)}}{2(1)}=\frac{4\pm\sqrt{16-4}}{2}$$

$$=\frac{4\pm\sqrt{12}}{2}=\frac{4\pm2\sqrt{3}}{2}=\frac{2(2\pm\sqrt{3})}{2}=2\pm\sqrt{3}$$

$x^2-4x=-1$

$x^2-4x+4=-1+4$

$(x-2)^2=3$

$x-2=\pm\sqrt{3}$

$x=2\pm\sqrt{3}$

$-4.9x^2+76.2x+9.67=0$

$x\approx-0.1, 15.7$

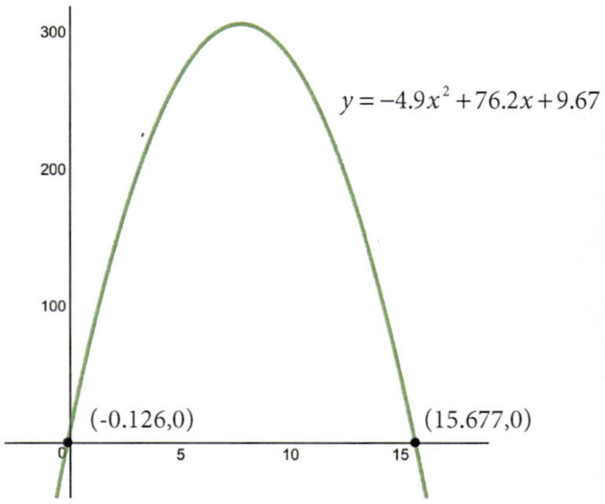

$x^2+8x=9$

$x=-9,1$

$x(x+8)=9$

$x=-9,1$

$(x+9)(x-1)=0$

$x+9=0 \quad x-1=0$

$x=-9,1$

Facilitation Notes

This version of the problem string lists short notes for important teacher moves during the string. After you've done the string yourself and studied the relationships involved, you might make similar notes for the things you want a reminder of or deem important.

$x^2+3x+2=0$	*Solve. Seek for efficiency. Share.* *What strategy do you think is the most efficient? Why?* *What other strategies could you use? Why not them?*
$3x^2+5x+20=0$	*Repeat.* *What about the numbers nudged you away from factoring?* *Could thinking about the graph influence your strategy choice?*
$x^2-4x+1=0$	*Repeat.* *What about the numbers nudged you away from factoring?* *Could thinking about the graph influence your strategy choice?*
$-4.9x^2+76.2x+9.67=0$	*What does this remind you of? What's going on?* *Repeat.* *What about the numbers influenced your strategy choice?*
$x^2+8x=9$	*Repeat.* *What about the structure influenced your strategy choice?* *What other strategies could you use? Why not them?*

PROBLEM STRINGS LESSON

5.8 Focus-Directrix: Solving for *f*

<table>
<tr>
<td rowspan="4">

At a Glance

$f = \frac{11}{5}$

$\frac{1}{f} = \frac{5}{11}$

$4f = 8$

$4f = \frac{1}{8}$

$\frac{1}{4f} = \frac{1}{8}$

$\frac{1}{4f} = 8$

$\frac{1}{4f} = -2$

</td>
<td>

Objectives

The goal of this problem string is to set up students for success when working with the Focus-directrix definition of a parabola. This string strengthens students' ability to solve simple rational equations using the "solve an equivalent reciprocal equation" strategy.

</td>
</tr>
<tr>
<td>

Placement

This string could come before or during student work on converting between the Focus-directrix definition and function notation for parabolas.

You could deliver this string before or as students are working on textbook section 5.8 Parabolas.

</td>
</tr>
<tr>
<td>

Guiding the Problem String

The first three problems should go quickly, just setting the stage for the rest. If students have a hard time reasoning about $4f = \frac{1}{8}$, you could model the relationships quickly using a double number line or a fraction bar model. Throughout the string, compare problems to each other and refer back to the reciprocal relationship between the first two problems, wondering aloud how that might be useful. Model the "using properties of equality by doing the same operation to both sides of the equation" strategy, the "solving an equivalent reciprocal equation" strategy, and the "numerators are the same so the denominators are equal" strategy.

This string works toward the same to solve an equivalent reciprocal equation strategy as the string in 6.9 Solving Rational Equations.

</td>
</tr>
<tr>
<td>

About the Mathematics

These equations are the type that students may solve to convert a quadratic function to the focus-directrix form of the parabola, where *f* is half of the distance between the focus and the vertex.

The solutions to an equation and that equation's reciprocal equation are the same, except for values that are outside of either equations' domain. For these equations, $f \neq 0$.

Refrain from using the word "over" to describe the rational expressions. "Over" describes position—it does not describe a mathematical relationship and is therefore, not helpful. Describe $\frac{1}{4f}$ as one divided by $4f$ and $\frac{1}{8}$ as one-eighth. Using this language potentially helps keep students making sense of expressions using relationships.

</td>
</tr>
</table>

Sample Interactions

Use the following as you plan how to elicit and model student strategies. This is not meant as a script, but as a view into the relationships involved and the intent of the problem string.

Teacher: *I'm going to put a problem on the board and ask you to think about some other ways to express f. We'll start with this problem, f is 11 divided by 5. What is 11 divided by 5? What's another way to express f?* **Student:** *Eleven-fifths?* **Teacher:** *Sure, that's a fine way to think about 11 divided by 5. What are other equivalent ways?* **Student:** $2\frac{1}{5}$. **Student:** 2.2. *Because 10 divided by 2 is 5. The one left over divided by 5 is like a dollar in 5 chunks. It's 20 cents.* **Teacher:** *I might model your thinking like this:*	$f=\frac{11}{5}$ $f=\frac{11}{5}=\frac{10}{5}+\frac{1}{5}=2\frac{1}{5}=2.2$
Teacher: *What about this next problem, 1 divided by f is equal to 5 divided by 11?* Students work and the teacher circulates, looking for students who multiply both sides by *f* or recognize the equality of the reciprocal first equation, and taking note of any other strategies. We choose not to model cross multiplication because we find that most students do not have any conceptual underpinning for it and it gets mixed up with any other rule students have memorized about rational expressions. **Teacher:** *I noticed you multiplied both sides by f. Tell us about that.* **Student:** *Yeah, so then I multiplied both sides by* $\frac{11}{5}$ *and so f is* $\frac{11}{5}$.	$\frac{1}{f}=\frac{5}{11}$ $f\cdot\frac{1}{f}=f\cdot\frac{5}{11}$ $1=\frac{5}{11}f$ $\frac{11}{5}\cdot 1=\frac{11}{5}\cdot\frac{5}{11}f$ $\frac{11}{5}=f$
Teacher: *I noticed that you did not do that. What were you thinking?* **Student:** *Well, it's the same thing as the first problem, so we already know the answer.* **Teacher:** *The same? It doesn't look the same to me.* **Student:** *Well, not the same, but if you flip it.* **Teacher:** *What do the rest of you think? Are the reciprocal equations, what you called flipping it, are those equations equivalent? Does the one imply the other?* **Student:** *It looks like it should. If you flip one, that's the same as flipping the other.* **Teacher:** *So you're saying that the reciprocal of one side should be equivalent to the reciprocal of the other? I wonder if that might be helpful in some of the equations we solve today.*	Does $f=\frac{5}{11}$ imply $\frac{1}{f}=\frac{11}{5}$?
Teacher: *The next problem is a quick one, 4f = 8. What is f? That's way too easy, that's just ...?* **Student:** 2	$4f=8,\ f=2$

(continued)

Teacher: *The next problem is 4 times f is one-eighth. What is f?* **Student:** *1⁄32. I divided both sides by 4.* **Student:** *I agree, but I just thought about what small pieces would I need 4 of to get 1⁄8. Half of 1⁄8 is 1⁄16 and half of that is 1⁄32.*	$4f=\frac{1}{8},\ f=\frac{1}{32}$
Teacher: *The next problem is 1 divided by 4f is equal to 1 divided by 8. What is f?* Students work briefly. **Teacher:** *What's f and what were you thinking?* **Student:** *It's 2. Since the 1s are the same, the denominators have to be the same too.* **Teacher:** *Does that make sense? If the numerators are the same, the denominators must be equal?* **Student:** *And if you do the reciprocal thing, you get the same answer.* **Student:** *I wish I would've thought of either of those. I multiplied both sides by 4f and 8 and got the same answer.* **Teacher:** *I bet you'll be looking for other strategies as we keep going!*	$\frac{1}{4f}=\frac{1}{8},\ f=2$ $\frac{1}{4f}=\frac{1}{8}$ $4f=8$
Teacher: *Next problem. What is f if 1 divided by 4f equals 8?* Students work briefly. **Teacher:** *What's f this time?* **Student:** *It's 1⁄32 again.* **Teacher:** *I noticed that you used our reciprocal equation idea. Tell us about that.* **Student:** *So, now 4f is one-eighth. We'd already done that one.*	$\frac{1}{4f}=8,\quad f=\frac{1}{32}$ $\frac{1}{4f}=8$ $4f=\frac{1}{8}$
Teacher: *Nice! I wonder if that influences this next one, 1 divided by 4f equals −2. Solve for f.* Students work. **Teacher:** *What did you find?* **Student:** *I found f is −1⁄8.* **Student:** *I did the reciprocal thing and got 4f is −1⁄2. So f is −1⁄8.* **Teacher:** *Any other strategies?* **Student:** *I ignored the negative at first. I knew I needed 1⁄2 in the denominator in order for it to flip to be 2. So I thought about what times 4 is 1⁄2. That's 1⁄8. Then I thought about the negative.*	$\frac{1}{4f}=-2$ $4f=-\frac{1}{2}$ $f=-\frac{1}{2}\cdot\frac{1}{4}=-\frac{1}{8}$

Teacher: *How would you summarize some of the things that came up in this string today?*

Elicit the following:

- *Reciprocal equations are equivalent.*
- *When the variable is in the denominator, you can use the reciprocal equation to solve.*

Sample Final Display

Your display could look like this at the end of the problem string:

$f=\frac{11}{5}$ $\qquad$ $f=\frac{11}{5}=\frac{10}{5}+\frac{1}{5}=2\tfrac{1}{5}=2.2$

$\frac{1}{f}=\frac{5}{11}$ $\qquad$ $f=\frac{11}{5}$ $\qquad$ $f\cdot\frac{1}{f}=f\cdot\frac{5}{11}$, $1=\frac{5}{11}f$, $\frac{11}{5}\cdot 1=\frac{11}{5}\cdot\frac{5}{11}f$, $\frac{11}{5}=f$ $\qquad$ Does $f=\frac{5}{11}$ imply $\frac{1}{f}=\frac{11}{5}$?

$4f=8,\ f=2$

$4f=\frac{1}{8},\ f=\frac{1}{32}$

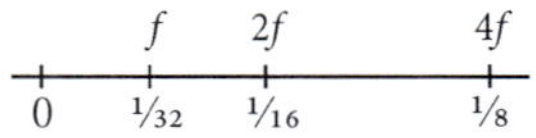

$\frac{1}{4f}=\frac{1}{8}$ $\qquad$ $f=2$ $\qquad$ $\frac{1}{4f}=\frac{1}{8}$, $4f=8$

$\frac{1}{4f}=8,\ f=\frac{1}{32}$ $\qquad$ $\frac{1}{4f}=8$, $4f=\frac{1}{8}$

$\frac{1}{4f}=-2$ $\qquad$ $f=-\frac{1}{8}$ $\qquad$ $4f=-\frac{1}{2}$, $f=-\frac{1}{2}\cdot\frac{1}{4}=-\frac{1}{8}$

Facilitation Notes

This version of the problem string lists short notes for important teacher moves during the string. After you've done the string yourself and studied the relationships involved, you might make similar notes for the things you want a reminder of or deem important.

$f=\frac{11}{5}$	What is 11 divided by 5? Decimal? Fraction?
$\frac{1}{f}=\frac{5}{11}$	(Say "divided by", not "over") Model multiplying both sides by f, 11. Is the reciprocal equation equivalent?
$4f=8$	Quick.
$4f=\frac{1}{8}$	Quick.
$\frac{1}{4f}=\frac{1}{8}$	Model multiplying both sides, equivalent numerators so = denominators, using reciprocal equations.
$\frac{1}{4f}=8$	Model multiplying both sides, using reciprocal equations.
$\frac{1}{4f}=-2$	Model using reciprocal equations.

PROBLEM STRINGS
LESSON

5.8 Using the Focus-Directrix Definition

At a Glance

Focus $(0, 0)$ directrix : $y = -2$

Focus $(0, 0)$ directrix : $y = -5$

Focus $(1, 2)$ directrix : $y = 10$

Focus $(1, 2)$ directrix : $y = 6$

Focus $(1, 2)$ directrix : $y = 4$

Focus $(1, 2)$ directrix : $y = 3$*

Focus $(1, 2)$ directrix : $y = 2.5$

Focus $(1, 2)$ directrix : $y = 2.1$

*optional problem

Objectives

The goal of this problem string is to give students practice and immediate feedback when finding the equation of a parabola in function notation given the focus and directrix. Students will look for patterns and begin to describe parameters that make a nice parabola using the focus-directrix definition versus using function notation.

Placement

This problem string comes after students have developed both the focus directrix definition and the function definition of a parabola.

You could deliver this string as students are working on 5.8 Parabolas in the textbook or after the section to help students solidify the two representations of parabolas.

Guiding the Problem String

The first two problems, with the focus at the origin, should go quickly, where students remind each other about the translation from the focus-directrix definition to the function notation for parabolas. The rest of the problems have the same focus, the point (1, 2), so students can attend to what happens as the directrix moves closer and closer to the focus. You might need to spend a bit more time on the last problem that deals with dividing by a decimal to find the vertical stretch in function notation.

About the Mathematics

Numbers that are pleasant to work with in the focus-directrix notation are not necessarily pleasant to work with when the parabola is written as a function of *x*. In the function world in math classes for $f(x) = ax^2 + bx + c$, we often deal with *a* values for vertical compressions of nice fractions like 1/2 or 1/4 and *a* values for vertical stretches like 2, 3, 4, 4.9, 16. Those parabolas do not have very nice focus-directrix definitions, often resulting in *f*-values of small fractions, where *f* is half of the distance between the focus and the vertex. And vice-versa. When the focus-directrix definition has pleasant, easy to graph values, like the first several problems of the string, the functions have more unwieldy *a*-values to graph in a typical [–10, 10] by [–10, 10] window.

Important Questions

Use the following as you plan how to elicit and model student strategies. This is not meant as a script, but as a view into the relationships involved and the intent of the problem string.

- *What does the parabola look like that has a focus at the origin and the directrix at $y = -2$? Sketch a graph and write the function definition.*
- *If the next problem has the same focus at the origin but the directrix is at $y = -5$, predict how you think that will change the graph? The function? Now actually find the function and sketch the graph on the same axes as the first problem. What happened to the graph and equation of the parabola as the directrix got closer to the focus?*
- *Here's a brand new parabola, with focus $(1, 2)$ and directrix $y = 10$. On a new graph, sketch the parabola and find the function definition. I'm going to make the display graph plenty big, so we can really see what's going on.*

- *How did you find the function definition? The graph?*
- *This next problem is going to have the same focus* (1, 2) *but directrix* $y = 6$. *Predict what you think will happen to the graph and the function definition. What is the function definition? The graph? Let's add it to the previous.*

Continue to ask each problem in turn, asking students to predict and then find the function definition and add a sketch of the graph to the previous ones. Display the graphs superimposed on each other.

- *For this last problem, with the same focus but the directrix is the line* $y = 2.1$, *how did you find the function?*
- *What is 1 divided by* ⅕*?*

Help students generalize at the end:

- *What part of the focus do you see in the function definition? (The x-value of the vertex is the same as the x-value of the focus.)*
- *What happens to the shape of the parabola as the directrix moves closer to the focus? Why?*
- *What do the functions of parabolas look like that have nice focus and directrix numbers? What do the focus and directrixes look like when the function definition has nice numbers?*

Sample Final Display

Your display could look like this at the end of the problem string:

Focus (0, 0) directrix : $y = -2$ $\quad \frac{1}{4(1)} = \frac{1}{4} \quad y = \frac{1}{4}x^2 - 1$

Focus (0, 0) directrix : $y = -5$ $\quad \frac{1}{4(2.5)} = \frac{1}{10} \quad y = \frac{1}{10}x^2 - 2.5$

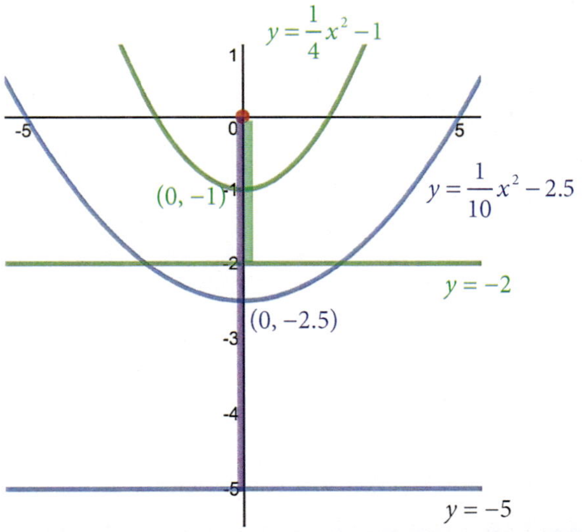

(continued)

Focus (0, 0) directrix : $y = 2$	$\frac{1}{4(1)} = \frac{1}{4}$	$y = \frac{1}{4}x^2 - 1$
Focus (0, 0) directrix : $y = 5$	$\frac{1}{4(2.5)} = \frac{1}{10}$	$y = \frac{1}{10}x^2 - 2.5$
Focus (1, 2) directrix : $y = 10$		$y = -\frac{1}{16}(x-1)^2 + 6$
Focus (1, 2) directrix : $y = 6$		$y = -\frac{1}{8}(x-1)^2 + 4$
Focus (1, 2) directrix : $y = 4$		$y = -\frac{1}{4}(x-1)^2 + 3$
Focus (1, 2) directrix : $y = 2.5$		$y = -1(x-1)^2 + 2.25$
Focus (1, 2) directrix : $y = 2.1$		$y = -5(x-1)^2 + 2.05$

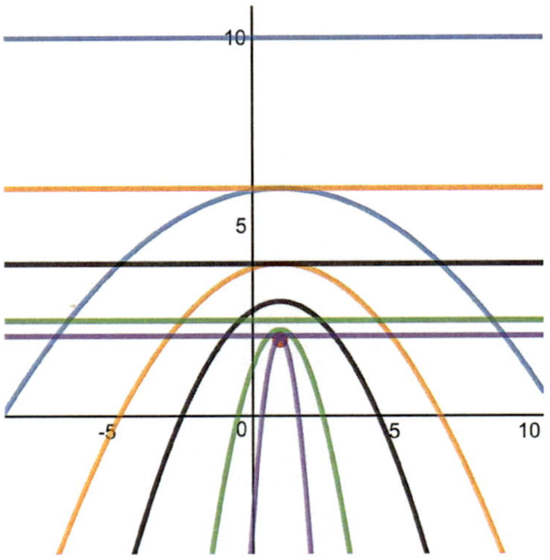

Facilitation Notes

This version of the problem string lists short notes for important teacher moves in the string. After you've done the string yourself and studied the relationships involved, you might make similar notes for the things you want a reminder of or deem important.

Focus (0, 0) directrix : $y = -2$	*Sketch graph, find function. Quickly graph with display calc.*
Focus (0, 0) directrix : $y = -5$	*How do you think this will relate to prior?* *Sketch graph, find function. Add to display calc.*
Focus (1, 2) directrix : $y = 10$	*How did you find graph, function with new focus?* *New display.*
Focus (1, 2) directrix : $y = 6$	*Predict based on previous. How did you find graph, function?* *How is this the same, different from previous? Add to display calc.*
Focus (1, 2) directrix : $y = 4$	*Repeat. Quick.*
(Focus (1, 2) directrix : $y = 3$)	*Optional. Quick.*
Focus (1, 2) directrix : $y = 2.5$	*Ugly directrix. Wonder what that will do? How did you solve for a?* *Nice function! Add to display calc.*
Focus (1, 2) directrix : $y = 2.1$	*Even uglier directrix. Wonder what that will do? How did you solve for a?* *Nice function! Add to display calc.*
	What makes for nice functions? Nice focus-directrix?

PROBLEM STRINGS LESSON

6.1 Polynomials

At a Glance

$y = x^3$

$y = 2x$

$y = x^3 + 2x$

$y = x^3 - 2x$

$y = x^2$

$y = x^3 + x^2$

$y = x^3 - x^2$

$y = x^3 + x^2 - 2x$

$y = x^3 - x^2 - 2x$

Objectives

The goal of this problem string is to help students develop a network of understandings about polynomial functions, specifically cubic functions, connecting and using multiple representations. This string continues to develop the graphing strategies of adding ordinates and factoring to find zeros that were developed in Problem String 5.1 Graphing Quadratic Functions.

Placement

This string is similar to the series of strings in chapter 5 and continues to help students develop notions about polynomial short and long run behavior, the graphing strategies of adding ordinates (combining a cubic, quadratic, and linear function), and factoring to find zeros. Students need prior experience with the quadratic parent function and zeros of functions.

You can use this string to introduce polynomial short and long run behavior in textbook Lesson 6.1 Polynomials.

Guiding the Problem String

If this is your students' first experience with these strategies, as you deliver the third problem wonder aloud about the combination of the first two problems: What does it even mean to add a cubic function and a line? The third and fourth problems have zeros at $x = 0, \pm\sqrt{2}$ so just estimate the zeros. The fifth, sixth, and seventh problems introduce the sum of a cubic and quadratic. The last two problems are built from previous functions and are an opportunity for you to assess students' depth of understanding. When working with a function like $y = x^3 - 2x$, encourage students to try both the difference of $y = x^3$ and $y = 2x$ and the sum of $y = x^3$ and $y = -2x$.

Throughout the string, encourage students to look at the short run behavior (what is happening around the zeros) and the long run behavior (in the extremes). Elicit that the long run behavior of the cubic increases and decreases faster than both the linear and the quadratic and therefore dominates the sum.

About the Mathematics

The strategy we are calling "adding ordinates" comes from the (abscissa, ordinate) language of ordered pairs. It simply means adding the y-values from the parts of a combination function. In this string, we look at combining cubic, quadratic, and linear functions. The idea therein, that the cubic function dominates the whole function thus making the combination function also a cubic function, portends the general meaning of polynomials, that is the term of highest degree dominates the polynomial.

This form, $y = x^3 + x^2 + x$, also begs to be factored into at least $y = x(x^2 + x + 1)$, where we know at least the root of $x = 0$. If the quadratic factor, $x^2 + x + 1$ can be further factored over the real numbers, then the parent cubic has more real roots. The y-value can be found by evaluating the function at the x-value.

(continued)

Important Questions

Use the following as you plan how to elicit and model student strategies.

- *Sketch a graph of* $y = x^3$. *What are some important points on the cubic parent function?*
- *Sketch a graph of* $y = 2x$. *What are some important points on this linear function?*
- *Sketch a graph of* $y = x^3 + 2x$. *I wonder what it means to add a cubic function to a linear function. Thoughts?*
- *What about those of you who factored,* $y = x(x^2 + 2)$*? What are you thinking?*
- *I wonder how everything we just discussed for* $y = x(x^2 + 2)$ *might impact how you will sketch a graph of* $y = x^3 - 2x$*?*
- *Did anyone think about subtracting* $y = 2x$ *from* $y = x^3$*? Let's model that thinking on the board.*
- *Did anyone think about adding* $y = -2x$ *to* $y = x^3$*? So, you're thinking about* $y = x^3 + (-2x)$*? Let's model that thinking on the board.*
- *Did anyone factor* $y = x(x^2 - 2)$*? Let's model that thinking on the board.*
- *Some of you are noticing a similarity between the graphs of the parent* $y = x^3$ *and* $y = x^3 + 2x$ *that is different from* $y = x^3 - 2x$. *What caused the wiggle, the dip, the local maximum and minimum in* $y = x^3 - 2x$. *What's similar between the three functions, long run or short run behavior? What's different between the three functions, long run or short run behavior? Why?*
- *Sketch a graph of* $y = x^2$. *What are some important points on the quadratic parent function?*
- *What about this function,* $y = x^3 + x^2$*? How would you graph it? What have we done today that might influence you?*
- *I wonder if that will affect how you graph the next problem,* $y = x^3 - x^2$*? You can sketch it any way you want. When you're done, step back and really think about the relationships and decide what you think is a really slick strategy that is efficient and sophisticated. What are some connections between all of the graphs so far?*
- *Short run behavior: Which of the last two functions has a wiggle (a dip, a local minimum and maximum)? Both? Why? Long run behavior: Which function,* $y = x^3$ *or* $y = x^2$, *"wins" over the long run, meaning it increases or decreases faster? Why? How does that affect the sum in the long run?*
- *What about this function,* $y = x^3 + x^2 - 2x$*? How will the linear term affect the long run behavior? The short run? What does the factored form tell you about the graph?*
- *Which term dominates the polynomial? Why? What does this have to do with the long run behavior of each of the terms? Why do we call a polynomial by its term of highest degree?*
- *How does adding the y-values (ordinates) of each of the terms of a polynomial help you think about the graph of the entire polynomial?*
- *How does factoring the polynomial help you think about the graph of the entire polynomial?*

How would you summarize some of the things that came up in this string today?

- *A cubic function that is a combination of a cubic term and a quadratic and/or linear term behaves like a cubic function. That is why the combination function is a cubic function.*
- *The long run behavior of the linear part pulls on the long run behavior of the cubic part but does not change the overall long run behavior of the combination function.*
- *The long run behavior of the quadratic part pulls on the long run behavior of the cubic part but does not change the overall long run behavior of the combination function.*
- *You can find the graph of a combination function by adding the y-values of the parts.*
- *If the polynomial factors nicely, you can use the factors to find the x-intercepts of the graph.*

Sample Final Display

Your display could look like this at the end of the problem string. Note that this final graph just shows the results of the last problem on the display grapher.

$y = x^3$

$y = 2x$

$y = x^3 + 2x$

$y = x(x^2 + 2)$

$y = x^3 - 2x$ $\quad y = -2x$

$y = x(x^2 - 2)$

$y = x^2$

$y = x^3 + x^2$

$y = x^2(x + 1)$

$y = x^3 - x^2$

$y = x^2(x - 1)$

$y = x^3 + x^2 - 2x$

$y = x(x^2 + x - 2) = x(x + 2)(x + 1)$

$y = x^3 - x^2 - 2x$

$y = x(x^2 - x - 2) = x(x - 2)(x + 1)$

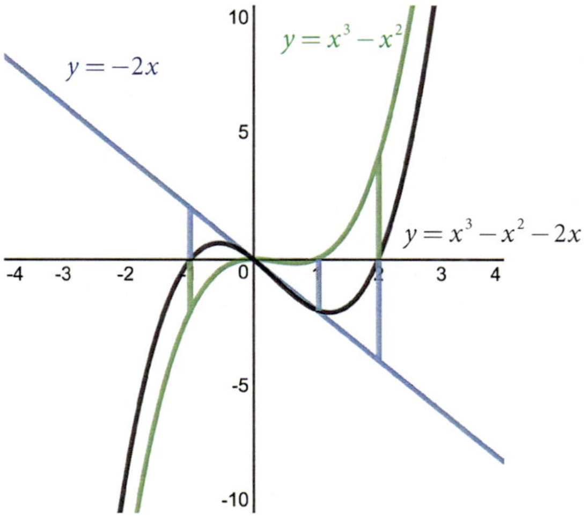

(continued)

Facilitation Notes

This version of the problem string lists short notes for important teacher moves during the string. After you've done the string yourself and studied the relationships involved, you might make similar notes for the things you want a reminder of or deem important.

$y=x^3$	Sketch a graph. What are some important points? Let's sketch it together on this display grid. Now I'll display a graph with the grapher. How does it compare?
$y=2x$	Repeat. Add to display graph. Quick.
$y=x^3+2x$	Sketch a graph. What does it mean to add a cubic and a line? Can you use the 2 previous? How would their important points help? How does the end behavior of the parts affect the end behavior of the combo? Who used factored form? How does that help?
$y=x^3-2x$	Repeat, including adding (subtracting) ordinates and factored form. Did anyone think about adding y=-2x instead of subtracting y=2x? Which do you like better? How does the previous problem compare to this one? Where did the wiggle come from?
$y=x^2$	Clear the graph except the cubic. Repeat as with previous parent functions. Add to display graph. Quick.
$y=x^3+x^2$	Sketch a graph. What does it mean to add a cubic and a quadratic? Who will win? Can you use the 2 terms? How does the end behavior of the parts affect the end behavior of the combo? Who used factored form? How does that help?
$y=x^3-x^2$	Repeat.
$y=x^3+x^2-2x$	Now what? Let's use the display grapher. How could we use some parts that we already have up? How will the linear term affect the long run behavior? Factored form?
$y=x^3-x^2-2x$	Repeat. Use the display grapher of the parts to reason about the combination. How does the long run behavior of the terms affect the long run behavior of the polynomial? Why do we call a polynomial by its term of highest degree?

PROBLEM STRINGS LESSON

6.2 Using Zeros

At a Glance

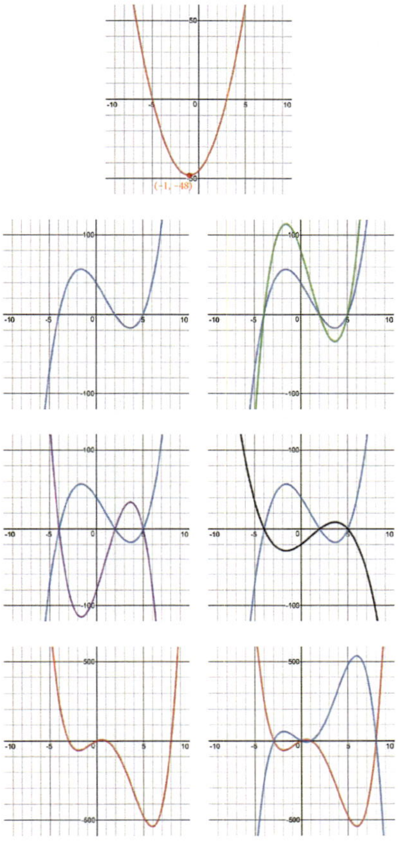

Objectives

The goal of this problem string is to firmly anchor students in the connection between the x-intercepts and the factored form of a polynomial. Students write the factored form of the equation given a graph and reason about transformations on the original.

Placement

Use this string before students begin the work of factoring polynomials to anchor them firmly in the relationship between those factors they will be finding, the roots of the equation, and the x-intercepts of the graph. Students should have prior experience with these relationships with quadratic functions.

You can use this problem string to start or support textbook Lesson 6.2 Factoring Polynomials.

Guiding the Problem String

This problem string is a set of visual questions, where you show students a graph of a polynomial and students work to find the equation. Make sure that you have hidden the "lists" of equations in your grapher so that students see only the graphs.

The first problem is a quadratic equation with the vertex noted. By giving students the vertex in a familiar quadratic problem, this should set the stage for them to work with the less familiar polynomials of higher degrees. Students may need paper and pencil for this problem but encourage them to put them down for the rest of the problems, emphasizing the relationships involved, rather than things to do. The second problem is a cubic function with a scale factor of 1. The third, fourth, and fifth problems are dilations of the second problem. Show the second problem along with each of them. If students ask, you can add any of the previous to help. The last two problems are similarly related, the last being a reflection of the previous across the x-axis.

Most of these problem should go quickly. Students shouldn't need much time to find the factored form. Instead, spend your time comparing the graphs and equations.

About the Mathematics

The fundamental theorem of algebra is that if $p(x)$ is an nth degree polynomial with complex coefficients, $p(x) = ax^n + bx^{n-1} + ... + k$, then $p(x)$ has exactly n complex roots. Therefore, when there are m x-intercepts visible in a graph, we can assume that the degree of the polynomial is at least m. In this string, students assume the minimum degree necessary and confirm by graphing.

(continued)

Sample Interactions

Use the following as you plan how to elicit and model student strategies. This is not meant as a script, but as a view into the relationships involved and the intent of the problem string.

Teacher: *This first problem in today's string is familiar. Here is a graph. What do you know about this polynomial?* **Student:** *It looks like a quadratic. It's opening up. It has zeros at* −5 *and* 3. **Teacher:** *Please write the factored form of this quadratic.* Students work and the teacher circulates, looking for students who are using the vertex to write the transformation form and who are using the x-intercepts to write the factored form.	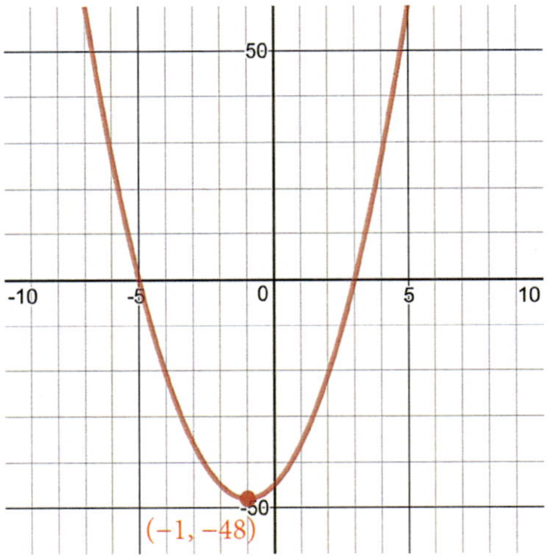
Teacher: *I saw you using the vertex first. Tell us about your thinking.* **Student:** *We shifted the quadratic over to the vertex, back* 1 *and down* 48. *But we know there might be a scale factor so we put in an a and then solved for a and got that a equals* 3. *So the equation is* $f(x)=3(x+1)^2-48$. **Teacher:** *Let's graph that and see if it matches. Yes.*	$f(x)=a(x+1)^2-48$ $f(3)=a(3+1)^2-48=0$ $16a=48$ $a=3$ $f(x)=3(x+1)^2-48$
Teacher: *You two had a different approach. Tell us about that, please.* **Student:** *Since it has x-intercepts at* −5 *and* 3, *it has to be x plus* 5 *times x minus* 3. If necessary, the teacher could press for justification about how the x-intercepts create the function. Since the teacher sees students working well with the relationship between zeros and x-intercepts, the teacher moves onto the vertex. **Teacher:** *How does that work with the vertex up there?* **Student:** *We know that* $f(-1)=-48$, *so we put* −1 *into the function we have and that is* −16, *so there must be a* 3. **Teacher:** *There needs to be a scale factor of* 3? *Like this? Great. Let's graph what we've found with the original. Yes, looks good.*	$f(x)=a(x+5)(x-3)$ $f(x)=a\left(x^2+2x-15\right),\ f(-1)=-48$ $f(-1)=a\left((-1)^2+2(-1)-15\right)=a(-16)=-48$ $a=3$ $f(x)=3(x+5)(x-3)$

Teacher: *The second problem in our string today is this graph. Please write the factored form of a function rule for this polynomial.*

Students work and the teacher circulates, looking for students who are using the zeros to write the factored form and students who are thinking about the cubic nature of the graph.

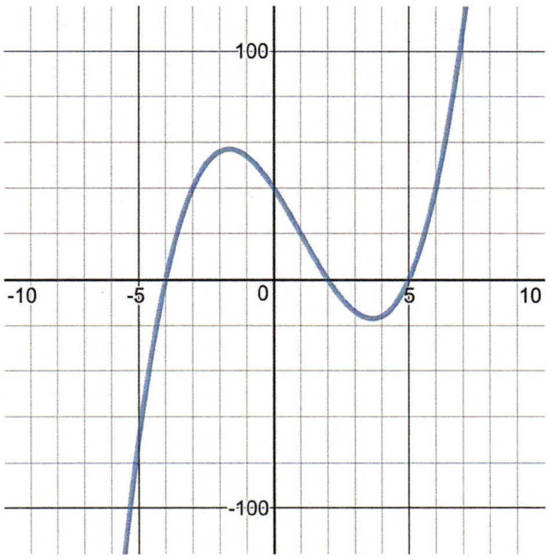

Teacher: *What do you think the factored form for this polynomial is?*

Student: *It's got three factors, one is x minus 5, another is x minus 2, and the last is x plus 4.*

Teacher: *Let's graph that and see if we need to stretch or compress it.*

The teacher adds the graph to the given and the graphs match. If students had given incorrect equations, they could adjust them at this time.

$f(x)=(x-5)(x-2)(x+4)$

Teacher: *That wasn't too bad. I am going to leave the graph of this blue function up and add the graph of the third problem. What is the factored form of this green function?*

We don't really have values for anything but the zeros on the graph, but I wonder if you could estimate a possible scale factor?

The teacher elicits students estimates, graphs the estimate(s), presses for justification, and the class decides that $f(x)=2(x-5)(x-2)(x+4)$ is a good fit.

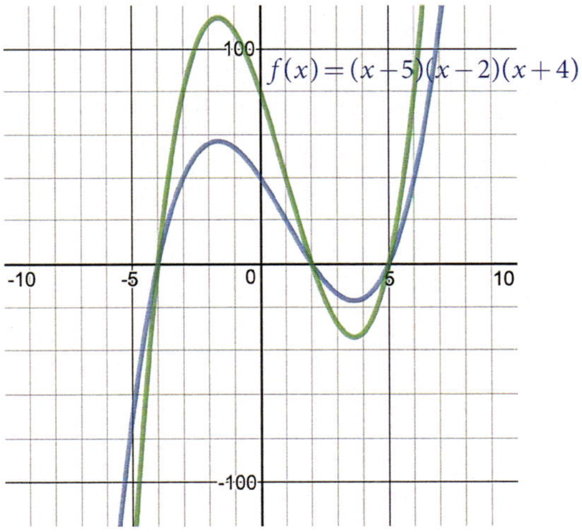

(continued)

Teacher: *I am going to leave the graph of the blue function up and add the graph of the fourth problem. What is the factored form of this purple function?*

The teacher elicits student suggestions, graphs the suggestion(s), presses students for justification, and the class decides if the functions are good fits.

The teacher then repeats with the next problem, the black graph.

$f(x) = -2(x-5)(x-2)(x+4)$

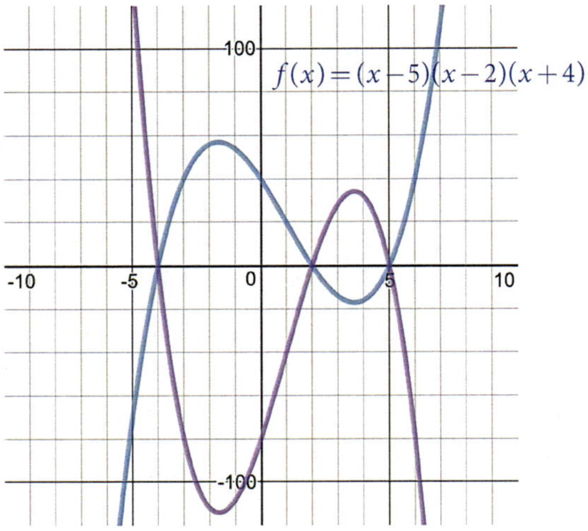

$f(x) = -\frac{1}{2}(x-5)(x-2)(x+4)$

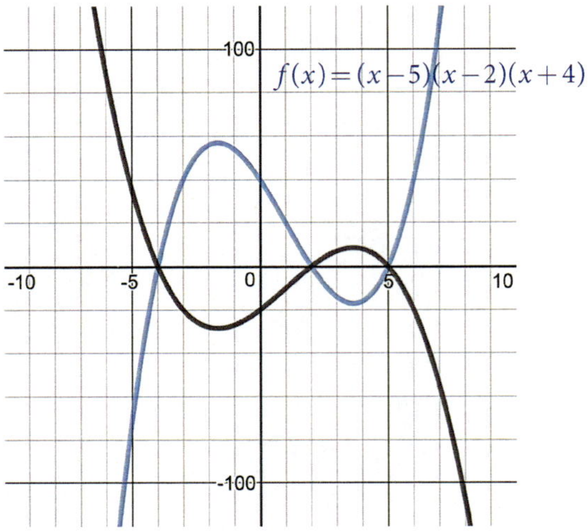

The teacher elicits generalizations.

Teacher: *So, what is happening here with the zeros of the function, the x-intercepts, and the factored form? Why are you only changing the scale factor and nothing else? What stayed the same and what was different?*

Student: *The x-intercepts were the same for all of those functions. Sometimes the graph was stretched or compressed and sometimes the graph was a reflection over the x-axis.*

The teacher presents the last two problems, one at time.

The teacher repeats the process of eliciting student suggestions, graphing the suggestion(s), pressing students for justification, and the class deciding if the functions are good fits.

$f(x) = x(x-1)(x-8)(x+3)$

$f(x) = -x(x-1)(x-8)(x+3)$

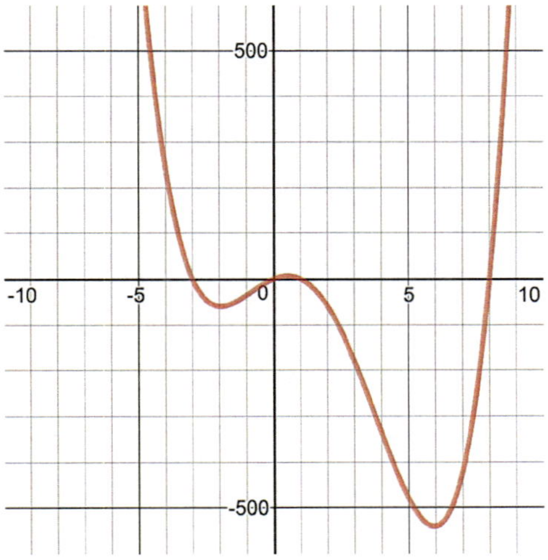

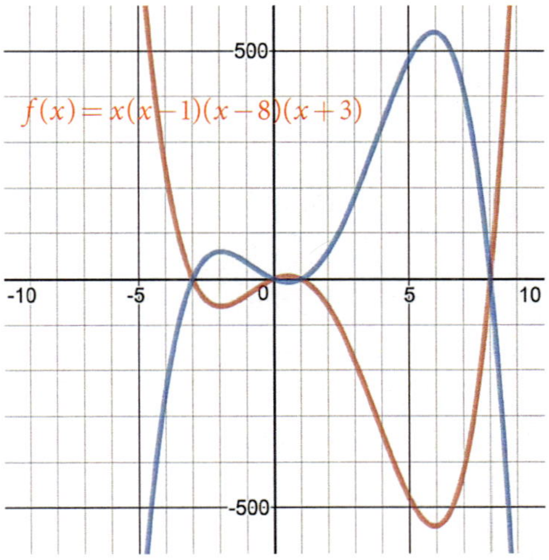

Teacher: *How would you summarize some of the things that came up in this string today?*

Elicit the following:

- *The x-intercepts of the graph of f(x) correspond to the factors of the expression, the zeros of the function, and the roots of the equation where f(x) = 0.*
- *If (x – a) is a factor of a polynomial, then x = a is an x-intercept of the graph, a is a zero of the function, and x = a is a solution to f(x) = 0.*

(continued)

Sample Final Display

Your display could look like this at the end of the problem string:

$f(x)=a(x+1)^2-48$

$f(3)=a(3+1)^2-48=0$

$16a=48$

$a=3$

$f(x)=3(x+1)^2-48$

$f(x)=a(x+5)(x-3)$

$f(x)=a(x^2+2x-15),\ f(-1)=-48$

$f(-1)=a((-1)^2+2(-1)-15)=a(-16)=-48$

$a=3$

$f(x)=3(x+5)(x-3)$

quadratic, stretched by scale factor 3

$f(x)=(x-5)(x-2)(x+4)$ cubic

$f(x)=2(x-5)(x-2)(x+4)$ stretched by scale factor 2

$f(x)=-2(x-5)(x-2)(x+4)$ reflected over the x-axis

$f(x)=-\frac{1}{2}(x-5)(x-2)(x+4)$ compressed by scale factor ½, reflected

$f(x)=x(x-1)(x-8)(x+3)$ fourth degree, quartic

$f(x)=-x(x-1)(x-8)(x+3)$ reflected over the x-axis

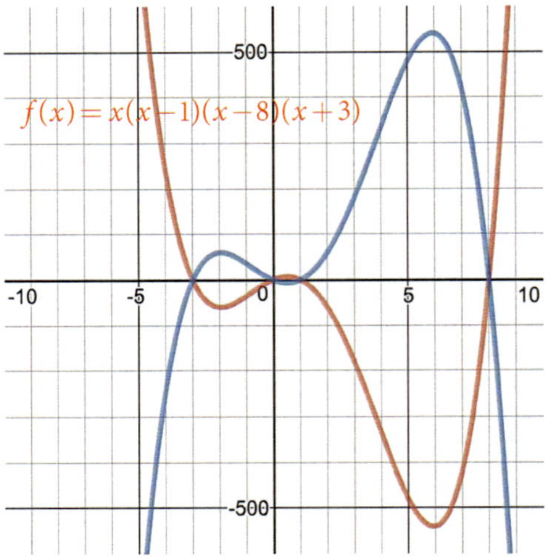

Facilitation Notes

This version of the problem string lists short notes for important teacher moves during the string. After you've done the string yourself and studied the relationships involved, you might make similar notes for the things you want a reminder of or deem important.

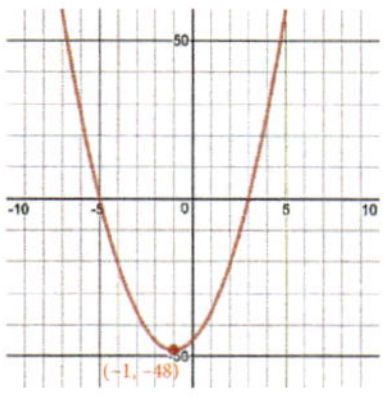

Here is a graph. What do you know about this polynomial?
Write the factored form of the polynomial.
How does the vertex help?
How do the x-intercepts help?
How can you find the scale factor?
Let's graph your equations and see if they match.

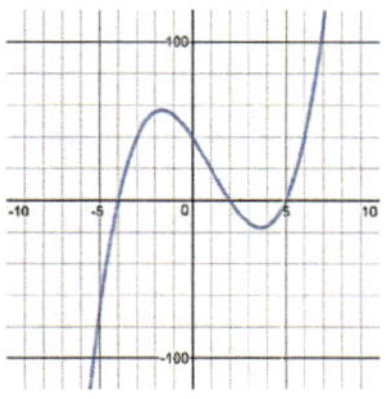

Here is the graph for the second problem.
Write the factored form.
How do you know?
Graph to see that they match.
So, what is the scale factor? 1.

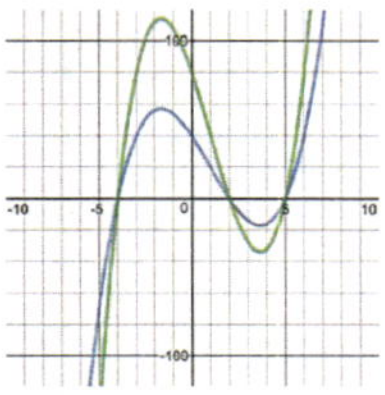

I'm going to leave that blue graph up and add the next problem's graph.
Write the factored form.
What do you estimate for the scale factor?
Add the estimated graph and adjust as needed.

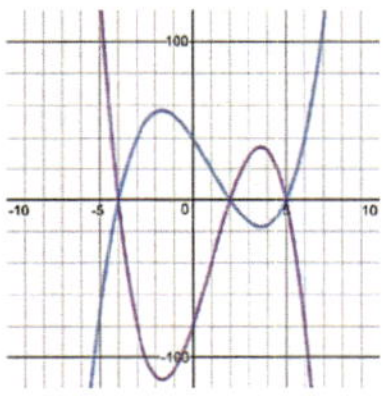

I'm going to leave that blue graph up and add the fourth problem's graph.
I wonder if they are related?
Write the factored form.
Add the green graph if requested.
Add the estimated graph and adjust as needed.

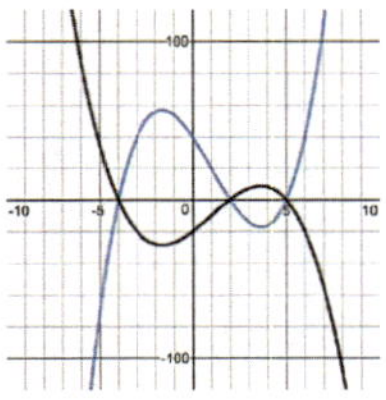

Repeat.
Add prior graphs if requested.
What is happening here with the zeros of the function, the x-intercepts, and the factored form? Why are you only changing the scale factor and nothing else?
What stayed the same and what was different?

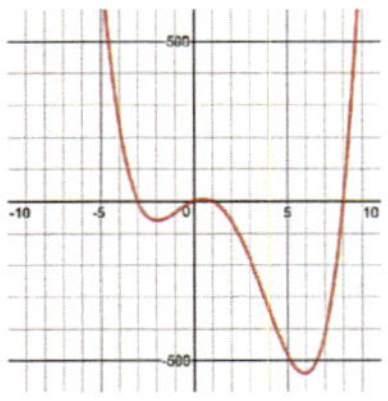

Here's a new graph. What do you notice? What do you predict?
Write the factored form.

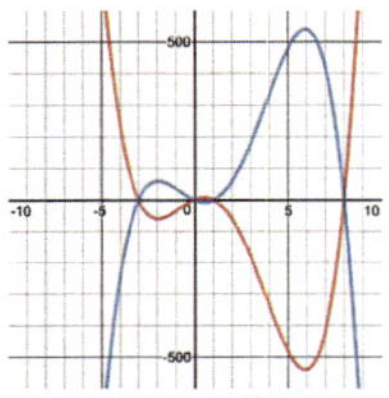

Repeat as before.
How are the factors of the expression related to the x-intercepts? Why?
How can you use the x-intercepts of a graph to find the factored form?

PROBLEM STRINGS
LESSON

6.3 Polynomial Viewing Windows

At a Glance

$f_1(x)=0.1(x+1)(x+1)(x-2)(x+20)$

$f_2(x)=-0.1(x-4)(x+4)(x-33)$

$f_3(x)=-0.1(x+2)(x+3)(x-3)^2(x-21)$

$f_4(x)=(x-3)(x-20)(x+22)$

$f_5(x)=0.5x^3(x+1)(x-1)(x+50)$

Objectives

The goal of this problem string is to develop students' connections between the zeros of a polynomial, the term of highest degree and the long run behavior of a polynomial. The vehicle to create or strengthen these relationships is finding viewing windows to see the important aspects of the graph.

Placement

This problem string can be used to introduce or strengthen students' understanding and facility with higher-degree polynomials, their zeros, the factored form, and long run behavior. Students should have some prior experience with the zero product property.

You can use this problem string with textbook Lesson 6.3 Higher-Degree Polynomials.

Guiding the Problem String

Each of these problems are designed so that they only partially appear in the Desmos home window. Therefore, when first graphed, a potentially confusing or deceiving graph appears.

If students are beginning to learn about polynomials, their zeros, factored form, and long run behavior, run the string by graphing the polynomial in the class grapher and asking questions to help students make connections between the equation and the graph. Once students see some zeros as x-intercepts but not others, nudge them to open the window to find the other x-intercepts. Use the zoom and swiping features at first but gradually press students to predict and have you change the window manually. Press them for justification. Help students generalize the long run behavior of even and odd degree polynomials.

If students are already familiar with the concepts, ask students to predict first, pressing for justification before showing the first graph. Continue to press for reasoning as you open up the window by manually choosing values based on student suggestions. Do not use the zoom or sliding features. Students must reason using what they know or are learning about. If students suggest incorrect moves, make them and let them learn from the missteps. The sample interaction is written as it might go with this group, to students who have some familiarity but need strengthening.

About the Mathematics

The fundamental theorem of algebra is that if $p(x)$ is an nth degree polynomial with complex coefficients, $p(x)=ax^n+bx^{n-1}+...+k$, then $p(x)$ has exactly n complex roots. The examples in this problem string each have real coefficients and real roots.

Sample Interactions

Use the following as you plan how to elicit and model student strategies. This is not meant as a script, but as a view into the relationships involved and the intent of the problem string.

Teacher: *Today's problem string is going to be a partner and group effort. We are going to predict the graphs of some functions and find good viewing windows together based on what we have been learning about polynomials.*
Here is the first problem. Study it please.

Think time.

Teacher: *What are some of your insights?*

Student: *It's long!; Lots of x's; Four x's; That* 0.1, *that's the vertical stretch, or actually I guess compression since it's so small.*

$f_1(x) = 0.1(x+1)(x+1)(x-2)(x+20)$

Teacher: *Well, here is what it looks like in the home window. What do you think?*

Student: *It looks like a cubic.*

Student: *But, wait, shouldn't it be a fourth degree? What is that called anyway?*

Student: *I see the two zeros at* 1 *and* –1. *But that x* + 1 *should be twice?*

Teacher: *There's lots of ideas floating around. Turn to your partner and talk about this window. Do you think it shows the important parts of the function? What do you know about the function from the equation? Do you see it in the graph?*

Students turn and talk while the teacher listens in, listening for students who are grappling with the function's four zeros, but only seeing 2.

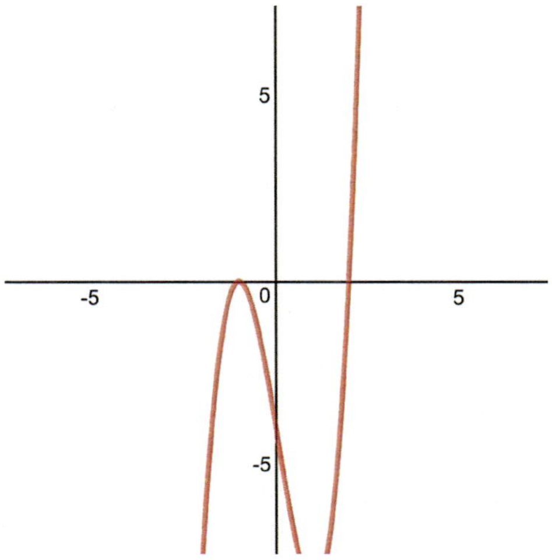

Teacher: *Let's come back together and start with you two. What were you talking about?*

Student: *We are pretty sure that there should be four x-intercepts but we only see two of them, so we think you should open up the window.*

Student: *Since we see the* –1 *and* 2, *we think you should zoom out.*

Teacher: *Why zoom out?*

(continued)

Student: *Because we need to see the −20. The one that's from the factor (x + 20).*

Teacher: *Is there any other reason you think we should zoom out, besides the −20 zero?*

Student: *There are four x's so when you multiply them all together, that's x^4. So, it should come back up on the left.*

Student: *Yeah, a fourth degree should be kind of like a parabola.*

Teacher: *You're saying that right now the end behavior is like a cubic, but it should be like a fourth degree? Okay, how should I change the window? Let's do it manually, using reasoning instead of just zooming.*

Student: *Go back on the x's to −21.*

Student: *Ahhh, yes, that looks better. Now make the y's lower. Change the y's to like −20. Better.*

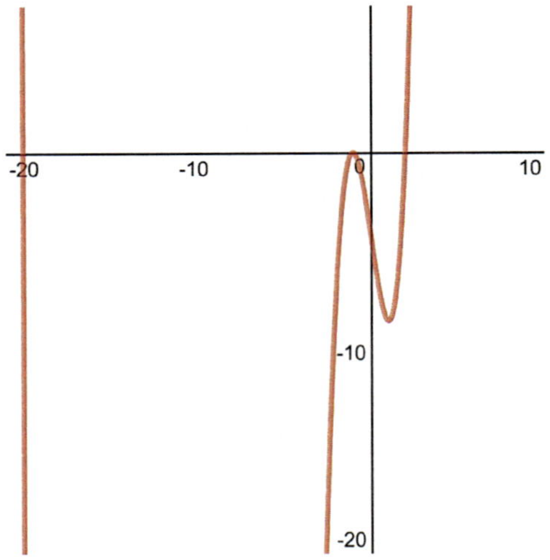

The students continue to give the teacher instructions to open up the window in different ways until they settle on a window similar to the following window [−21, 8] by [−1700, 500].

Teacher: *What makes this a suitable viewing window?*

Student: *We can see the long run behavior of a fourth degree. Up and up, as you go to the left and to the right.*

Student: *And we can at least sort of see the funny stuff that's happening near the origin.*

Teacher: *But didn't you two earlier have a question about the number of x-intercepts and the number of factors in the equation? Didn't you say we should see four x-intercepts?*

Student: *Oh yeah. At first we thought that there were four x-intercepts, but then we realized that the function is bouncing at x = −1. And (x − 1) appears twice, so we think it's like a parabola that only intersects right at the y-axis.*

Student: *What are you talking about? A parabola?*

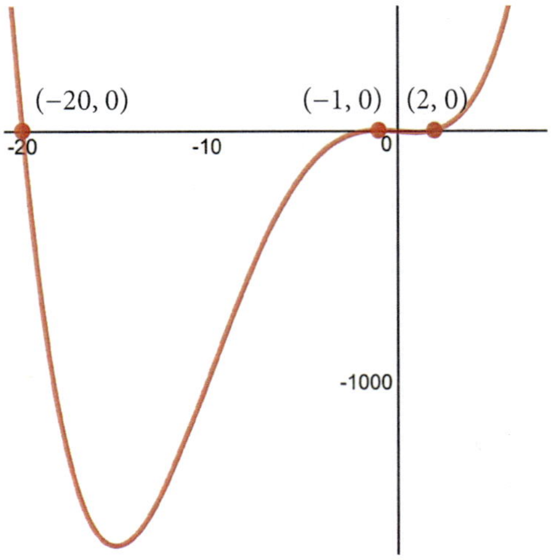

Student: *If we had the parabola $y=(x+1)^2$, it would intersect the x-axis at (0, −1) right?*

Student: *Ahhh, so since there are two (x + 1) factors, the function just touches there and doesn't go through?*

Teacher: *Let's zoom in and look at it. And by the way, we call a factor that appears twice a "double root."*

The teacher hits the home button again and the students note that the function intersects the x-axis at $x = -1$.

The teacher notes on the display, "3 x-intercepts (1 double root), 4th degree"

Before graphing the next function, the teacher makes sure that the grapher is back in the home window.

Teacher: *Nice work. Let's predict a bit for this function. What do you think?*

Think time.

$$f_2(x)=-0.1(x-4)(x+4)(x-33)$$

Teacher: *Before I put the graph up, make sure you have predicted first. Does everyone have a picture in your mind? Or have sketched a quick graph? Yes? Tell us about your predictions please. Each of you just give us one prediction.*

Student: *It is a cubic; It's been reflected across the x-axis; It should have three x-intercepts; One of them is far away.*

The teacher displays the graph in the home window.

Teacher: *What do you think? Do your predictions match this?*

Student: *Well, there are the zeros at −4 and 4.*

Student: *It looks like a quadratic. That's not right. It should be cubic.*

Student: *We need to make the window bigger so we can see the rest.*

Teacher: *The rest of what? Which way will it go? How do you know?*

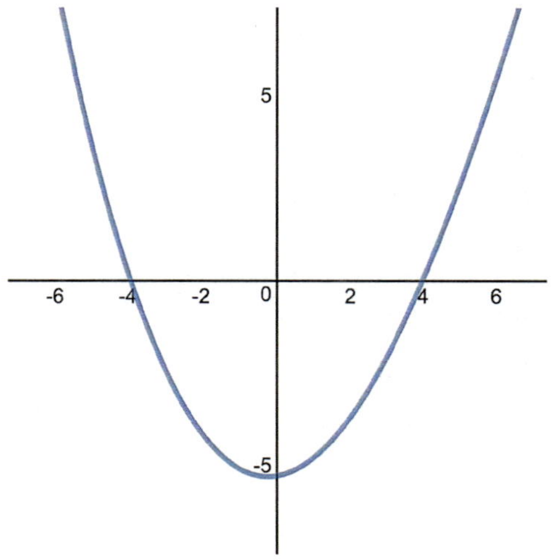

Student: *I think that it will be decreasing, because the coefficient is negative. So it should be like $-x^3$ in the long run.*

The students continue to give the teacher instructions to open up the window in different ways until they settle on a window similar to the following window [−15, 36] by [−30, 100].

Teacher: *What makes you think this is a decent viewing window? How do you know we found it all, that the graph isn't going to turn around again? Convince us!*

Student: *Since it's a cubic, it should have that end behavior. We know it's not going to turn around again because we already see three x-intercepts. There aren't any more.*

Student: *If it turned one more time, then the end behavior wouldn't be right. It'd be like the one we had before, a fourth degree.*

The teacher notes on the display "3 x-intercepts, 3rd degree, reflected over the x-axis"

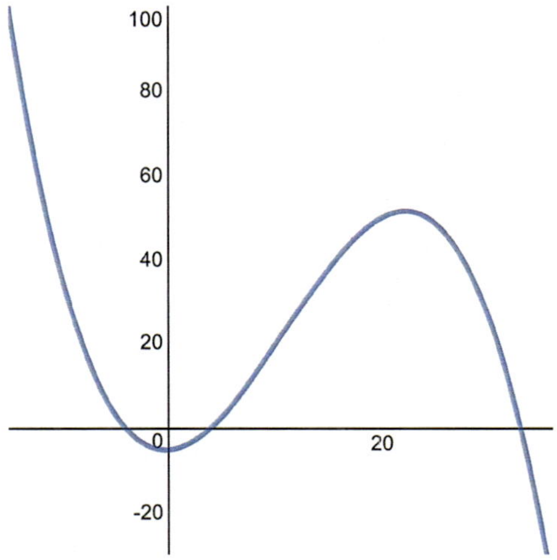

Teacher: *Here is our third polynomial. What do you think? Predict what you will see. Be prepared to justify why.*

The teacher elicits predictions.

$$f_3(x) = -0.1(x+2)(x+3)(x-3)^2(x-21)$$

(continued)

The teacher displays the graph in the home window.

Teacher: *It looks like two parabolas! Did I type it in wrong? Weird. How did your predictions compare?*

The teacher elicits student insights and conjectures and presses for justification. They work together to find a good viewing window.

As they find the window on the left to show the long run behavior, they realize that they might need two different windows to see all of the important behaviors, one to see long run and one to see the zeros of the function.

After the discussion, the teacher notes on the display, "4 x-intercepts (one double root), 5th degree"

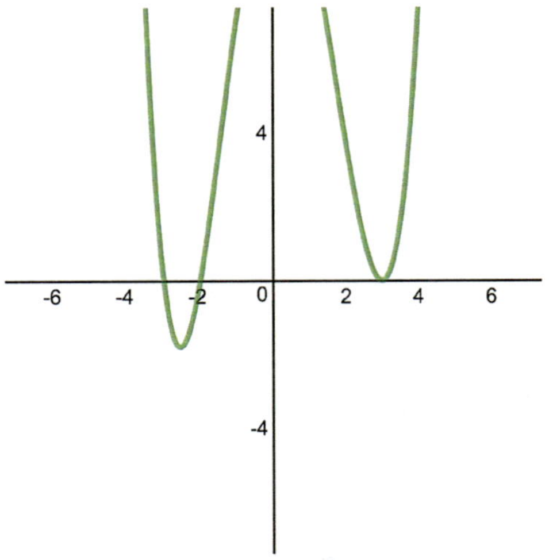

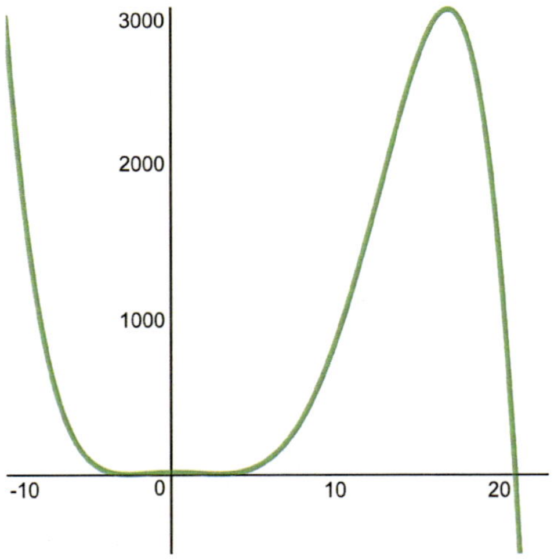

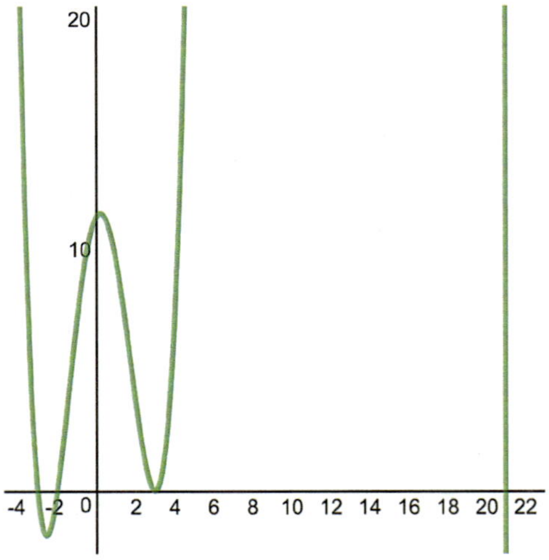

The teacher repeats with the last two problems, each in turn, presenting the next problem, asking students to predict and defend their reasoning. They look at the graph in the home window and then work to find a viewing window that shows the important features of the polynomial. The teacher works to focus the conversation on end behavior, factors, and x-intercepts.

$$f_4(x) = (x-3)(x-20)(x+22)$$

3 x-intercepts, 3rd degree

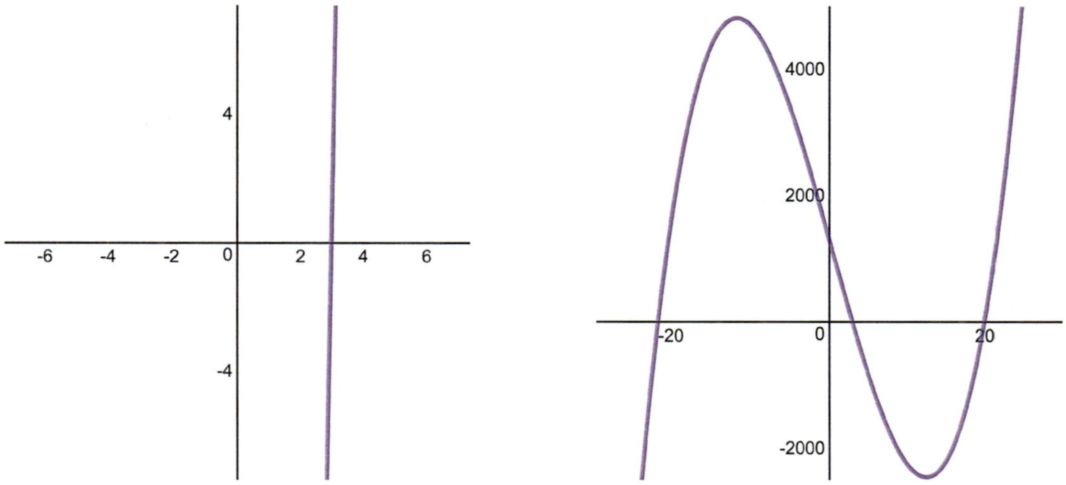

$$f_5(x) = 0.5x^3(x+1)(x-1)(x+50)$$

4 x-intercepts, one triple root, 6th degree

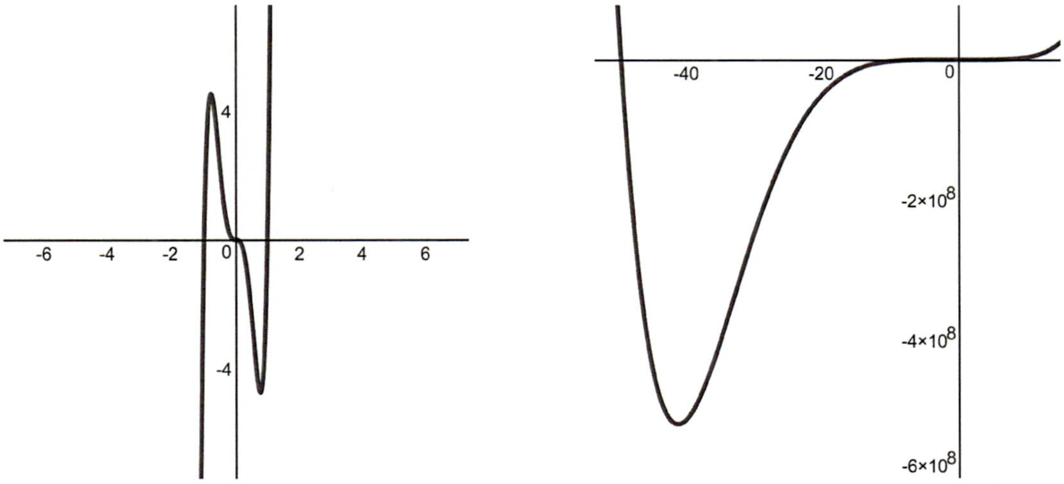

Teacher: *How would you summarize some of the things that came up in this string today?*

Elicit the following:

- *You can determine the long run behavior based on the sign and exponent of the term of the highest degree.*
- *The factors of the polynomial determine the zeros of the function and x-intercepts of the graph.*
- *Double roots intersect but do not cross the x-axis.*

(continued)

Sample Final Display

Your display could look like this at the end of the problem string:

$f_1(x) = 0.1(x+1)(x+1)(x-2)(x+20)$

3 x-intercepts (1 double root), 4th degree

$f_2(x) = -0.1(x-4)(x+4)(x-33)$

3 x-intercepts, 3rd degree, reflected over the x-axis

$f_3(x) = -0.1(x+2)(x+3)(x-3)^2(x-21)$

4 x-intercepts (one double root), 5th degree

$f_4(x) = (x-3)(x-20)(x+22)$

3 x-intercepts, 3rd degree

$f_5(x) = 0.5x^3(x+1)(x-1)(x+50)$

4 x-intercepts, one triple root, 6th degree

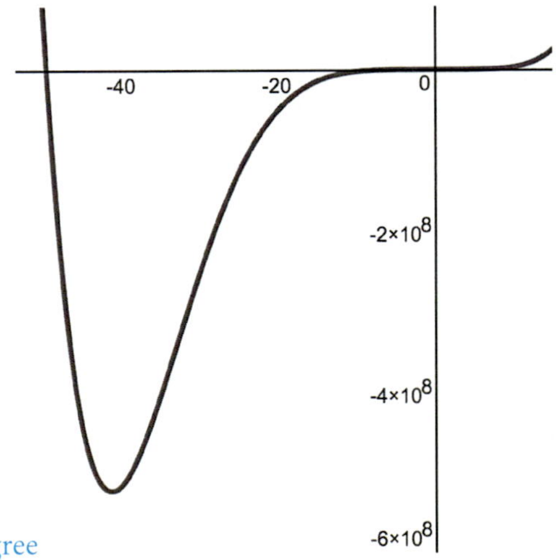

end behavior—sign and degree
x-intercepts—factors of polynomial

Facilitation Notes

This version of the problem string lists short notes for important teacher moves during the string. After you've done the string yourself and studied the relationships involved, you might make similar notes for the things you want a reminder of or deem important.

$f_1(x) = 0.1(x+1)(x+1)(x-2)(x+20)$	*Show all graphs first in the home window.* *Study this problem. What do you notice?* *Here's what it looks like in the home window. What do you think?* *Find a good viewing window (show important features) manually.* *No zoom or swiping.* *Double root.*
$f_2(x) = -0.1(x-4)(x+4)(x-33)$	*Predict graph. Why?* *Show in home window. How does this graph compare to your prediction?* *Find a good window together.*
$f_3(x) = -0.1(x+2)(x+3)(x-3)^2(x-21)$	*Repeat. Don't forget home window first!* *Double root.* *Two viewing windows necessary to show all important features.*
$f_4(x) = (x-3)(x-20)(x+22)$	*Repeat.*
$f_5(x) = 0.5x^3(x+1)(x-1)(x+50)$	*Repeat. Don't forget home window first!* *How do you make sense of the x^3?* *How low can you go?* *Two viewing windows necessary to show all important features.* *How can you determine the long run behavior?* *How can you determine the x-intercepts of the graph?*

PROBLEM STRINGS
LESSON

6.4 Polynomial Division

At a Glance

	$2x^2$	___	4
___	$2x^3$		
-2		$6x$	

	$3x$	$+1$
___	$3x^3$	
___	$6x^2$	
___		-5

	___	___	___
x	$4x^3$	$-5x^2$	$3x$
$+1$	$4x^2$	$-5x$	3

$(4x^3-13x^2+5x-6)\div(x-3)$

$(6x^3+11x^2-x-6)\div(2x+3)$

Objectives

The goal of this problem string is to help students make connections between multiplication and division of polynomials and to apply these relationships to polynomial division.

Placement

This problem string can be used to introduce students to using the rectangle diagram as a tool for computing polynomial division. Students should have prior experience using the rectangle diagram to multiply polynomials.

You can use this problem string to introduce or support textbook Lesson 6.4 More About Finding Solutions.

Guiding the Problem String

The first two problems are in the form of puzzles, where enough of the values are given to figure out the others. Present these and ask students which values they found first. Emphasize the multiplication and division relationships that are represented. After the second problem, look back at patterns in the diagonals. The third problem is a bridge from filling in puzzles to using the rectangle diagram as a tool for polynomial division. Encourage students to use the patterns to help them divide the next two problems.

About the Mathematics

In a multiplication equation, the factors multiply to a product. In a division equation, the dividend divided by the divisor is the quotient.

Division notation is social knowledge. Help students by writing the multiplicative relationships involved as both multiplication with missing factors equations and division using the division symbol ÷, as a ratio, and with the traditional long division symbol. Understanding and using the multiplicative relationships is logical-mathematical. Students need experience multiplying and dividing polynomials to build this facility.

Sample Interactions

Use the following as you plan how to elicit and model student strategies. This is not meant as a script, but as a view into the relationships involved and the intent of the problem string.

Teacher: *Here is the first problem of today's string. We have worked with rectangular diagrams before. This one is a puzzle because some of the values are given and your job is to find the missing ones. What belongs in the blanks and in the empty spots inside the rectangle?*

Students work and the teacher circulates, noting where students have started.

	$2x^2$	___	4
___	$2x^3$		
-2		$6x$	

(continued)

Teacher: *Please tell us where you started.*

Student: *I saw the −2 and 4 on the outside and knew that meant their product was −8.*

Student: *We started at the left side. We knew that the blank on the top left is x, because x times $2x^2$ is $2x^3$.*

Student: *And I think we started below that, −2 times $2x^2$ is $-4x^2$.*

	$2x^2$	___	4
$\underline{x}$	$2x^3$		
-2	$-4x^2$	$6x$	-8

Teacher: *And then you could get the rest of these? Help me fill them in.*

Students suggest the rest of the terms while the teachers records.

Teacher: *So you could start in a few different places and still get it all filled in.*

	$2x^2$	$-\underline{3x}$	4
$\underline{x}$	$2x^3$	$-3x^2$	$4x$
-2	$-4x^2$	$6x$	-8

Teacher: *As you were telling me what goes where, you kept saying "times." There are many multiplicative relationships going on in this diagram. Let's write down the main ones. If this is one of the factors, (x – 2), represented by the dimensions on the side of the rectangle, what is the other factor?*

The teacher circles the $(x - 2)$ factor on the left of the rectangle in blue and begins to write a multiplication sentence off to the right also in blue.

	$2x^2$	$-\underline{3x}$	4
$\underline{x}$	$2x^3$	$-3x^2$	$4x$
-2	$-4x^2$	$6x$	-8

$(x-2)$

Student: $2x^2 - 3x + 4$.

The teacher circles the $2x^2 - 3x + 4$ factor on the top of the rectangle and writes this factor in the multiplication sentence in magenta.

	$2x^2$	$-\underline{3x}$	4
$\underline{x}$	$2x^3$	$-3x^2$	$4x$
-2	$-4x^2$	$6x$	-8

$(x-2)(2x^2-3x+4)$

Teacher: *And what is the product of these two factors?*

Student: *You have to add up the stuff inside,* $2x^3 - 7x^2 + 10x - 8$.

Teacher: *I'll finish this multiplication sentence by writing the product. What are other ways we can write this relationship? Using division?*

	$2x^2$	$-\underline{3x}$	4
$\underline{x}$	$2x^3$	$-3x^2$	$4x$
-2	$-4x^2$	$6x$	-8

$$(x-2)(2x^2-3x+4) = 2x^3 - 7x^2 + 10x - 8$$

$$\frac{2x^3 - 7x^2 + 10x - 8}{x-2} = 2x^2 - 3x + 4$$

$$\frac{2x^3 - 7x^2 + 10x - 8}{2x^2 - 3x + 4} = x - 2$$

$$x-2\overline{\smash{)}2x^3 - 7x^2 + 10x - 8}\quad \text{quotient: } 2x^2 - 3x + 4$$

Teacher: *Here's the next problem, another puzzle for us to fill in. First thing, decide where you want to start.*

Brief think time.

Teacher: *Turn to your partner, choose a place to start, and decide what should go where.*

	$3x$	$+1$
___	$3x^3$	
___	$6x^2$	
___		-5

Teacher: *Let's all fill this in together. We could start in a few different places, but since you've already talked about that with your partners, let's just get these filled in and talk about some other relationships you're seeing.*

The teacher records as students suggest terms.

	$3x$	$+1$
x^2	$3x^3$	x^2
$2x$	$6x^2$	$2x$
-5	$-15x$	-5

Teacher: *What multiplication equation can we write? And a division equation?*

With student input, the teacher records a multiplication and division equation.

	$3x$	$+1$
x^2	$3x^3$	x^2
$2x$	$6x^2$	$2x$
-5	$-15x$	-5

$$(x^2+2x-5)(3x+1)=3x^3+7x^2-13x-5$$

$$\frac{3x^3+7x^2-13x-5}{(x^2+2x-5)}=3x+1$$

(continued)

Teacher: *Some of you have been noticing a pattern in the diagrams, about where the terms are. Turn and talk to your partner about where the term of highest degree is, where the constant term is, and what is happening in the diagonals.*

Partners talk while the teacher listens in.

Teacher: *What patterns are you noticing?*

Student: *The highest term is in the top left corner. The number is in the bottom right spot. The diagonals have to be added together. They make the middle terms. That all is true if the factors on the outside are in descending order from top to bottom and left to right.*

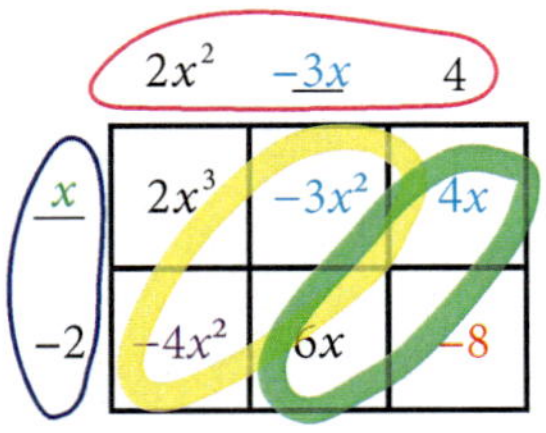

$(x-2)(2x^2-3x+4)=2x^3-7x^2+10x-8$

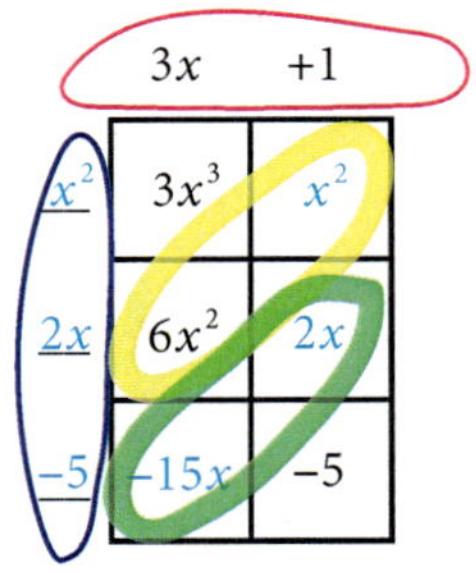

$(x^2+2x-5)(3x+1)=3x^3+7x^2-13x-5$

Teacher: *Here's the next problem. The puzzle is kind of straight forward; just fill in the top factor. But first please write the accompanying problem. Is this a multiplication problem? A division problem?*

Brief pause.

	__	__	__
x	$4x^3$	$-5x^2$	$3x$
$+1$	$4x^2$	$-5x$	3

Teacher: *What is the problem that we could write to represent this rectangle, before we found the missing factor? What do you think?*

Student: *I think it's multiplication.*

As a student says the multiplication equation, the teacher writes it.

Student: *I think it's division.*

As a student says the division equation, the teacher writes it.

Teacher: *Do both of these work? Yes?*

$(x+1)(\underline{\qquad\qquad})=4x^3-x^2-2x+3$

$(4x^3-x^2-2x+3)\div(x+1)=\underline{\qquad\quad}$

Teacher: *How did you find the product in the multiplication equations? Which is the dividend in the division equation?*

Student: *It's all the stuff on the inside. You have to add the stuff in the diagonals. There's only one $4x^3$ and 3, then you put together the x squareds and the x's.*

Teacher: *Everyone agree? Yes?*

	___	___	___
x	$4x^3$	$-5x^2$	$3x$
$+1$	$4x^2$	$-5x$	3

Teacher: *What you think the top factor is and why?*

A student explains as the teacher writes the top factor.

	$4x^2$	$-5x$	$+3$
x	$4x^3$	$-5x^2$	$3x$
$+1$	$4x^2$	$-5x$	3

Teacher: *So, let's fill in the blanks in the problems we wrote.*

$$(x+1)(\underline{4x^2-5x+3}) = 4x^3 - x^2 - 2x + 3$$

$$(4x^3 - x^2 - 2x + 3) \div (x+1) = \underline{4x^2 - 5x + 3}$$

Teacher: *The next problem today is just a division problem. Where would these parts go on a rectangular diagram?*

$$(4x^3 - 13x^2 + 5x - 6) \div (x-3)$$

Student: *Well, we know the $4x^3$ goes in the top left.*

Student: *And we know that the x + 3 goes along the left side.*

Student: *And I think the negative 6 goes in the bottom right.*

Teacher: *Do you know where any of the rest goes? Can you fill in the puzzle? How?*

As students work, the teacher circulates looking for students who are grappling with the diagonals.

x	$4x^3$		
-3			-6

Teacher: *I saw you two put something on top. What and why?*

Student: *We know that x times something is $4x^3$. So that has to be $4x^2$.*

$$(4x^3 - 13x^2 + 5x - 6) \div (x-3)$$

	$4x^2$		
x	$4x^3$		
-3			-6

Student: *And that means that we can fill in the bottom left part because -3 times $4x^2$ is $-12x^2$.*

Teacher: *Does all of this make sense? Yes? So, now what?*

$$(4x^3 - 13x^2 + 5x - 6) \div (x-3)$$

	$4x^2$		
x	$4x^3$		
-3	$-12x^2$		-6

(continued)

Teacher: *I heard you guys talking about the diagonals. Can you tell us what you are thinking?*

Student: *We know we need to have a total of* $-13x^2$*. Right now, we've got* $-12x^2$*. So, the empty part of the diagonal has to have* $-1x^2$*.*

Student: *And then that forces the top to be* $-x$*.*

Teacher: *Wait, what's going on? Why?*

Student: *It's like we are undoing the diagonals. Every time when we were finding the total for the inside, the diagonals added to the middle terms. So, since we have the total and a part, the* $-13x^2$ *and* $-12x^2$*, we need the other part,* $-1x^2$*.*

Student: *And that makes the top be* $-1x$*.*

Teacher: *Turn to your partner and explain what you think is happening here. How are you using the diagonals?*

$(4x^3 - 13x^2 + 5x - 6) \div (x - 3)$

	$4x^2$	$-x$	
x	$4x^3$	$-x^2$	
-3	$-12x^2$		-6

Teacher: *Since it sounds like you're all making sense of how the diagonals are playing a part, let's fill in the rest. What are you thinking?*

Student: *First, we know the bottom part of the middle. It has to be* $3x$ *because of the* -3 *and the* $-x$*.*

Student: *Now, if you look at the problem, the long part...*

Teacher: *The dividend.*

Student: *Yeah, that has* $5x$*, so we need a* $2x$ *to go with the* $3x$ *in the diagonal to add to the* $5x$*. So put* $2x$ *in the top right. That means that the top part of the factor has to be 2, because 2 times* x *is* $2x$*. And 2 times* -3 *is* -6*, so that fits.*

$(4x^3 - 13x^2 + 5x - 6) \div (x - 3)$

	$4x^2$	$-x$	2
x	$4x^3$	$-x^2$	$2x$
-3	$-12x^2$	$3x$	-6

Teacher: *So, what is* $(4x^3 - 13x^2 + 5x - 6) \div (x - 3)$*?*

Student: *It's that stuff on the top,* $4x^2 - x + 2$*.*

$(5x^3 - 13x^2 + 5x - 6) \div (x - 3)$

$= 4x^2 - x + 2$

Teacher: *The last problem today is* $(6x^3 + 11x^2 - x - 6) \div (2x + 3)$*. I wonder what this would look like in a rectangle diagram?*

Students work and the teacher brings students together, emphasizing helping students verbalize using the diagonals to find the dividend.

$(6x^3 + 11x^2 - x - 6) \div (2x + 3)$

Teacher: *How would you summarize some of the things that came up in this string today?*

Elicit the following:

- *Multiplication and division are related.*
- *Division can be written many ways.*
- *You can find a missing factor using division and thinking about multiplication.*
- *In a rectangular diagram, the middle terms in the quotient are created using the diagonals.*

Sample Final Display

Your display could look like this at the end of the problem string:

	$2x^2$	$-3x$	4
x	$2x^3$	$-3x^2$	$4x$
-2	$-4x^2$	$6x$	-8

$(x-2)(2x^2-3x+4)=2x^3-7x^2+10x-8$

$\dfrac{2x^3-7x^2+10x-8}{x-2}=2x^2-3x+4$

$\dfrac{2x^3-7x^2+10x-8}{2x^2-3x+4}=x-2$

$$x-2\overline{)2x^3-7x^2+10x-8}\quad \text{quotient: } 2x^2-3x+4$$

	$3x$	$+1$
x^2	$3x^3$	x^2
$2x$	$6x^2$	$2x$
-5	$-15x$	-5

$(x^2+2x-5)(3x+1)=3x^3+7x^2-13x-5$

$\dfrac{3x^3+7x^2-13x-5}{(x^2+2x-5)}=3x+1$

	$4x^2$	$-5x$	$+3$
x	$4x^3$	$-5x^2$	$3x$
$+1$	$4x^2$	$-5x$	3

$(x+1)(\underline{\ 4x^2-5x+3\ })=4x^3-x^2-2x+3$

$(4x^3-x^2-2x+3)\div(x+1)=\underline{4x^2-5x+3}$

$(4x^3-13x^2+5x-6)\div(x-3)$

$=4x^2-x+2$

	$4x^2$	$-x$	2
x	$5x^3$	$-x^2$	$2x$
-3	$-12x^2$	$3x$	-6

$(6x^3+11x^2-x-6)\div(2x+3)$

	$3x^2$	x	-2
$2x$	$6x^3$	$2x^2$	$-4x$
3	$9x^2$	$3x$	-6

(continued)

Facilitation Notes

This version of the problem string lists short notes for important teacher moves during the string. After you've done the string yourself and studied the relationships involved, you might make similar notes for the things you want a reminder of or deem important.

	$2x^2$	__	4
__	$2x^3$		
-2		$6x$	

Here is a puzzle. What belongs in the blanks and in the empty spots inside?
Where did you start?
How would you write the multiplicative relationship? Division equations?

	$3x$	$+1$
__	$3x^3$	
__	$6x^2$	
__		-5

Turn to your partner, choose a place to start, decide what goes where.
Fill in together.
Write multiplication and division equations.

Let's look at both problems, what patterns do you see?
What's happening in the diagonals?

	__	__	__
x	$4x^3$	$-5x^2$	$3x$
$+1$	$4x^2$	$-5x$	3

Is this a mulitplication problem? Division? Write equations with blanks.
How did you find middle terms in quotient?
Fill in the puzzle.
Fill in the equation blanks.
Connect factor, factor, product to divisor, quotient, dividend.

$(4x^3 - 13x^2 + 5x - 6) \div (x - 3)$ *Where would these go on a rectangle diagram?*
So, we know where the first and last terms go, what else can we figure out?
How do you know what goes in the diagonals?

$(6x^3 + 11x^2 - x - 6) \div (2x + 3)$ *I wonder what this would look like in a rectangle diagram?*

PROBLEM STRINGS LESSON 6.5

The Rational Parent Functions

At a Glance

$1 \div 2$

$1 \div 4$

$1 \div 10$

$1 \div 1{,}000$

$1 \div 1$

$1 \div \frac{1}{2}$

$1 \div \frac{1}{4}$

$1 \div \frac{1}{10}$

$1 \div \frac{1}{1{,}000}$

$1 \div -1$

$1 \div -3$

$1 \div -9$

$1 \div -1{,}000$

$1 \div -\frac{1}{4}$

$1 \div -\frac{1}{8}$

$1 \div \frac{1}{1{,}000}$

$y = \frac{1}{x}$

$y = \frac{1}{x^2}$

Objectives

The goal of this problem string is to give students a quick experience constructing the parent rational functions, $y=\frac{1}{x}$ and $y=\frac{1}{x^2}$, reasoning about both the horizontal and vertical asymptotes.

Placement

This problem string can be used to introduce students to the rational parent functions by having students think briefly about some well-chosen points and reasoning about the relationships which create horizontal and vertical asymptotes.

You can use this problem string to begin textbook Lesson 6.5 Introduction to Rational Functions. You can also use this string to cinch the learning from the Breaking Point Investigation in Lesson 6.5.

Guiding the Problem String

This problem string should go very quickly, with strategic pauses at places where the reasoning changes. Students will not need paper and pencil.

Most of the problems of this string are given as a division problem. When students find the quotient, then you will record the results of each problem as an ordered pair and plotted on a graph. The ordered pair is (divisor, quotient).

Say each problem using division language, "What is 1 divided by 2?" and simultaneously write $1 \div 2 = \frac{1}{2}$. Then, plot the point $\left(2, \frac{1}{2}\right)$. As you get to the extreme points, ask students to envision where that point would be (next classroom over) and use that to help them extend to larger values. At the end of the first string, ask, "What is 1 divided by any number *x*?"

About the Mathematics

A rational function is the ratio of polynomials.

$$r(x)=\frac{p(x)}{q(x)}=\frac{a_n x^n + a_{n-1}x^{n-1}+\ldots+a_0}{b_m x^m + b_{m-1}x^{m-1}+\ldots b_0}$$

We call these two functions, $y=\frac{1}{x}$ and $y=\frac{1}{x^2}$, the rational parent functions. All rational functions have long run behavior models that are either a polynomial (the ratio of the term of highest degree of the polynomial in the numerator to the term of highest degree of the polynomial in the denominator) or one of the two rational parent functions.

(continued)

Important Questions

Use the following as you plan how to elicit and model student strategies.

- *We can't plot the* (1, 1,000) *because given the scale we are using, it would be out the door (across the hall, next classroom over, the gym), but where would it be? What about a point even further out? Would we ever be able to go far enough out that* 1 *divided by that number would be equal to* 0, *or less than* 0?
- *We can't plot* (0.001, 1,000) *because it would be way higher than the ceiling, but where would it be? What about a point even further up? What is happening to the graph the closer to the y-axis,* 0, *we are getting? What's the closest we can get? Is there a "closest" we can get? If you give me a really small, close to* 0, *number, like one-millionth, could you find a number closer to* 0? *And where would that point be?*
- *What's happening if we continue looking at this relationship,* $\left(x, \frac{1}{x}\right)$ *but in the second and third quadrants?*
- *If we connected all of these points, meaning we are considering all of the x-values, what is the behavior of the graph?*
- *So, what is the graph of* $y = \frac{1}{x}$?
- *Given everything we've just done, what do you think the graph of* $y = \frac{1}{x^2}$ *looks like? Why? Justify your reasoning.*

How would you summarize some of the things that came up in this string today?

- *The function* $y = \frac{1}{x}$ *has long run behavior that as x increases, y decreases, approaching* 0 *but never getting to* 0; *and as x decreases, y increases, approaching* 0 *but never getting to* 0. *This means that the function has a horizontal asymptote at* $y = 0$, the *x*-axis.
- *The function* $y = \frac{1}{x}$ *has short run behavior that as x approaches* 0 *from the positive side, y keeps getting bigger and bigger and we can say that y approaches infinity. As x approaches* 0 *from the negative side, y keeps getting smaller and smaller and we can say that y approaches negative infinity. We call this a vertical asymptote. This vertical asymptote is at* $x = 0$, *the y-axis.*
- *The function* $y = \frac{1}{x^2}$ *has similar behavior in the first quadrant as* $y = \frac{1}{x}$, *with a horizontal asymptote* $y = 0$ *and vertical asymptote at* $x = 0$, *but it is steeper. However, since all x-values are squared, there are no negative y-values. The second quadrant contains the reflection of the first quadrant over the y-axis.*

Sample Final Display

Your display could look like this at the end of the problem string:

$1 \div 2 = \frac{1}{2} \quad (2, \frac{1}{2})$

$1 \div 4 = \frac{1}{4} \quad (4, \frac{1}{4})$

$1 \div 10 = \frac{1}{10} \quad (10, \frac{1}{10})$

$1 \div 1{,}000 = \frac{1}{1{,}000} \quad (1000, \frac{1}{1000})$

$1 \div 1 = \frac{1}{1} = 1 \quad (1, 1)$

$1 \div \frac{1}{2} = \frac{1}{\frac{1}{2}} = 2 \quad (\frac{1}{2}, 2)$

$1 \div \frac{1}{4} = \frac{1}{\frac{1}{4}} = 4 \quad (\frac{1}{4}, 4)$

$1 \div \frac{1}{10} = \frac{1}{\frac{1}{10}} = 10 \quad (\frac{1}{10}, 10)$

$1 \div \frac{1}{1{,}000} = \frac{1}{\frac{1}{1{,}000}} = 1{,}000 \quad (\frac{1}{1000}, 1000)$

$1 \div -1 = \frac{1}{-1} = -1 \quad (-1, -1)$

$1 \div -3 = \frac{1}{-3} = -\frac{1}{3} \quad (-3, -\frac{1}{3})$

$1 \div -9 = \frac{1}{-9} = -\frac{1}{9} \quad (9, -\frac{1}{9})$

$1 \div -1{,}000 = \frac{1}{-1{,}000} = -\frac{1}{1{,}000} \quad (-1000, -\frac{1}{1000})$

$1 \div -\frac{1}{4} = \frac{1}{-\frac{1}{4}} = -4 \quad (-\frac{1}{4}, -4)$

$1 \div -\frac{1}{8} = \frac{1}{-\frac{1}{8}} = -8 \quad (-\frac{1}{8}, -8)$

$1 \div -\frac{1}{1{,}000} = \frac{1}{\frac{1}{-1{,}000}} = -1{,}000 \quad (\frac{-1}{1000}, -1000)$

$\frac{1}{x} \quad (x, \frac{1}{x})$

$\frac{1}{x^2} \quad (x, \frac{1}{x^2})$

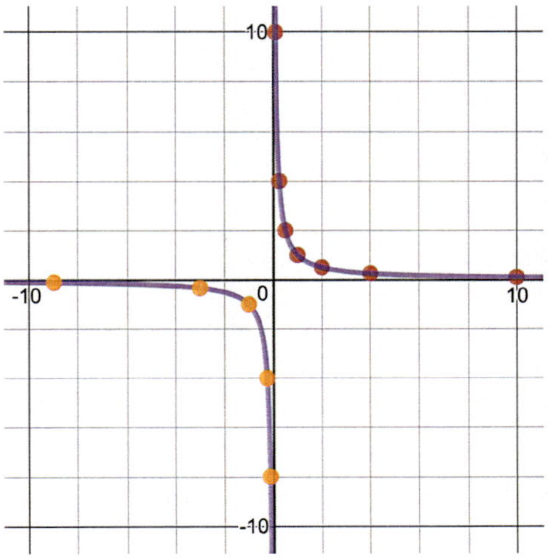

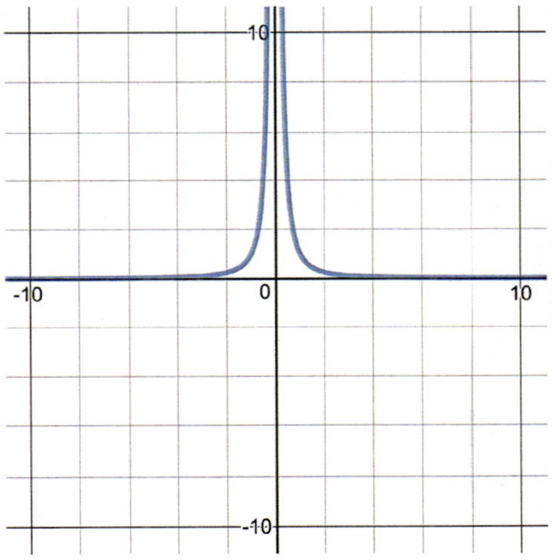

(continued)

Facilitation Notes

This version of the problem string lists short notes for important teacher moves during the string. After you've done the string yourself and studied the relationships involved, you might make similar notes for the things you want a reminder of or deem important.

$1 \div 2$	Today's string will go fast. Just sit and reason together. I'll just display this grid up here. What is 1 divided by 2? Write $1 \div 2$ and $\frac{1}{2}$. Plot (2, ½).
$1 \div 4$	Repeat for $1 \div 4$, $1 \div 10$.
$1 \div 10$	
$1 \div 1{,}000$	For $1 \div 1{,}000$, ask students where it would be if you could reach it? (Classroom next door?)
$1 \div 1$	Let's back up a bit. What's $1 \div 1$. Easy, I know, but where would I plot the corresponding point? What's $1 \div ½$? What does that mean? One way to think of it is, how many ½s are in 1? 2. Plot (½, 2).
$1 \div \frac{1}{2}$	
$1 \div \frac{1}{4}$	Repeat for ¼, 1/10.
$1 \div \frac{1}{10}$	
$1 \div \frac{1}{1{,}000}$	For this one, ask students where it would be if you could reach it? (Outside?)
$1 \div -1$	Let's back up even more, what's $1 \div -1$? Plot (−1, −1) Repeat for $1 \div -3$, $1 \div -9$.
$1 \div -3$	
$1 \div -9$	
$1 \div -1{,}000$	Where would it be if you could reach it?
$1 \div -\frac{1}{4}$	Let's head back toward zero, what's $1 \div -¼$? Repeat.
$1 \div -\frac{1}{8}$	
$1 \div \frac{1}{1{,}000}$	What's 1 divided by any number, $1 \div x$? What does the graph look like? The behavior, approaching but never reaching, is called an asymptote. There is a horizontal asymptote at $y = 0$ and a vertical asymptote at $x = 0$. Graph on the display grapher.
$y = \frac{1}{x}$	
$y = \frac{1}{x^2}$	Given everything we just did, what do you think this function will look like? What's 1 divided by any number squared? Graph on the display grapher.

PROBLEM STRINGS
LESSON

6.6 Rational Functions

At a Glance	
$\frac{x^2+x}{x+2}$ $y = x - 1$ $\frac{x^3+x}{x+2}$ $\frac{x^4+x}{x+2}$ $\frac{2x}{x+2}$ $\frac{2x}{x^2-4}$ $\frac{2x}{x^3-1}$	**Objectives** The goal of this problem string is to bridge students from dividing polynomials where the result is another polynomial (as they did when factoring polynomials) to dividing polynomials where the result is a rational function. **Placement** Students should have experience with multiplying polynomials and using division to factor polynomials. They should have some sense of the long run behavior of the parent rational functions (see Problem String 6.5). Use this problem string to help students learn to describe the long run behavior of rational functions and differentiate it from their short run behavior. You can use this problem string after textbook Lesson 6.5 Introduction to Rational Functions or during textbook Lesson 6.6 Graphs of Rational Functions. **Guiding the Problem String** The first two problems are a pair, where the second problem is the long run model for the first. After students do the division and deal with the remainder, graph the rational function and the line together and discuss the long run behavior. The rest of the functions do not have their long run behavior models listed, but they will come out of the division work. The first four rational functions have long run behavior that is a polynomial. The last two rational functions have long run behavior that is a rational function. As the string progresses help students gradually realize that the long run behavior of the rational function is the long run of the ratio of the terms of highest degree of the numerator to the denominator. In other words, the long run behavior of the rational function is based on the quotient of its polynomials' highest degree terms. **About the Mathematics** Rational functions are ratios of polynomials so the long run behavior of the polynomials determine the long run behavior of the rational function. Just as the long run behavior of a polynomial is based on its term of highest degree, the long run behavior of the rational function is based on the quotient of its polynomials' highest degree terms. If the degree of the numerator is higher or equal to the degree of the denominator, the long run behavior is a polynomial. If the degree of the numerator of a rational expression is less than the denominator, the long run behavior is dilation of either $y=\frac{1}{x}$ or $y=\frac{1}{x^2}$.

(continued)

Sample Interactions

Use the following as you plan how to elicit and model student strategies. This is not meant as a script, but as a view into the relationships involved and the intent of the problem string.

Teacher: *We have been factoring polynomials using division. We also explored the long run behavior of the rational parent functions. Let's connect those two things today. Here is the first problem for today's problem string—a rational expression. This fraction bar means division, an expression divided by an expression, in this case a polynomial divided by a polynomial. It can be read "the quantity x squared plus x all divided by the quantity x + 2." So, let's divide! What is this numerator divided by this denominator?*

Students work and the teacher circulates, looking for students who are grappling with the remainder. The teacher helps students with the beginning of the division as necessary.

$$\frac{x^2+x}{x+2}$$

Teacher: *You two had a great start. Will you start us off please?*

Student: *We know that the x plus 2 goes on the side, it's one of the factors. And the x squared goes in the top left corner. Then that has to be x up there, so now we get 2 times x is 2x. Then we know that the total has to have one positive x, so the diagonal must be negative x because 2x and −x is x.*

Student: *So that forces the bottom right to be −2. But that's not right. There shouldn't be a − 2.*

	x	-1
x	x^2	$-x$
2	$2x$	-2

Teacher: *What are you thinking about that?*

Student: *There will be a remainder. Keep going and put a 2 in the top right corner and the 2 is a remainder.*

Teacher: *Why did you put a 2 in the top right corner?*

Student: *There are no numbers in the* x^2+x, *so we need the 2 to add to the −2 to get 0. So then the 2 is left over. It's a remainder.*

	x	-1	
x	x^2	$-x$	2
2	$2x$	-2	

Teacher: *So, we could write the other factor, the dividend, as* $\frac{x^2+x}{x+2} = x-1\ R2$. *How could we write the dividend without a remainder?*

Student: *I think it's x minus 1 plus 2 over x + 2.*

Teacher: *How do you know?*

$$\frac{x^2+x}{x+2} = x-1\ R2$$

$$\frac{x^2+x}{x+2} = x-1\ R2 = x-1+\frac{2}{x+2}$$

Student: *If you multiply the two factors together, x + 2 times x − 1, you get* x^2+x-2, *but if you multiply x + 2 times 2 divided by x + 2 you get 2. The −2 and 2 add to zero.*

$$(x+2)\left(\frac{2}{x+2}\right)$$
$$=2$$

Teacher: *Let's take a look at a graph of this as a function. What do you think the graph might look like?*

Here it is. How would you describe it?

Student: *It kind of looks like two weird parabolas.*

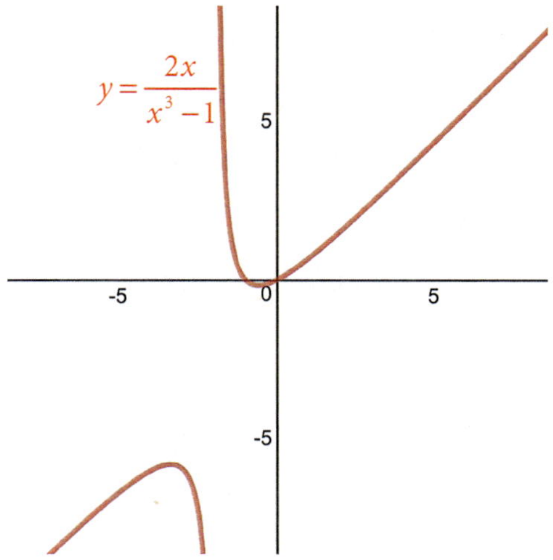

Teacher: *The second problem in today's string is* $y = x - 1$*? Predict in your mind the graph of* $y = x - 1$*. Got it?*

$y = x - 1$

Teacher: *Let's look at it at the display grapher along with the first problem.*

What do you see? How do the two graphs compare?

Why do you think it might be that way?

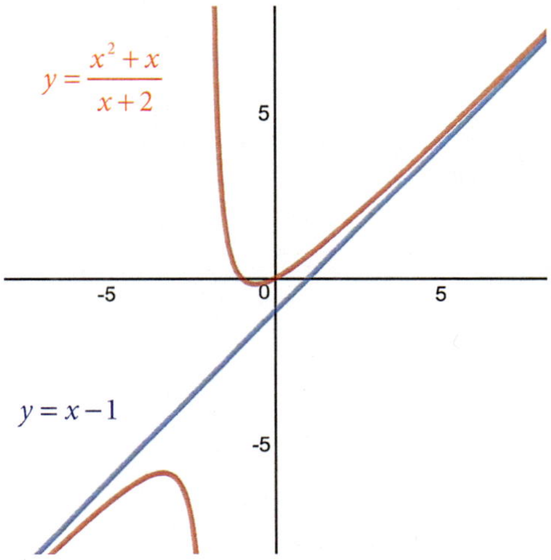

(continued)

Teacher: *Let's zoom out a bit. What are you seeing? Why? What are you thinking?*

Student: *I notice that they look really similar when you zoom out.*

Teacher: *Did anyone notice that the second problem, the line, is part of the quotient for the division in the first problem?*

Student: *It's almost like the division is saying the original rational function is equal to a line plus a rational function, that* $\frac{x^2+x}{x+2} = x-1+\frac{2}{x+2}$.

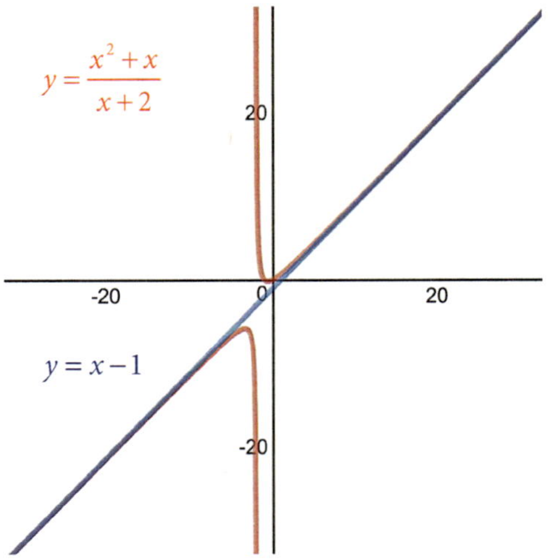

Teacher: *That the rational function behaves like the line* $y = x-1$ *in the long run sounds like a helpful thing.*

Student: *Yeah, it helps me understand the rational function better. It's basically like the line as you zoom out.*

Student: *But it's not like the line zoomed in.*

Teacher: *What is happening in the short run? Where does the short run stuff, that's not like the line, happen?*

Student: *It looks like it's around* $x = -2$.

Student: *That's where the denominator is zero, at* $x = -2$.

Teacher: *Why is that important?*

Student: *You can't divide by zero.*

Student: *This is like the rational parent functions, where they freak out at zero. They do that asymptote thing, where the numbers get either really big or really small but never get to zero.*

Student: *Except in this case, it looks like it's happening at* $x = -2$.

Teacher: *Let's add in the graph of* $x = -2$.

Student: *Sure enough, that looks like an asymptote.*

Teacher: *So, we have a rational function that behaves a lot like a line except where it has that asymptote. Nice.*

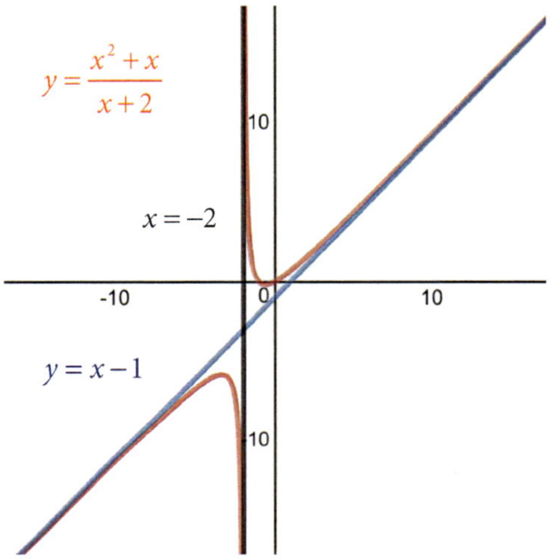

Teacher: *The next problem in the string is this rational expression. What do you think it would behave like in the long run? Just predict quietly for now.*

Do the division please. What is this numerator divided by this denominator?

$$\frac{x^3+x}{x+2}$$

The students work while the teacher circulates. Then the teacher has students describe the division as the teacher represents their thinking on the board. They discuss the remainder as needed.

	x^2	$-2x$	5	
x	x^3	$-2x^2$	$5x$	-10
2	$2x^2$	$-4x$	10	

$$\frac{x^3+x}{x+2} = x^2 - 2x + 5 + \frac{-10}{x+2}$$

Teacher: *What will this function look like? Let's graph it.*

What do you think? What does it look like? How does it compare to what you predicted? Turn to your partner and discuss.

Partners turn and talk while the teacher listens in, looking for students who are grappling with the long and short run behavior.

Student: *It makes sense that it looks cubic. There is that x cubed on top.*

Student: *But I'm not sure. It seems like the x on the bottom might be affecting it.*

Student: *I am wondering if it's like the parabola in the answer.*

Teacher: *The parabola in the quotient? Yes? Does anyone want to see the graph of the quotient?*

Student: *Yes!*

Student: *Maybe?*

Student: *Why?*

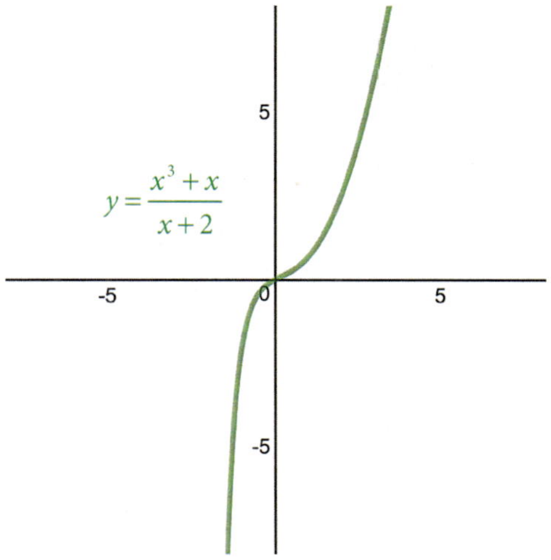

Teacher: *What do you think?*

Student: *To the right, it looks like they will have the same long run behavior. But to the left...*

Teacher: *Should the end behavior of the rational function be like the end behavior of the parabola?*

Student: *Will you zoom the graph out?*

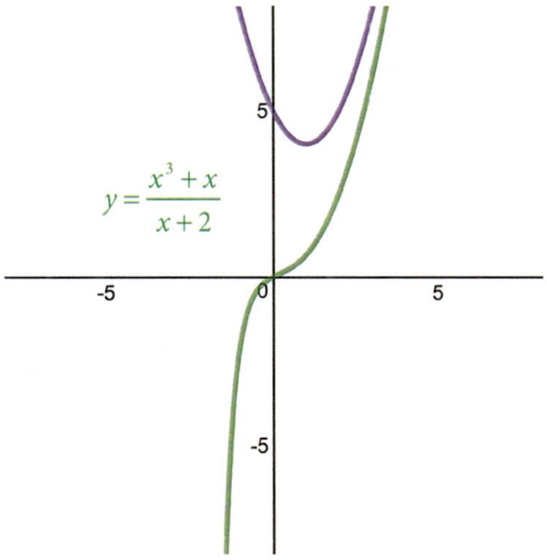

(continued)

Teacher: *Talk to me!*

Student: *Ahhh, it does act like a parabola. Interesting.*

Student: *In the long run it does. It the short run it has another asymptote.*

Student: *Yes, it has an asymptote at $x = -2$ again. But most of the time, it acts just like the parabola.*

Student: *I'm noticing that the first problem had x^2 divided by x to get an x and it acted like a line. This problem had an x^3 divided by an x and it acted like a parabola.*

Teacher: *Might be a pattern worth noting.*

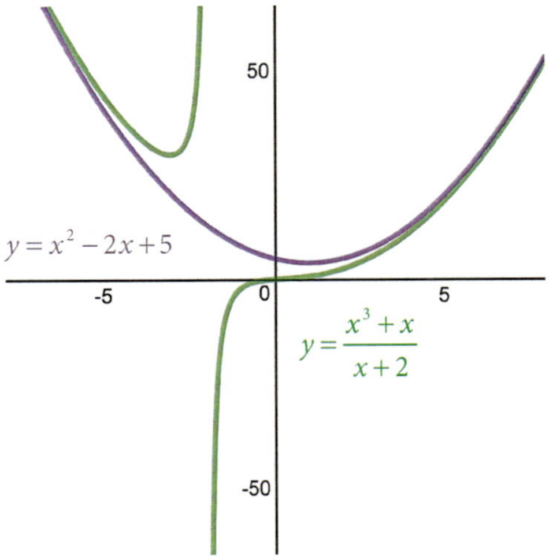

The teacher gives the next two problems, one at a time, $\frac{x^4 + x}{x + 2}$ and then $\frac{2x}{x + 2}$ and crafts a similar conversation, where students divide, predict, graph the rational function and the quotient, and compare the long run behavior.

	x^3	$-2x^2$	$+4x$	-7	
x	x^4	$-2x^3$	$4x^2$	$-7x$	14
2	$2x^3$	$-4x^2$	$8x$	-14	

$$\frac{x^4 + x}{x + 2} = x^3 - 2x^2 + 4x - 7 + \frac{14}{x + 2}$$

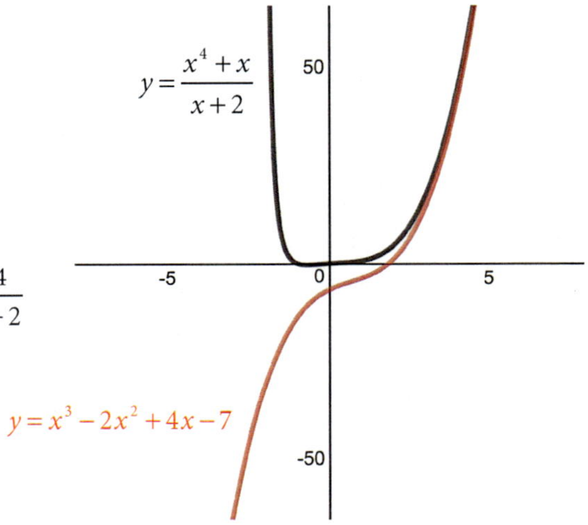

$$\frac{2x}{x+2}$$

	2	
x	$2x$	-4
2	4	

a horizontal line

$$\frac{2x}{x+2} = 2 + \frac{-4}{x+2}$$

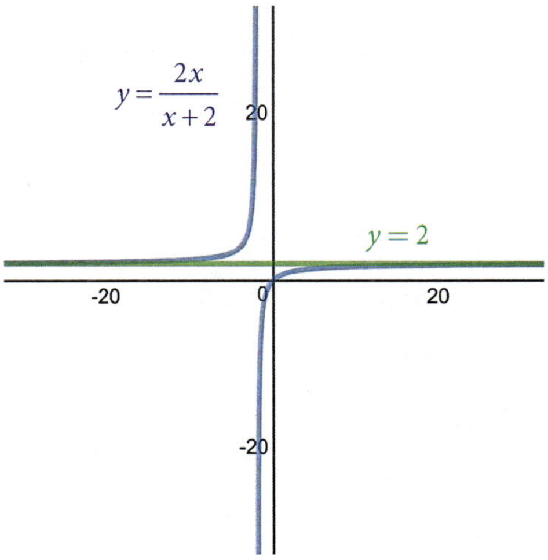

Teacher: *The next problem in the string is this rational function, another polynomial divided by a polynomial. What are you thinking about this one? Should we divide?*

Student: *No. I don't think so. It's like it's already done.*

Student: *Yeah, I agree. The top is less than the bottom.*

Teacher: *Do you mean that the numerator has an exponent that is less than the exponent in the denominator?*

Student: *Yes. Like a fraction where the denominator is greater than the numerator, like ⅘ then it's already simplified.*

$$\frac{2x}{x^2-4}$$

Student: *I tried to divide and I got to a weird place. I put the x^2 and -4 on the side and then the 2x in the top left corner and I tried to find what would go on top. What times x^2 is 2x? I think it's 2 divided by x. But then we are getting another rational expression so I agree to just leave it.*

$$\frac{2x}{x^2-4}$$

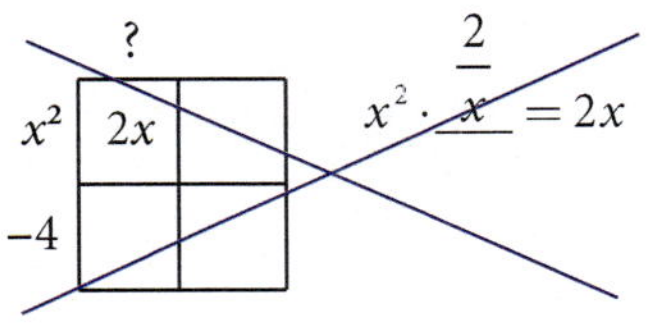

(continued)

Teacher: *Let's take a look at the graph. Before we do, predict. What do you think it might look like? What kind of long run behavior might it have? Short run behavior? Turn to your partner and predict out loud.*

The students turn and talk while the teacher gets the graph ready to display.

Teacher: *Here it is. How does it compare to your predictions? What do you think?*

Student: *It looks different than the other ones.*

Student: *It's like a cubic in the middle of a rational.*

Teacher: *Let's focus on the short run behavior. What's happening at $x = -2$?*

Student: *A vertical asymptote?*

Student: *Yes, and I think there is one at $x = 2$ also.*

Teacher: *What is going on at those values in the function?*

Student: *That's where the denominator is zero so it's doing that thing where you are dividing by values closer and closer to zero so you get an asymptote.*

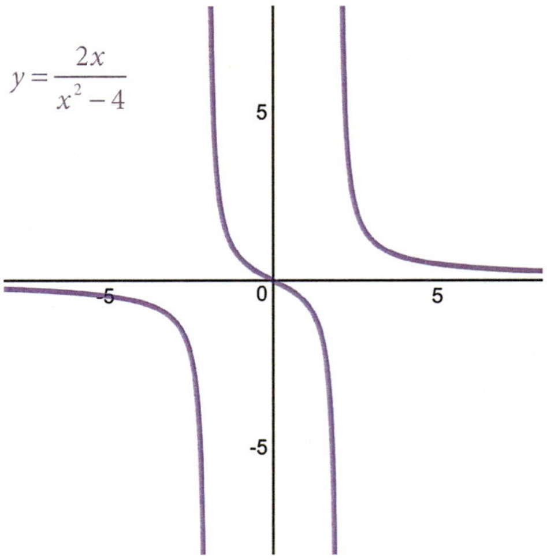

Teacher: *What's going on with the long run behavior? Since we didn't divide, is there something that we can say about what's happening in the function in the long run? What if we considered that first part when we started dividing, what is $2x$ divided by x^2?*

Student: *It's 2 divided by x.*

Teacher: *Let's graph that.*

The teacher displays the graph of both functions.

Teacher: *What do you think? Do they have the same long run behavior?*

Student: *It looks like it. Can you zoom out?*

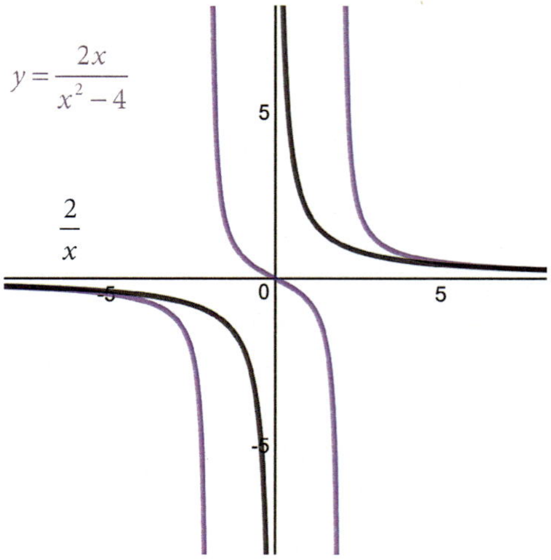

Teacher: *Sure thing. And?*

Student: *Yes, and it makes sense that they would have the same long run behavior because they are both like the one over x function.*

Teacher: *You think they are both acting like that parent function, one divided by x? Or maybe in this case, two divided by x?*

Student: *Yes, as you go out either direction, it's like 2 divided by bigger and bigger numbers which makes it get closer and closer to zero.*

Student: *And 2 divided by smaller and smaller negative numbers also gets closer and closer to zero.*

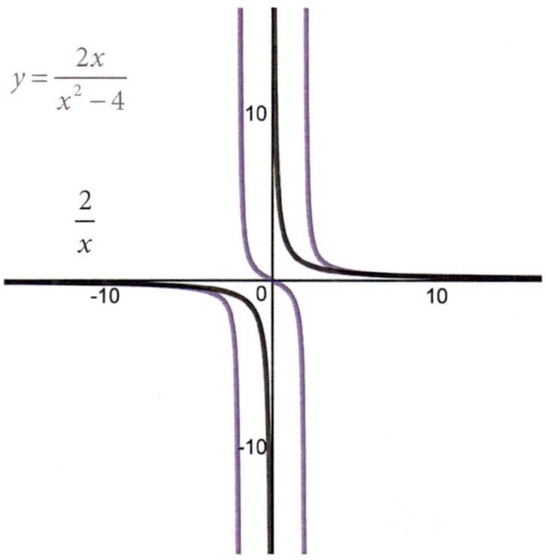

Teacher: *The last problem today is this rational expression. What do you think?*

The teacher repeats a similar experience with the last problem, asking students to predict, graphing, looking at and discussing the long run behavior. The display could look like the Sample Final Display that follows.

Teacher: *How would you summarize some of the things that came up in this string today?*

Elicit the following:

- *Rational functions are related to its polynomials.*
- *The long run behavior of the rational function is based on the quotient of its polynomials highest degree terms.*
- *Vertical asymptotes happen when the denominator of a rational function is zero.*

(continued)

Sample Final Display

Your display could look like this at the end of the problem string:

$\frac{x^2+x}{x+2}$

	x	-1	
x	x^2	$-x$	2
2	$2x$	-2	

$y=x-1$

$(x+2)\left(\frac{2}{x+2}\right)$
$=2$

a line

a rational function

$\frac{x^2+x}{x+2}=x-1\ R2 \neq x-1+\frac{2}{x+2}$

$\frac{x^3+x}{x+2}$

	x^2	$-2x$	5	
x	x^3	$-2x^2$	$5x$	-10
2	$2x^2$	$-4x$	10	

a parabola

$\frac{x^3+x}{x+2}=x^2-2x+5+\frac{-10}{x+2}$

$\frac{x^4+x}{x+2}$

	x^3	$-2x^2$	$+4x$	-7	
x	x^4	$-2x^3$	$4x^2$	$-7x$	14
2	$2x^3$	$-4x^2$	$8x$	-14	

a cubic

$\frac{x^4+x}{x+2}=x^3-2x^2+4x-7+\frac{14}{x+2}$

$\frac{2x}{x+2}$

	2	
x	$2x$	-4
2	4	

a horizontal line

$\frac{2x}{x+2}=2+\frac{-4}{x+2}$

$\frac{2x}{x^2-4}$

	?	
x^2	$2x$	
-4		

$x^2 \cdot \frac{2}{x}=2x$

end behavior like $y=\frac{1}{x}$

$\frac{2x}{x^3-1}$

$\frac{2x}{x^3}=\frac{2}{x^2}$

end behavior like $y=\frac{1}{x^2}$

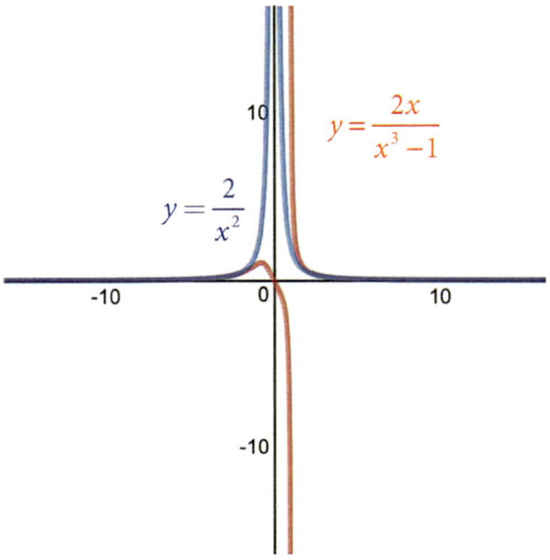

Facilitation Notes

This version of the problem string lists short notes for important teacher moves during the string. After you've done the string yourself and studied the relationships involved, you might make similar notes for the things you want a reminder of or deem important.

$\dfrac{x^2+x}{x+2}$	*Let's connect some of this polynomial division to rational parent functions.* *The fraction bar means division. Let's divide!* *Let's graph it! How would you describe it?*
$y=x-1$	*Predict this graph quickly.* *Let's take a look at both this and the previous graphs together.* *What do you see? How do the graphs compare? Why might it be that way?* *Let's zoom out. What are you seeing?* *Did anyone notice the relationship between the rational expression, the divisor, and the second problem?* *That the rational function behaves in the long run like the part of the quotient sounds like a helpful thing.*
$\dfrac{x^3+x}{x+2}$	*Predict how you think this one will behave in the long run.* *Divide.* *What does it look like? Why do you think that is happening in the short run? The long run?*
$\dfrac{x^4+x}{x+2}$	*Repeat for the next two problems.* *Notice the relationship between the quotient of the terms of highest degree of the numerator to the denominator.*
$\dfrac{2x}{x+2}$	
$\dfrac{2x}{x^2-4}$	*Should we divide? Elicit finding the ratio of the terms of highest degree.* *Graph that ratio. Zoom out. The end behavior is similar to the parent $y = 1/x$.*
$\dfrac{2x}{x^3-1}$	*Repeat.* *Discuss the long run behavior of rational functions behaving like the ratio of the terms of highest degrees of the numerator to the denominater. The end behavior is similar to the parent $y = 1/x^2$.*

PROBLEM STRINGS LESSON

6.7 Rational Expressions

At a Glance

True or False?

$$\frac{36}{60}=\frac{30+6}{30+30}=\frac{6}{30}$$

$$\frac{36}{60}=\frac{2^2\cdot 3^2}{2^2\cdot 3\cdot 5}=\frac{3}{5}$$

$$\frac{x+2}{x+5}=\frac{2}{5}, x\neq 0$$

$$\frac{2(x+15)}{5(x+15)}=\frac{2}{5}$$

$$\frac{x^2+5x+6}{x^2-4}=\frac{x+3}{x-2}$$

$$\frac{x^2+5x+6}{x^2-4}=\frac{5x+6}{-4}$$

$$\frac{x+3}{x-2}+\frac{x-2}{x+3}=1$$

Objectives

The goal of this problem string is to help students confront and reason about common mistakes made with rational expressions. Rather than telling students to stop making commonly made errors, we can help students use technology and properties to reason and justify the relationships.

Placement

This problem string can help students parse out the different properties at work when dealing with rational expressions so that they can more correctly manipulate them. You can use this problem string before or during students' work with operations with rational expressions.

You can use this problem string to support the work in textbook Lesson 6.7 Adding and Subtracting Rational Expressions and Lesson 6.8 Multiplying and Dividing Rational Expressions.

Guiding the Problem String

Students often try to "cancel" everything because they do not understand what is really happening when others have colloquially used the word "cancel." Refrain from using that often misused term. Instead, describe what is happening: in these cases factors are dividing to one. Refrain from using the positional word "over" to describe the rational expressions. Instead say "divide by" or "the ratio of".

The string is designed in partner problems, where one is true and the false one demonstrates an often seen misstep. The first two problems use numbers to help students parse out the difference between decomposing numerators and denominators by multiplication versus addition because equivalent factors divide to 1 but equivalent addends do not. The rest of the sets demonstrate the same additive mistakes but with variables in different ways. The last problem can segue into discussing the addition of rational expressions.

Encourage the use of technology to graph both sides of the equation for students to support their assertions. Encourage the use of properties to justify their assertions.

About the Mathematics

Equivalent factors may seem to "cancel" but in reality they divide to 1. This is an example of the associate property of multiplication at work, reassociating as needed to achieve an expression of a factor divided by itself.

$$\frac{ab}{ac}=\frac{a}{a}\cdot\frac{b}{c}=1\cdot\frac{b}{c}=\frac{b}{c} \text{ but } \frac{a+b}{a+c}\neq\frac{a}{a}+\frac{b}{c}$$

Important Questions

Use the following as you plan how to elicit and model student strategies.

- *What does "cancel" mean in your mind? What is a mathematical way to say what you mean?*
- *What properties and operations are involved when things divide to 1?*
- *How does a graph of both sides of the equation help support your argument?*

How would you summarize some of the things that came up in this string today?

- *You can use the associative property of multiplication to rearrange factors in a rational expression in order to rewrite the expression including a factor divided by itself. Something divided by itself is 1. And 1 times anything is that thing. This can be handy when finding equivalencies.*
- *It may not make sense to decompose rational expressions into addition or subtraction because then you cannot create a factor divided by itself. The expression is still composed of addition or subtraction, not multiplication which is needed if you want to use the fact that anything times 1 is that thing.*

Sample Final Display

Your display could look like this at the end of the problem string:

$\frac{36}{60}=\frac{30+6}{30+30}=\frac{6}{30}$ false $\quad \frac{36}{60}=\frac{3}{5}\neq\frac{6}{30}$

$\frac{36}{60}=\frac{2^2\cdot 3^2}{2^2\cdot 3\cdot 5}=\frac{3}{5}$ true $\quad \frac{2^2\cdot 3^2}{2^2\cdot 3\cdot 5}=\frac{2^2}{2^2}\cdot\frac{3}{3}\cdot\frac{3}{5}=1\cdot 1\cdot\frac{3}{5}=\frac{3}{5}$

$\frac{x+2}{x+5}=\frac{2}{5}$ false $\quad \xcancel{\frac{\cancel{x}+2}{\cancel{x}+5}} \quad \frac{x+2}{x+5}\neq\frac{x}{x}+\frac{2}{5}\neq\frac{2}{5}$

$\frac{2(x+15)}{5(x+15)}=\frac{2}{5}$ true $\quad \frac{2(x+15)}{5(x+15)}=\frac{2}{5}\cdot\frac{x+15}{x+15}=\frac{2}{5}\cdot 1=\frac{2}{5}$

$\frac{x^2+5x+6}{x^2-4}=\frac{x+3}{x-2}$ true $\quad \frac{(x+3)(x+2)}{(x-2)(x+2)}=\frac{(x+3)}{(x-2)}\cdot\frac{(x+2)}{(x+2)}=\frac{(x+3)}{(x-2)}\cdot 1=\frac{(x+3)}{(x-2)}$.

$\frac{x^2+5x+6}{x^2-4}=\frac{5x+6}{-4}$ false $\quad \xcancel{\frac{\cancel{x^2}+5x+6}{\cancel{x^2}-4}} \quad \frac{x^2+5x+6}{x^2-4}\neq\frac{x^2}{x^2}+5x+\frac{6}{-4}\neq\frac{5x+6}{-4}$

$\frac{x+3}{x-2}+\frac{x-2}{x+3}=1$ false $\quad \xcancel{\frac{\cancel{x+3}}{\cancel{x-2}}+\frac{\cancel{x-2}}{\cancel{x+3}}} \quad \frac{x+3}{x-2}+\frac{x-2}{x+3}\neq\frac{x+3}{x+3}+\frac{x-2}{x-2}\neq 1$

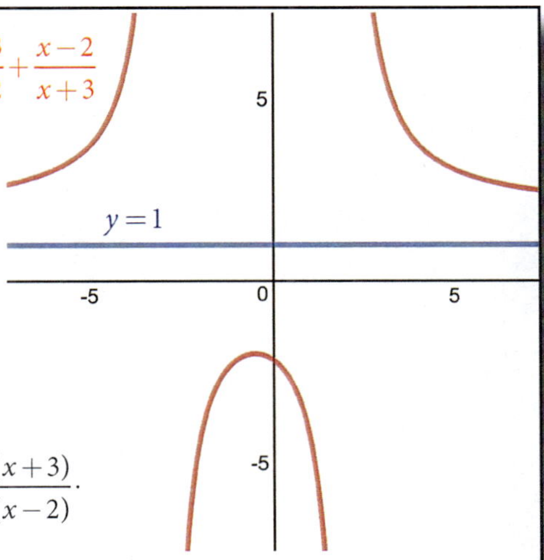

(continued)

Facilitation Notes

This version of the problem string lists short notes for important teacher moves during the string. After you've done the string yourself and studied the relationships involved, you might make similar notes for the things you want a reminder of or deem important.

$\frac{36}{60}=\frac{30+6}{30+30}=\frac{6}{30}$	*True or False? How do you know?* *What is 36/60 equivalent to?* *What does "cancel" really mean? When does it make sense? When not?*
$\frac{36}{60}=\frac{2^2\cdot3^2}{2^2\cdot3\cdot5}=\frac{3}{5}$	*True or False?* *But I thought you said we couldn't "cancel?"* *What does make sense? Why?*
$\frac{x+2}{x+5}=\frac{2}{5}$	*True or False?* *Why? What would a graph of both sides look like?* *Does that support your argument?* *What operation is involved? What property is involved?*
$\frac{2(x+15)}{5(x+15)}=\frac{2}{5}$	*Do these divide to 1? How do you know?* *What operation is involved? What property is involved?*
$\frac{x^2+5x+6}{x^2-4}=\frac{x+3}{x-2}$	*What would a graph of both sides look like?* *Does that support your argument?* *Would factoring help? What operation is involved?*
$\frac{x^2+5x+6}{x^2-4}=\frac{5x+6}{-4}$	*Do these x^2s divide to 1? How do you know?* *What property is involved? What operation is involved?* *Would a graph support your arguement?*
$\frac{x+3}{x-2}+\frac{x-2}{x+3}=1$	*All of these divide to 1? How do you know?* *What operation is involved? What property is involved?* *Graph?*

PROBLEM STRINGS LESSON

6.8 Dividing Rational Expressions

<table>
<tr>
<td>

At a Glance

cheese (c.)	3
pizza	2

cheese (c.)	2
pizza	$\frac{1}{5}$

cheese (c.)	8
pizza	$\frac{4}{5}$

cheese (c.)	2
pizza	$\frac{a}{b}$

x^2	
$\frac{x+1}{x}$	

$x^2 - 1$	
$\frac{x+1}{x}$	

$$\frac{x^2 - 5x + 6}{x^2 - 5x - 6} \div \frac{x^2 - 4}{x + 1}$$

</td>
<td>

Objectives

The goal of this problem string is to develop students' strategy for dividing rational expressions based on finding equivalent ratios where the divisor is 1.

Placement

This problem string can help students develop dividing rational expressions after students have worked with simplifying and multiplying rational expressions.

You can use this problem string during textbook Lesson 6.8 Multiplying and Dividing Rational Expressions to help students develop strategies for dividing rational expressions.

Guiding the Problem String

The first four problems are set in the context of the number of cups of grated cheese needed for parts of pizzas. Keep the problems in the context, with the goal for each question of finding how many cups of cheese are needed for one pizza. The second and third problem are related where both the cheese and the pizza are quadrupled. Since the goal is to find the amount of cheese for one pizza, students are reasoning about making the divisor one, which creates an easy division problem. The fourth problem is meant to help students generalize the notion of creating an equivalent rational expression where the divisor is one. The next two problems are given in a ratio table format to help students transition to rational expressions with variables. The last problem is given in a more traditional format.

About the Mathematics

This ratio context of cheese to pizza is a helpful, partitive division perspective that can help students realize how they are creating a divisor of one. This can be generalized to any division problem which can be solved by finding an equivalent ratio where the divisor is 1. This is colloquially known as "ours is not to reason why, just invert and multiply" but without the underpinnings of why and how it works, this oblique rule becomes confused, mixed up, and implemented incorrectly in any expression that remotely resembles a rational expression. Ours is to reason why!

</td>
</tr>
</table>

Sample Interactions

Use the following as you plan how to elicit and model student strategies. This is not meant as a script, but as a view into the relationships involved and the intent of the problem string.

<table>
<tr>
<td>**Teacher:** *Today's problem string is yummy. What if we were going to talk about making cheesy pizza and I told you that the recipe is for 3 cups of cheese for 2 pizzas.*</td>
<td>

cheese (c.)	3
pizza	2

</td>
</tr>
<tr>
<td>**Teacher:** *How many cups of cheese would we need for 1 pizza?*

The teacher draws in the extra part of the ratio table and the 1.</td>
<td>

cheese (c.)	3	
pizza	2	1

</td>
</tr>
</table>

(continued)

Student: *You just need half as much, 1.5 cups of cheese.*

The teacher writes 1.5 and models taking half of the cheese and the pizza on the ratio table. You can model this with ÷2 or ×½.

Teacher: *So, if you need half as much pizza, then you need half as much cheese? Makes sense.*

	÷ 2 →	
cheese (c.)	3	1.5
pizza	2	1
	÷ 2 →	

Teacher: *Another way we can represent this problem is that we have 3 cups of cheese divided over 2 pizzas and you just told me that's equivalent to 1.5 cups of cheese per 1 pizza.*

	÷ 2 →	
cheese (c.)	3	1.5
pizza	2	1
	÷ 2 →	

$3 \div 2 = 1.5 \div 1 = 1.5$

Teacher: *The next problem is a different recipe. This time we have 2 cups of cheese for one-fifth of a pizza. More cheesy or not as cheesy as the first?*

Student: *Lots more cheesy!*

cheese (c.)	2
pizza	$\frac{1}{5}$

Teacher: *So, how many cups of cheese would we need for one pizza?*

The teacher draws in the extra part of the ratio table and the 1.

cheese (c.)	2	
pizza	$\frac{1}{5}$	1

Student: *More!*

Teacher: *How much more?*

Student: *If you only have one-fifth of a pizza, you need five times as much to get a whole pizza. So 5 times 2 is 10 cups of cheese. That's a ton of cheese!*

	× 5 →	
cheese (c.)	2	10
pizza	$\frac{1}{5}$	1
	× 5 →	

Teacher: *Another way we can represent this problem is that we have 2 cups of cheese divided over one-fifth of a pizza and you just told me that's equivalent to 10 cups of cheese per 1 pizza.*

	× 5 →	
cheese (c.)	2	10
pizza	$\frac{1}{5}$	1
	× 5 →	

$2 \div \frac{1}{5} = 10 \div 1 = 10$

Teacher: *The next problem is a different recipe. This time we have 8 cups of cheese for four one-fifths, or four-fifths, of a pizza. How much cheese for one pizza?*

I wonder if the previous problems might help us think about this one. How are they related?

cheese (c.)	8	
pizza	$\frac{4}{5}$	1

Student: *It looks like this is just four times as much, cheese and pizza.*

Teacher: *How can that help?*

Student: *You still have just as much cheese per pizza as before.*

Teacher: *Is there some way that we can get this problem back to the previous problem?*

Student: *You can divide both the cheese and the pizza by 4.*

	÷ 4 →		
cheese (c.)	8	2	
pizza	$\frac{4}{5}$	$\frac{1}{5}$	1
	÷ 4 →		

Teacher: *And?*

Student: *Now we are back to the previous recipe.*

Teacher: *So we can just divide by 4 to get back to the previous case of 2 cups of cheese for one-fifth of a pizza, which we decided we could scale up by 5 to get 10 cups for 1 pizza.*

		÷ 4	× 5
cheese (c.)	8	2	10
pizza	$\frac{4}{5}$	$\frac{1}{5}$	1

Teacher: *When we had this case of cups of cheese for a fractional part, four-fifths, of a pizza, how did we get that four-fifths to one pizza?*

Student: *It's like we scaled down to get to one-fifth and then back up to get to one.*

Student: *Yeah, we got to a unit fraction and then once you know that it's 2 cups for one-fifth of a pizza, you will always need 5 of those one-fifths to make a whole pizza.*

Teacher: *So, if we have 8 cups of cheese divided over four-fifths of pizza, that's equivalent to 2 cups of cheese divided over one-fifth of a pizza which is equivalent to 10 cups of cheese for 1 pizza? We can record that this way. We can also record that as the ratio of 8 to four-fifths.*

		÷ 4	× 5
cheese (c.)	8	2	10
pizza	$\frac{4}{5}$	$\frac{1}{5}$	1

$8 \div \frac{4}{5} = 2 \div \frac{1}{5} = 10 \div 1 = 10$

Teacher: *I'm going to write this ratio of 8 to four-fifths. Looking back at the ratio table of cups of cheese to pizzas, you said that we divided by 4 and then multiplied by 5. Can anyone think of another way to write divided by 4 and multiplied by 5? Could we write it as multiplying by five-fourths? And if we did write it that way, what happens to the denominator? Yep, there's that one pizza. And the numerator? There's the 8 divided by 4 times 5. What do you think of this?*

		÷ 4	× 5
cheese (c.)	8	2	10
pizza	$\frac{4}{5}$	$\frac{1}{5}$	1

$8 \div \frac{4}{5} = 2 \div \frac{1}{5} = 10 \div 1 = 10$

$$\frac{8}{\frac{4}{5}} = \frac{8\cdot\frac{5}{4}}{\frac{4}{5}\cdot\frac{5}{4}} = \frac{8\cdot\frac{5}{4}}{1} = 8\cdot\frac{5}{4}$$

Student: *Right, because multiplying the four-fifths by five-fourths will always be one. So then you're just dividing by 1.*

Teacher: *With this next problem, let's generalize that. What if we have 2 cups of cheese for some part of a pizza that is a divided by b parts? What is an equivalent amount of cheese for 1 pizza?*

cheese (c.)	2
pizza	$\frac{a}{b}$

The teacher records students thinking as they generalize how they could find equivalent expressions with a divisor of 1.

		÷ a	× b
cheese (c.)	2	$\frac{2}{a}$	$\frac{2b}{a}$
pizza	$\frac{a}{b}$	$\frac{1}{b}$	1

$2 \div \frac{a}{b} = \frac{2}{a} \div \frac{1}{b} = \frac{2b}{a} \div 1$

$$\frac{2}{\frac{a}{b}} = \frac{2\cdot\frac{b}{a}}{\frac{a}{b}\cdot\frac{b}{a}} = \frac{2\cdot\frac{b}{a}}{1} = \frac{2b}{a}$$

Teacher: *What do you think about 2 divided by any fraction?*

Student: *Division is easy when the divisor is one! You are just multiplying by the reciprocal.*

(continued)

Teacher: *The next problem today is no longer about cheese and pizza but it's still in a ratio table. I wonder if we can use the same kind of reasoning to help us find an equivalent expression where the divisor is one?*	x^2 over $\frac{x+1}{x}$ (ratio table)

Students work and the teacher records students' thinking in the ratio table step-by-step, multiplying by the reciprocal, and as an expression with the division symbol, each time emphasizing equivalency and finding a divisor of 1.

Ratio table (steps: $\div(x+1)$, then $\cdot x$):

x^2	$\frac{x^2}{x+1}$	$\frac{x^3}{x+1}$
$\frac{x+1}{x}$	$\frac{1}{x}$	1

Ratio table (step: $\cdot \frac{x}{x+1}$):

x^2	$x^2 \cdot \frac{x}{x+1}$
$\frac{x+1}{x}$	1

$$x^2 \div \frac{x+1}{x} = \frac{x^2}{\frac{x+1}{x}} = \frac{x^2 \cdot \frac{x}{x+1}}{\frac{x+1}{x} \cdot \frac{x}{x+1}} = \frac{\frac{x^3}{x+1}}{1} = \frac{x^3}{x+1}$$

The teacher repeats with the next problem, where students work and the teacher records students' thinking as an expression with the division symbol, emphasizing equivalency and finding a divisor of 1.

Ratio table: $x^2 - 1$ over $\frac{x+1}{x}$

$$(x^2-1) \div \frac{x+1}{x} = \frac{(x^2-1)}{\frac{x+1}{x}} = \frac{(x^2-1) \cdot \frac{x}{x+1}}{\frac{x+1}{x} \cdot \frac{x}{x+1}} = \frac{\frac{(x-1)(x+1)x}{(x+1)}}{1} = \frac{(x-1)x}{1} = x^2 - x$$

Teacher: *What are you noticing?*

Student: *Each time it's about getting that dividing number.*

Teacher: *The divisor.*

Student: *Yeah, getting the divisor to be 1. When you divide by 1, then it's just whatever is in the numerator.*

Student: *And the numerator turns out to be the original numerator times the reciprocal of the denominator.*

Teacher: *Every time? Why?*

Student: *Because you're always multiplying by the reciprocal of the denominator times the denominator to get it to be one. Because division by one is easy!*

Teacher: *The last problem today just looks like a division problem. I wonder if we can still try to find an equivalent problem where the divisor is one?*	$\frac{x^2-5x+6}{x^2-5x-6} \div \frac{x^2-4}{x+1}$

Students work and the teacher records students thinking, emphasizing equivalency and finding a divisor of 1.

$$\frac{x^2-5x+6}{x^2-5x-6} \div \frac{x^2-4}{x+1} \qquad \frac{x^2-5x+6}{x^2-5x-6} \cdot \frac{x+1}{x^2-4} \div \frac{x^2-4}{x+1} \cdot \frac{x+1}{x^2-4}$$

$$= \frac{(x-3)(x-2)}{(x-6)(x+1)} \cdot \frac{x+1}{(x+2)(x-2)} \div 1 = \frac{x-3}{(x-6)(x+2)}$$

Teacher: *How would you summarize some of the things that came up in this string today?*

Elicit the following:

- *Division can be thought of as a ratio.*
- *We can write equivalent expressions where the divisor is one. That makes the division easy.*

Sample Final Display

Your display could look like this at the end of the problem string:

cheese (c.)	3	1.5
pizza	2	1

(÷ 2 applied to both rows)

$3 \div 2 = 1.5 \div 1 = 1.5$

cheese (c.)	2	10
pizza	$\frac{1}{5}$	1

(× 5 applied to both rows)

$2 \div \frac{1}{5} = 10 \div 1 = 10$

cheese (c.)	8	2	10
pizza	$\frac{4}{5}$	$\frac{1}{5}$	1

(÷ 4, then × 5 applied to both rows)

$8 \div \frac{4}{5} = 2 \div \frac{1}{5} = 10 \div 1 = 10$ $\qquad \frac{8}{\frac{4}{5}} = \frac{8 \cdot \frac{5}{4}}{\frac{4}{5} \cdot \frac{5}{4}} = \frac{8 \cdot \frac{5}{4}}{1} = 8 \cdot \frac{5}{4}$

cheese (c.)	2	$\frac{2}{a}$	$\frac{2b}{a}$
pizza	$\frac{a}{b}$	$\frac{1}{b}$	1

(÷ a, then × b applied to both rows)

$2 \div \frac{a}{b} = \frac{2}{a} \div \frac{1}{b} = \frac{2b}{a} \div 1$ $\qquad \frac{2}{\frac{a}{b}} = \frac{2 \cdot \frac{b}{a}}{\frac{a}{b} \cdot \frac{b}{a}} = \frac{2 \cdot \frac{b}{a}}{1} = \frac{2b}{a}$

Division is easy when the divisor is 1!

x^2		$x^2 \cdot \frac{x}{x+1}$
$\frac{x+1}{x}$		1

($\cdot \frac{x}{x+1}$ applied to both rows)

$$x^2 \div \frac{x+1}{x} = \frac{x^2}{\frac{x+1}{x}} = \frac{x^2 \cdot \frac{x}{x+1}}{\frac{x+1}{x} \cdot \frac{x}{x+1}} = \frac{\frac{x^3}{x+1}}{1} = \frac{x^3}{x+1}$$

$x^2 - 1$	
$\frac{x+1}{x}$	

$$(x^2-1) \div \frac{x+1}{x} = \frac{(x^2-1)}{\frac{x+1}{x}} = \frac{(x^2-1) \cdot \frac{x}{x+1}}{\frac{x+1}{x} \cdot \frac{x}{x+1}} = \frac{\frac{(x-1)(x+1)x}{(x+1)}}{1} = \frac{(x-1)x}{1} = x^2 - x$$

$$\frac{x^2-5x+6}{x^2-5x-6} \div \frac{x^2-4}{x+1} \qquad \frac{x^2-5x+6}{x^2-5x-6} \cdot \frac{x+1}{x^2-4} \div \frac{x^2-4}{x+1} \cdot \frac{x+1}{x^2-4}$$

$$= \frac{(x-3)(x-2)}{(x-6)(x+1)} \cdot \frac{x+1}{(x+2)(x-2)} \div 1 = \frac{x-3}{(x-6)(x+2)}$$

(continued)

Facilitation Notes

This version of the problem string lists short notes for important teacher moves during the string. After you've done the string yourself and studied the relationships involved, you might make similar notes for the things you want a reminder of or deem important.

$$\begin{array}{l|c} \text{cheese (c.)} & 3 \\ \hline \text{pizza} & 2 \end{array}$$

Keep the context of pizzas and cups of cheese.
How many cups of cheese do we need for 1 pizza?
Represent as using the division symbol:
3 cups of cheese divided over 2 pizzas is 1.5 c. cheese per 1 pizza.

$$\begin{array}{l|c} \text{cheese (c.)} & 2 \\ \hline \text{pizza} & \frac{1}{5} \end{array}$$

Different recipe. More cheesy or less cheesy?
How many cups of cheese do we need for 1 pizza?
Represent as using the division symbol.

$$\begin{array}{l|c} \text{cheese (c.)} & 8 \\ \hline \text{pizza} & \frac{4}{5} \end{array}$$

Different recipe. Related to previous? How?
How can we get back to previous? Divide by 4?
Represent as using the division symbol.
How can we represent dividing by 4 and multiplying by 5?
Represent as equivalent ratio out of context.

$$\begin{array}{l|c} \text{cheese (c.)} & 2 \\ \hline \text{pizza} & \frac{a}{b} \end{array}$$

Let's get a little general.
How can we get to 1 pizza? What can we multiply by?
Represent as using the division symbol and as equivalent ratios out of context.
Division is easy when the divisor is 1!

$$\begin{array}{c|c} x^2 & \\ \hline \frac{x+1}{x} & \end{array}$$

No longer about cheese and pizza, but still in a ratio table.
How can we get the divisor (denominator) to be one multiplicatively?

$$\begin{array}{c|c} x^2 - 1 & \\ \hline \frac{x+1}{x} & \end{array}$$

This time model with equations (not in ratio table).
How can we get the divisor (denominator) to be one multiplicatively?
Keep the language about equivalency and division.

$$\frac{x^2 - 5x + 6}{x^2 - x + 6} \div \frac{x^2 - 4}{x + 1}$$

This question looks like a division problem.
Can we find an equivalent expression where the divisor is 1?

PROBLEM STRINGS
LESSON
6.9 Solving Rational Equations

At a Glance

$$\frac{17}{4} = x$$

$$\frac{4}{17} = \frac{1}{x}$$

$$\frac{x}{2} = x + 1$$

$$\frac{2}{x} = \frac{1}{x+1}$$

$$\frac{4}{2x+3} = \frac{1}{x}$$

Objectives

The goal of this problem string is to strengthen students' ability to solve simple rational equations using the "solve an equivalent reciprocal equation" strategy

Placement

This problem string can be used to help students solve simple rational equations. Students should have some prior experience with solving for a variable that is in the denominator.

You can use this problem string in textbook Lesson 6.9 Solving Rational Equations.

Guiding the Problem String

The first three problems should go quickly, setting the stage for the rest. The problems are in sets of equivalent partners, where the solution to the first equation can potentially help students develop the notion that it can be useful to consider solving the equivalent reciprocal equation. Throughout the string, compare problems to each other and refer back to the reciprocal relationship between the first two problems, wondering aloud how that might be useful. Model the "using properties of equality by doing the same operation to both sides of the equation" strategy, the "solving an equivalent reciprocal equation" strategy, and the "numerators are the same so the denominators are equal" strategy.

This string works toward the same "solve an equivalent reciprocal equation strategy" as the string in Lesson 5.8. See that problem string for an example of a full dialog and modeling of a similar string. Even though students may have begun to build this strategy before, students will benefit from multiple exposures to it as they construct it for themselves.

About the Mathematics

The solutions to an equation and that equation's reciprocal equation are the same, except for values that are outside of either equations' domain.

Refrain from using the word "over" to describe the rational expressions. "Over" describes position—it does not describe a mathematical relationship and is therefore not helpful. Describe $\frac{1}{x}$ as one divided by x and $17/4$ as seventeen-fourths. Using this language potentially helps students make sense of expressions using relationships.

(continued)

Important Questions

Use the following as you plan how to elicit and model student strategies.

- *What is equivalent to seventeen fourths?*
- *Is the reciprocal equation equivalent? Does that mean that x is the same in equivalent equations?*
- *Can you use an equivalent equation that is easier to solve?*

How would you summarize some of the things that came up in this string today?

- *Reciprocal equations are equivalent.*
- *When the variable is in the denominator, you can use the reciprocal equation to solve.*
- *Consider looking for equivalent equations that might be easier to solve.*

Sample Final Display

Your display could look like this at the end of the problem string:

$\frac{17}{4} = x$

$x = \frac{17}{4} = \frac{16}{4} + \frac{1}{4} = 4\frac{1}{4} = 4.25$

$\frac{4}{17} = \frac{1}{x}$

$x = \frac{17}{4}$

$x \cdot \frac{4}{17} = \frac{1}{x} \cdot x$

$x \cdot \frac{4}{17} \cdot \frac{17}{4} = 1 \cdot \frac{17}{4}$

$x = \frac{17}{4}$

Does $\frac{4}{17} = \frac{1}{x}$ imply $\frac{17}{4} = x$?

$\frac{x}{2} = x + 1$

$x = -2$

$\frac{x}{2} = x + 1$

$-1 = 0.5x$

$-2 = x$

$\frac{x}{2} = x + 1$

$x = 2x + 2$

$-2 = x$

$\frac{2}{x} = \frac{1}{x+1}$

$x = -2$

$(x)(x+1)\frac{2}{x} = \frac{1}{x+1}(x)(x+1)$

$2x + 2 = x$

$x = -2$

$\frac{2}{x} = \frac{1}{x+1}$ implies $\frac{x}{2} = x + 1$ so $x = -2$ from previous.

$\frac{4}{2x+3} = \frac{1}{x}$

$x = 1.5$

$\frac{4}{2x+3} = \frac{1}{x} \Leftrightarrow \frac{2x+3}{4} = x$

$2x + 3 = 4x$

$-2x = -3$

$x = 1.5$

Facilitation Notes

This version of the problem string lists short notes for important teacher moves during the string. After you've done the string yourself and studied the relationships involved, you might make similar notes for the things you want a reminder of or deem important.

$\frac{17}{4} = x$	*What is 17 divided by 4? Fraction? Decimal?*
$\frac{4}{17} = \frac{1}{x}$	*Say "divided by" not over. Model multiplying both sides by x.* *Same solution as previous. Hm...* *Is the reciprocal equation equivalent?*
$\frac{x}{2} = x + 1$	*Model multiplying both sides by 2, subtracting 0.5x.* *Quick.*
$\frac{2}{x} = \frac{1}{x+1}$	*Model multiplying both sides by x(x+1).* *Use equivalent reciprocal equation, already know the answer to previous.* *That might be a handy strategy!*
$\frac{4}{2x+3} = \frac{1}{x}$	*Could multiply both sides by (2x + 3)(x) but why not consider reciprocal?* *Model solving reciprocal.*

PROBLEM STRINGS
LESSON
7.1 Special Right Triangles

At a Glance

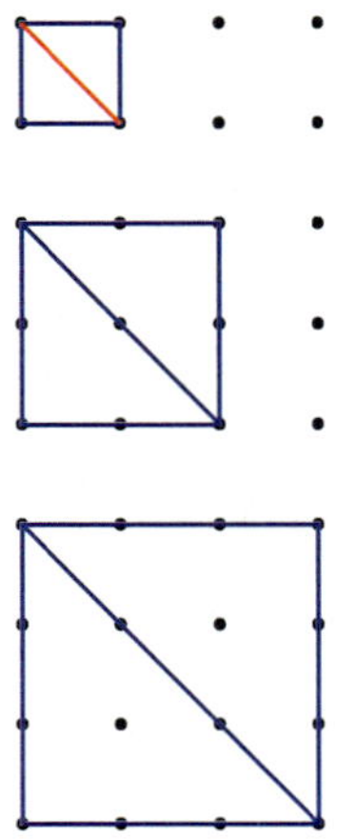

In a 45°, 45°, 90°, what if the:
side length is 6, hypotenuse?
side length is 21, hypotenuse?
side length is x, hypotenuse?

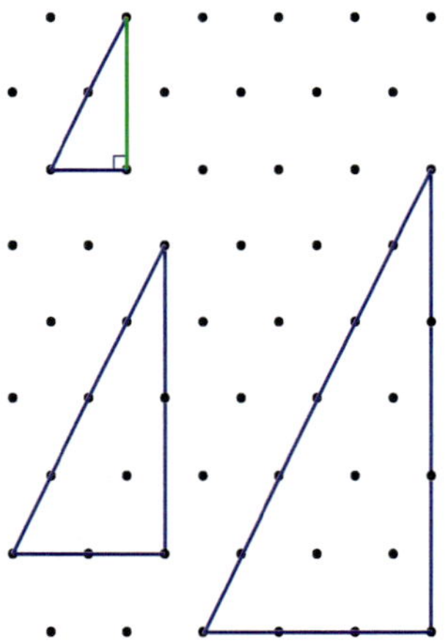

In a 30°, 60°, 90°, what if the:
short side is 6, hypotenuse? long side?
short side is 21, hypotenuse? long side?
short side is x, hypotenuse? long side?

Objectives
The goal of this problem string is to build students' understanding of the relationships of special right triangle side lengths and the idea that each angle has a specific constant ratio of side lengths.

Placement
This problem string is for students to review or build the relationships in special right triangles before students draw on those relationships in a study of trigonometry. Students should have prior experience with the Pythagorean Theorem.

You can use this problem string to introduce textbook Chapter 7 Trigonometry and Trigonometric Functions.

Guiding the Problem String
The first six problems are based on 45°, 45°, 90° triangles. The first three problems are given as triangles that are half of a square with unit side lengths. Students find the hypotenuse of the triangle (diagonal of the square) using the Pythagorean Theorem. The second and third problems' hypotenuses can also be found by finding the first triangle inside the bigger triangles. Have students describe the pattern they are seeing and apply it in the fourth and fifth problems, and generalize it in the sixth problem. The next six problems use the same sequence but with the relationships in 30°, 60°, 90° triangles that are built in isometric dots.

One way to sketch the triangles as you ask each question is to project dot paper on a board and write on the projection. Another way is to write on dot paper under a document camera. A display page of both kinds of dot paper is provided at the end of the problem string. Do not prepare the figures ahead of time. There is value in having students watch you quickly sketch them. Students get a sense of the figure, just by watching you sketch.

About the Mathematics
Both 45°, 45°, 90° triangles and 30°, 60°, 90° triangles are often referred to as special right triangles. Historically, these triangles have been used when working with trigonometry as important touch points because the side length relationships are easily derived and the trigonometric relationships are workable without arduous calculations. With the advent of calculators and technology, they are gradually receiving less attention. They can still play a critical role in constructing the idea that individual angle measurements correspond to unique ratios of side lengths that remain constant, regardless of the dilation of the triangle.

Important Questions

Use the following as you plan how to elicit and model student strategies.

- *What are the angle measures in this triangle? How do you know? How does the triangle relate to the square? How do the dots help?*
- *What are the relationships in a right triangle between the side lengths and the length of the hypotenuse (the Pythagorean Theorem)?*
- *How can you use the smaller triangle to help you find the missing length(s) in the bigger triangles?*
- *Given a side length of x, what is the length of the hypotenuse in a* 45°, 45°, 90° *triangle?*
- *Which angle in this triangle is* 30°? *Which is* 60°? *How do you know?*
- *Given a short side length of x, what is the length of the hypotenuse and the long side in a* 30°, 60°, 90° *triangle?*

How would you summarize some of the things that came up in this string today?

- *You can figure out the missing side and/or hypotenuse lengths in a special right triangle.*
- *The shortest side of a triangle is opposite the smallest angle. The longest side of a triangle is opposite the largest angle. There are* 180° *in a triangle.*
- *The relationships in a* 45°, 45°, 90° *triangle are* x, x, $x\sqrt{2}$.
- *The relationships in a* 30°, 60°, 90° *triangle are* x, $2x$, $x\sqrt{3}$.
- *The relationships of side lengths of a triangle stay the same regardless of the size of the triangle.*
- *You can use the Pythagorean Theorem or visualize the smaller unit triangles inside the triangle to reason about the hypotenuse length.*

Sample Final Display

Your display could look like this at the end of the problem string:

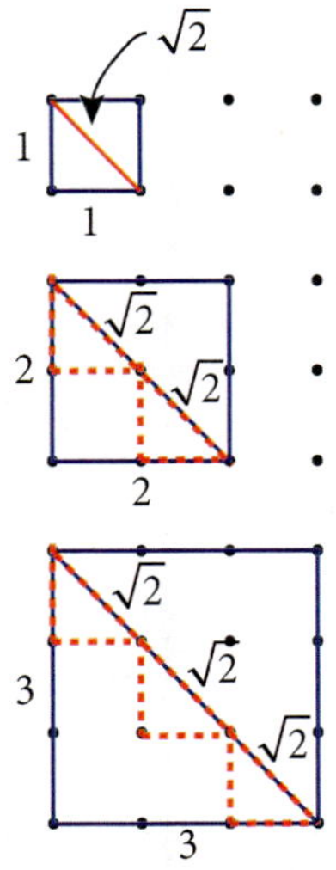

$1^2 + 1^2 = c^2$

$2 = c^2$

$c = \sqrt{2}$

$2^2 + 2^2 = c^2$

$8 = c^2$

$c = \sqrt{8} = 2\sqrt{2}$

$3^2 + 3^2 = c^2$

$18 = c^2$

$c = \sqrt{18} = 3\sqrt{2}$

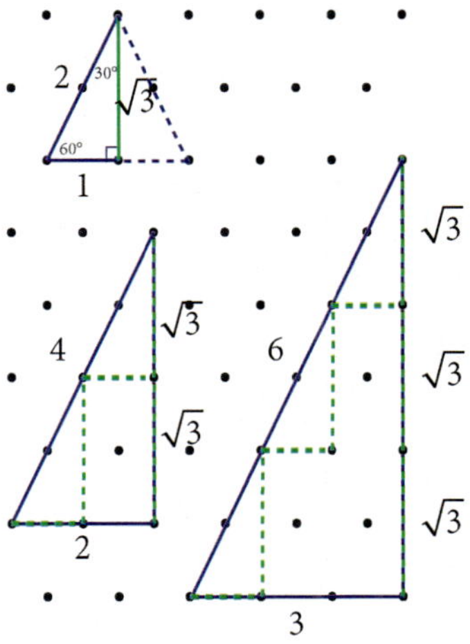

$1^2 + b^2 = 2^2$

$b^2 = 3$

$b = \sqrt{3}$

$2^2 + b^2 = 4^2$

$b^2 = 12$

$b = \sqrt{12} = 2\sqrt{3}$

$3^2 + b^2 = 6^2$

$b^2 = 27$

$b = \sqrt{27} = 3\sqrt{3}$

In a 45°, 45°, 90°:
What if the side length is 6? Hypotenuse is $6\sqrt{2}$
What if the side length is 21? Hypotenuse is $21\sqrt{2}$
What if the side length is x? Hypotenuse is $x\sqrt{2}$

In a 30°, 60°, 90°:
What if the short side is 6? Hypotenuse is 12 and side is $6\sqrt{3}$
What if the short side is 21? Hypotenuse is 42 and side is $21\sqrt{3}$
What if the short side is x? Hypotenuse is $2x$ and side is $x\sqrt{3}$

(continued)

Facilitation Notes

This version of the problem string lists short notes for important teacher moves during the string. After you've done the string yourself and studied the relationships involved, you might make similar notes for the things you want a reminder of or deem important.

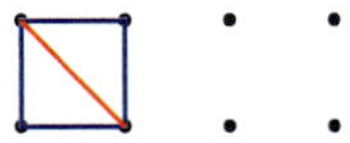

I'm going to draw a unit square using these dots. How long is this diagonal?
What do you know about the diagonal? This triangle? Label 45°, 45°, 90°.
Record Pythagorean theorem.
Label the side and diagonal lengths.

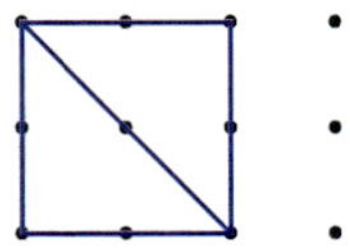

Side length 2. Hypotenuse of the triangle?
Compare Pythagorean Theorem results to previous.
Draw the similar interior unit triangles.

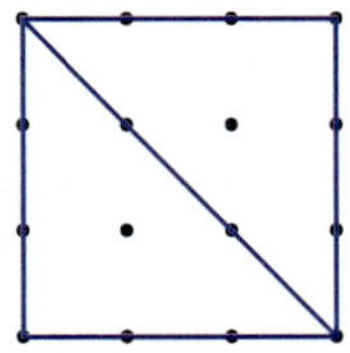

What about a 45°, 45°, 90° triangle that has a side length of 3?
Compare Pythagorean Theorem results to the 1 x 1 square.
Draw the similar interior unit triangles.

In a 45°, 45°, 90°, what if the:
side length is 6, hypotenuse?
side length is 21, hypotenuse?
side length is x, hypotenuse?

What pattern are you seeing?
In a 45°, 45°, 90° triangle:
Without drawing, if the side length is 6, what is the hypotenuse? Quick.
Repeat w 21. Quick.

Let's get a little general, if the side length of a 45°, 45°, 90° triangle is x, what is the length of the hypotenuse?

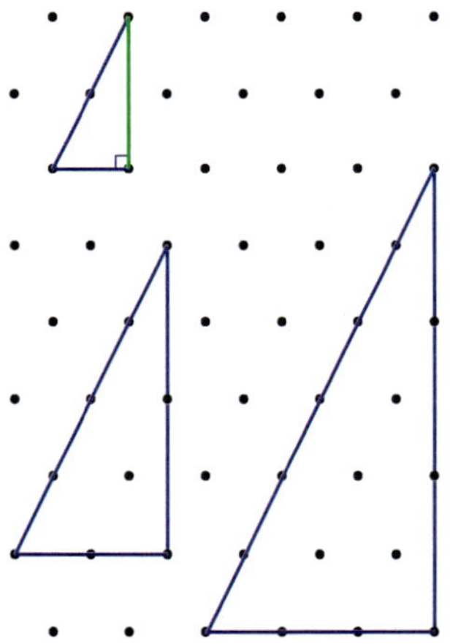

Different dot paper allows us to draw triangles with different angles.
What are the angle measures? How do you know? What lengths do you know?
Would drawing a copy of this triangle next to it and reflected help?
Missing side lengths?
Pythagorean Theorem.
What is the shortest side? Longest side?
How do they compare to the hypotenuse in length?

What lengths do you know in this second triangle (short side 2)?
Missing side length?
Compare the Pythagorean Theorem results to using the previous triangle.
Draw the similar interior unit triangles.

What about a 30°, 60°, 90° triangle that has a short side length of 3?
Missing side length?
Compare the Pythagorean Theorem results to triangle w short side 1.
Draw the similar interior unit triangles.

In a 30°, 60°, 90°, what if the:
short side is 6, hypotenuse? long side?
short side is 21, hypotenuse? long side?
short side is x, hypotenuse? long side?

What pattern are you seeing?
In a 30°, 60°, 90° triangle:
Without drawing, if the side length is 6, what is the hypotenuse? Quick.
Repeat w 21. Quick.

Let's get a little general, if the short side length of a 30°, 60°, 90° triangle is x, what is the length of the hypotenuse, the long side?
What if you know the hypotenuse is x, what are the short and long sides?

Dot Paper to Display

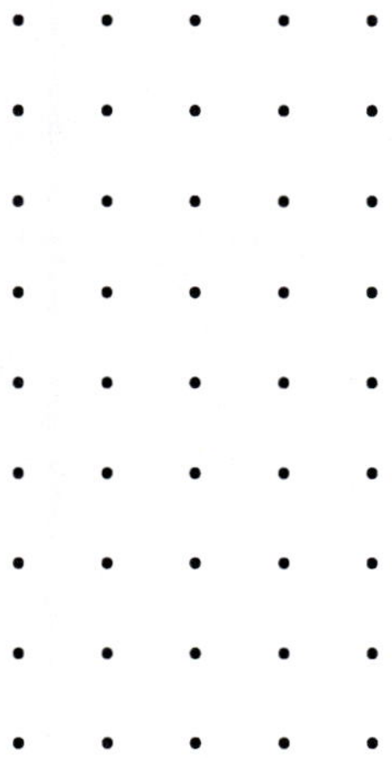

PROBLEM STRINGS LESSON

7.2 Using Reference Angles

At a Glance

Find the sine, cosine, and tangent of the angle created by these points. What is the angle?

(3, 4)

(−3, 4)

(−3, −4)

(3, −4)

Objectives

The goal of this problem string is for students to use ordered pairs on the coordinate grid to describe angles, formed by rotating a ray along the positive x-axis about the origin, and three trigonometric values of those angles. This provides a bridge from right triangle trigonometry to the unit circle and trigonometric functions in later lessons.

Placement

This problem string takes what students know about right triangle trigonometry and uses it to bridge to angles in the coordinate grid. Students should have prior experience finding right triangle trigonometry values and the Pythagorean Theorem to find the distance between points.

You can use this problem string to introduce or support textbook Lesson 7.2 Extending Trigonometry.

Guiding the Problem String

This problem string consists of four points, where students find the sine, cosine, and tangent for the angle formed by rotating a ray along the positive x-axis about the origin until the ray intersects the given point. Students also find the measure of the angle formed using arc functions with technology. If students have never been introduced to reflex angles, be prepared to have students describe spinning or rotating more than 180 degrees.

Since the absolute value of the coordinates are the same, the absolute value of the sine, cosine, and tangent will be the same. Don't tell students that. Help them realize it as they work through the string by asking them to notice patterns. After the first problem, ask students to predict and then check their predictions. Press them for justification about why their predictions were correct or not.

As the string progresses, ask students to generalize the relationships they are developing. Create an anchor chart to help students organize their thoughts and solidify the learning.

About the Mathematics

Reference angles are the smallest angle between the terminal side and the x-axis. The absolute value of the sine, cosine, and tangent of an angle are equivalent to the absolute value of the same trigonometric relationship of the reference angle.

Sample Interactions

Use the following as you plan how to elicit and model student strategies. This is not meant as a script, but as a view into the relationships involved and the intent of the problem string.

Teacher: *We've just been reviewing right triangle trigonometry. Let's do a little more of that with a twist. In today's string, we'll start with the point (3, 4). Envision a ray at the origin, along the x-axis to the right. If that ray rotated, keeping the end point at the origin, until it intersected the point (3, 4), can you picture a triangle formed?*

The teacher is pointing and using motions to simulate the angle opening up to intersect (3, 4).

Can you see how it could form a triangle with part of that ray, one side is the x distance, the other side is the y distance, and the hypotenuse is the length? Basically, we just created a triangle on the coordinate plane by opening up an angle until it intersected the point. We'll call that angle θ.

What can you tell me about this triangle? Can you tell what any of the angles are?

Student: *It's a right triangle. That angle on the right is a right angle because that's how points work. The four means it's four units straight up.*

Student: *We know the angle that you opened up is less than 90 degrees because the axes meet at 90 degrees.*

Student: *The angle has to be more than 45 degrees because it would be at 45 degrees if the point was (3, 3), but it's not, it's higher.*

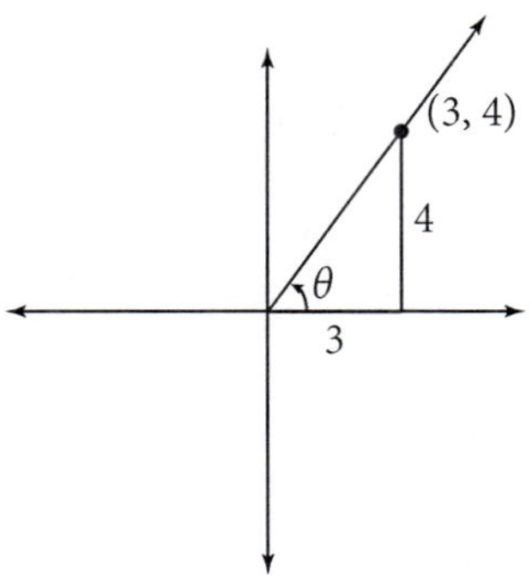

Teacher: *We're going to call the angle that opened up θ. You've got four tasks now. You can do them in any order. Find the $\sin\theta$, $\cos\theta$, $\tan\theta$ and the measure of the angle θ.*

Students work and the teacher circulates, looking for students who have chosen to start in different places and who found the angle differently.

(3, 4) $\quad \sin\theta \quad \cos\theta \quad \tan\theta \quad \theta$

Teacher: *Did anyone not start with the sine or cosine? Please tell us about that.*

Student: *I started with the tangent because we already know the opposite and adjacent, so the tangent is four-thirds.*

Teacher: *The tangent of θ is the ratio 4 to 3? Did anyone get anything different? No? You guys are groaning. What?*

Student: *We didn't even think that you could just find the tangent without having to do anything.*

Teacher: *Nice insight. And did I see that you found the angle then?*

Student: *Yes, we did the arctan of four-thirds in our calculator and got 53.13 degrees.*

Teacher: *So the* $\arctan\left(\frac{4}{3}\right) = 53.13°$.

$$\tan\theta = \frac{\text{opposite}}{\text{adjacent}} = \frac{4}{3}$$

$$\arctan\left(\frac{4}{3}\right) = 53.13°$$

(continued)

Teacher: *Who started with finding the sine?* **Student:** *I started with the sine and knew I needed the hypotenuse. So I did 3 squared plus 4 squared and that's 9 and 16 is 25. So that's 5 squared.* **Teacher:** *You used the Pythagorean Theorem. So the hypotenuse is 5? How does that help?*	$3^2+4^2=9+16=25=5^2$
Student: *The sine of the angle is opposite over hypotenuse so it's four-fifths.* **Teacher:** *The sine is the ratio of 4 to 5 or four-fifths to one? Everyone agree?*	$\sin\theta=\dfrac{\text{opposite}}{\text{hypotenuse}}=\dfrac{4}{5}$
Teacher: *Did anyone find the hypotenuse a different way? I'm just curious.* **Student:** *I just know that there's a Pythagorean triple that's 3, 4, 5.* **Teacher:** *That sounds handy. How about the cosine?* **Student:** *It's just adjacent over hypotenuse, so three-fifths.* **Teacher:** *So, the cosine of θ is the ratio 3 to 5 or three-fifths to 1. You are noticing that I am always saying the sine, cosine, and tangent as ratios. That's what they are. Sometimes we get lazy and just say three-fifths instead of three-fifths to one.*	$\cos\theta=\dfrac{\text{adjacent}}{\text{hypotenuse}}=\dfrac{3}{5}$

Teacher: *So, the students who found the tangent first, used arctangent to find the angle. Did those of you who found sine and cosine first use those to find the angle? Could we?*

Student: *I used arcsine.*

Student: *I used arccosine.*

Teacher: *Did you get the same result? The same angle measure? Yes? I wonder if that will always be true?*

(3, 4) $\quad 3^2+4^2=9+16=25=5^2$

$\sin\theta=\dfrac{\text{opposite}}{\text{hypotenuse}}=\dfrac{4}{5} \qquad \cos\theta=\dfrac{\text{adjacent}}{\text{hypotenuse}}=\dfrac{3}{5}$

$\tan\theta=\dfrac{\text{opposite}}{\text{adjacent}}=\dfrac{4}{3}$

$\arctan\left(\tfrac{4}{3}\right)=53.13^\circ \qquad \arcsin\left(\tfrac{4}{5}\right)=53.13^\circ \qquad \arccos\left(\tfrac{3}{5}\right)=53.13^\circ$

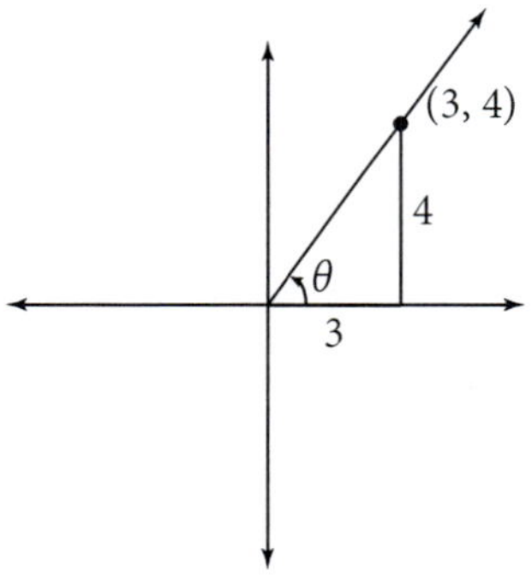

Teacher: *The next problem in our string is (−3, 4). Let's plot that and chat about the angle briefly.* *What quadrant is the point in? Yep, the second quadrant.* *So, if the point is over here at (−3, 4), where is the angle? It still starts over here anchored at the origin and laying on the x-axis toward the right. Tell me about this angle θ.* **Student:** *It's bigger than the last one. Its' more than 90°. It's less than 180°* **Teacher:** *Find the same things, sine, cosine, tangent, and the measure of the angle formed if we start a ray at the origin along the x-axis and open it to the point (−3, 4)?* Students work while the teacher circulates, looking for students who are starting to make connections between this and the previous problem, especially the location and use of the reference angle. While circulating, the teacher helps to clear up any confusion about the signs, that the −3 does not represent length, but position.	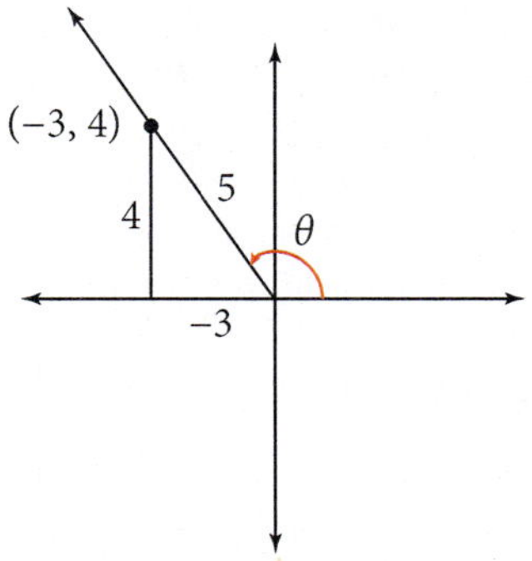
Teacher: *Who thinks they have a nice place to start. Even if it's not where you started, but you want your brain to go there next time you hit a problem like this.* **Student:** *We think it would be easy—no work—to start with the sine, because it's going to be the same, 4 to 5. Oh, I could've said that we already knew that it is the same triangle, 3, 4, 5.*	$\sin\theta = \frac{4}{5}$
Teacher: *That could be a fine place to start. Let's quickly get up on the board the other two trig values and then talk about that angle measure that seems to have you talking.*	$\cos\theta = \frac{-3}{5}$ $\tan\theta = \frac{4}{-3} = -\frac{4}{3}$
Teacher: *Okay, tell me about the angle measure. What's going on?* **Student:** *We did the arc things and got mixed results.* **Student:** *What do you mean? We just knew that because we could use exactly the same arcsine as the first problem, that the angle would be, oh, wait. That doesn't make sense. It's not the same angle as the first one. This angle is much bigger. Okay, I'm listening.* **Student:** *This is what we got. We agree that arcsine would give the same result, arccosine is 126.87° and arctangent is -53.13! What does that even mean?*	$\arcsin\left(\frac{4}{5}\right) = 53.13°$ $\arccos\left(\frac{-3}{5}\right) = 126.87°$ $\arctan\left(\frac{4}{-3}\right) = -53.13°$

(continued)

Teacher: *Anyone have any ideas on what that negative angle means? Mathematicians say that if you open up the angle from the x-axis like we have been doing by rotating the ray counter clockwise, that is a positive angle measure, but if you open it clockwise, that is a negative angle measure. So, an angle of measure −53.13° would be here, in the fourth quadrant.*

Does that help?

Student: *Well, that's interesting.*

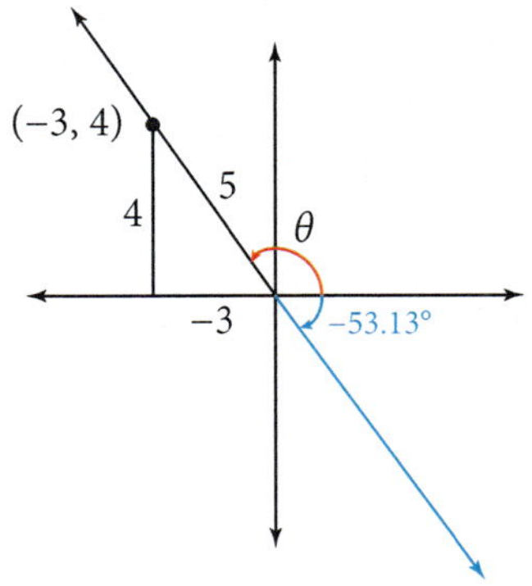

Teacher: *Is there anything we can say using vertical angles here?*

Student: *Yes, the angle across from the −53.13°, the one in the second quadrant, the one in the triangle, that is the same size.*

Teacher: *We call that angle in the second quadrant the reference angle.*

Student: *Oh, and then that agrees with the arccosine.*

Student: *What do you mean?*

Student: *The arccosine gave us 126.87°. That's like the other part to the 180°. Yeah, 180° – 53.13° is 126.87°.*

Student: *And I can see that 90° – 53.13° is 36.87°. So the first quadrant 90° plus the 36.87°*

Student: *I can see the connections but why would the calculator do that?*

Teacher: *That's a really good question. Let's leave that hanging for a few minutes. What did we decide θ is?*

Student: 126.87°.

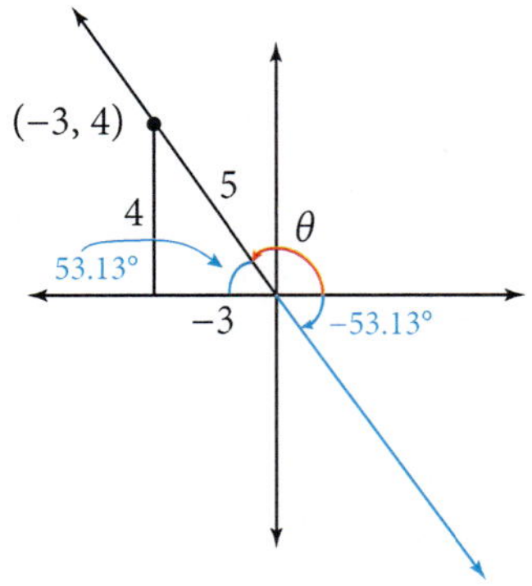

Teacher: *Take a look at the work we have up so far. What patterns are you noticing?*

Student: *I am noticing that the values for the opposite side of the triangle are the same for both angles so far, so the arcsine was the same. But the values for both arccosine and arctangent are different than the first angle. That's because the 5 hypotenuse is the same, but the adjacent side is 3 in the first problem and −3 in the second problem.*

Student: *So, any trig thing that deals with the adjacent side will be different.*

Student: *The trig things will be negatives. And the angles will be different.*

Teacher: *Let's keep all of those ideas in mind.*

(−3, 4) still a 3, 4, 5 triangle

$\sin\theta = \frac{4}{5}$ $\cos\theta = \frac{-3}{5}$ $\tan\theta = \frac{4}{-3} = -\frac{4}{3}$

$\arcsin\left(\frac{4}{5}\right) = 53.13°$ $\arccos\left(\frac{-3}{5}\right) = 126.87°$ $\arctan\left(\frac{4}{-3}\right) = -53.13°$

reference angle is 53.13°

$\theta = 180° - 53.13° = 126.87°$ $90 - 53.13° = 36.87°$ $\theta = 90° + 36.87° = 126.87°$

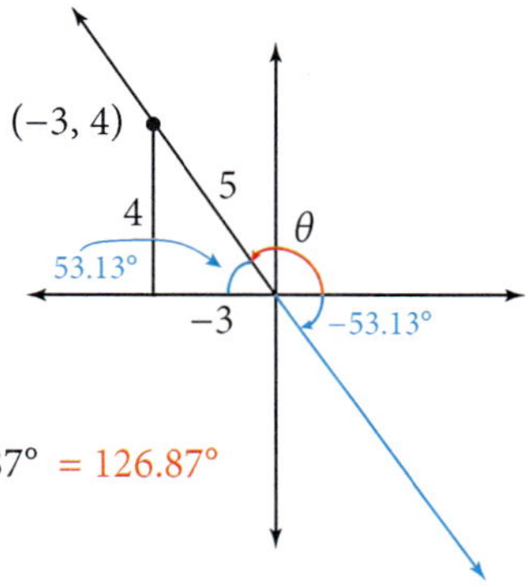

Teacher: *The next problem in our string is (−3, −4). Let's plot that and chat about the angle briefly.*

What quadrant is the point in? Yep, the third quadrant.

So, if the point is over here at (−3, −4), where is the angle? It still starts over here anchored at the origin and laying on the x-axis toward the right. Tell me about this angle θ.

Student: *It's bigger than the last one. It's more than 180°. It's less than 270°.*

Teacher: *We are going to find the same things, sine, cosine, tangent, and the measure of the angle formed if we start a ray at the origin along the x-axis and open it to the point (−3, −4).*

First, let's do some predicting. Turn to your partner and without actually figuring anything or using your calculator, what might stay the same and what might be different and why?

Students talk while the teacher listens in.

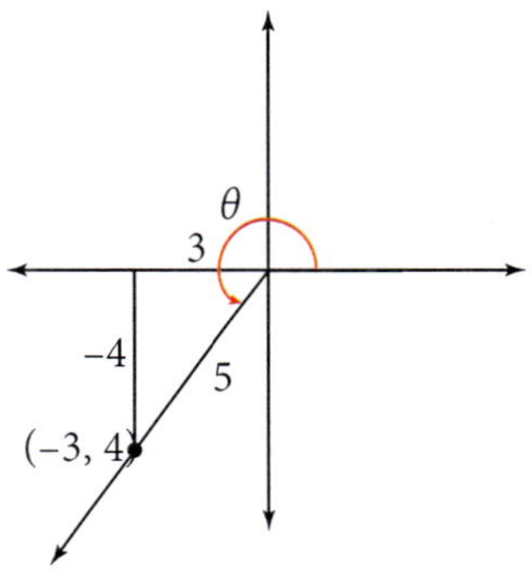

(continued)

Teacher: *If you haven't already, go ahead and start finding the sine, cosine, tangent, and the measure of the angle. Then we'll talk about your predictions and what you find.*

Students work while the teacher circulates. Then the teacher crafts a conversation about the signs of the trigonometric values based on the third quadrant point and how the reference angle is used.

$(-3, -4)$ still a 3, 4, 5 triangle

$\sin\theta = \frac{-4}{5}$ $\cos\theta = \frac{-3}{5}$ $\tan\theta = \frac{-4}{-3} = \frac{4}{3}$

reference angle is 53.13°

$\arcsin\left(\frac{-4}{5}\right) = -53.13$ $\arccos\left(\frac{-3}{5}\right) = 126.87°$ $\arctan\left(\frac{4}{3}\right) = 53.13°$

$\theta = 180° + 53.13° = 133.13°$

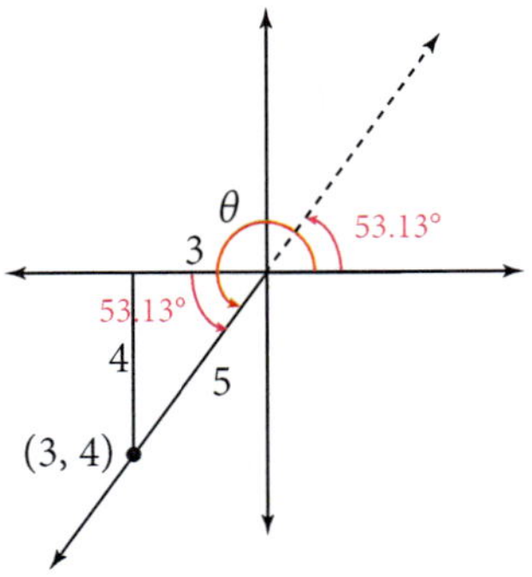

The teacher helps students begin or continue to generalize some important relationships.

Teacher: *Let's look back a bit. We seem to be using the x- and y-coordinates a bunch in our trig values. If we are putting points in the coordinate grid, tell me about the coordinates and the sides on the triangles.*

As students suggest the relationships, the teacher starts to make an anchor chart like the following.

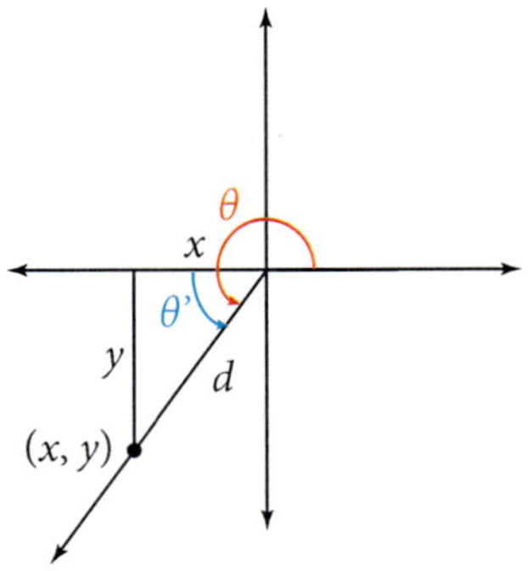

$\cos\theta = \frac{x}{d}$

$\sin\theta = \frac{y}{d}$

$\tan\theta = \frac{y}{x}$

The teacher gives the last problem and students predict, use patterns to find the sign of the trig values, and use the reference angle to find the angle.

Teacher: *Anyone want to guess what the last problem is in today's string? Yes! The point* (3, −4). *What do you know that can help you? Try to reason through all of the trig values and the angle by using what we have already found today.*

Students work briefly and then the teacher crafts a conversation about using the previous problems and records student thinking.

Where can you find a cosine that is equivalent? Why?

Where can you find a sine that is equivalent? Why?

Tangent? Why?

How can you use the reference angle to help you?

(3, −4) still a 3, 4, 5 triangle

$\sin\theta = \frac{-4}{5}$ $\cos\theta = \frac{3}{5}$ $\tan\theta = \frac{-4}{3}$

reference angle is 53.13°

$\arcsin\left(\frac{-4}{5}\right) = -53.13$ $\arccos\left(\frac{3}{5}\right) = 53.13°$ $\arctan\left(\frac{-4}{3}\right) = -53.13°$

$\theta = 360° - 53.13° = 306.87°$ $\theta = 270° + 36.87° = 306.87°$

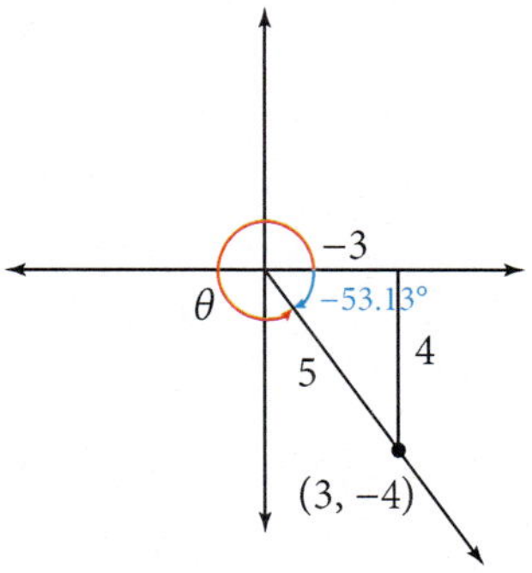

The teacher asks students to generalize about the signs, positive or negative, of the trigonometric values based on the quadrant the point is in.

Teacher: *Where is cosine positive? Why?*

Where is sine positive? Why?

Where is tangent positive? Why?

How could we represent this thinking on our anchor chart?

The teacher continues to make an anchor chart to help organize and solidify the learning as students reason through it.

$\cos\theta = \frac{x}{d}$

$\sin\theta = \frac{y}{d}$

$\tan\theta = \frac{y}{x}$

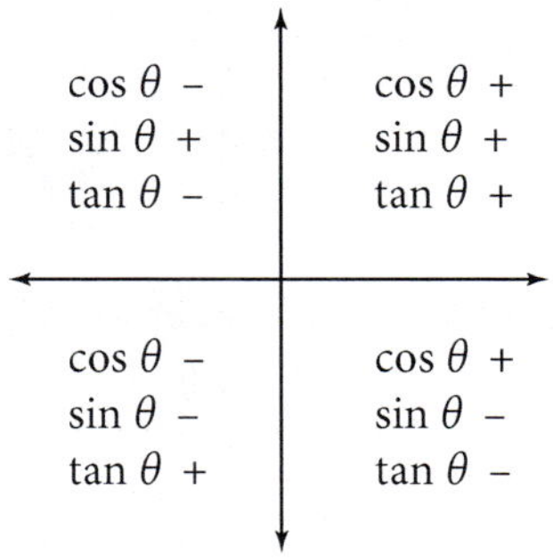

Teacher: *How would you summarize some of the things that came up in this string today?*

Elicit the following:

- *Points can correspond to angles that result from rotating a ray along the positive x-axis about the origin.*
- *The absolute value of trigonometric values are the same for points with coordinates that have the same absolute value.*
- *The sign of the trigonometric values depends on which quadrant the point is in.*
- *The calculator gives values for reference angles. We need to use the reference angles to find the angle measure.*
- *We do not need to memorize signs and quadrants because we can reason about it.*

(continued)

Sample Final Display

Your display could look like this at the end of the problem string:

(3, 4) $3^2+4^2=9+16=25=5^2$

$\tan\theta=\frac{\text{opposite}}{\text{adjacent}}=\frac{4}{3}$ $\sin\theta=\frac{\text{opposite}}{\text{hypotenuse}}=\frac{4}{5}$ $\cos\theta=\frac{\text{adjacent}}{\text{hypotenuse}}=\frac{3}{5}$

$\arctan\left(\frac{4}{3}\right)=53.13°$ $\arcsin\left(\frac{4}{5}\right)=53.13°$ $\arccos\left(\frac{3}{5}\right)=53.13°$

$\theta=53.13°$

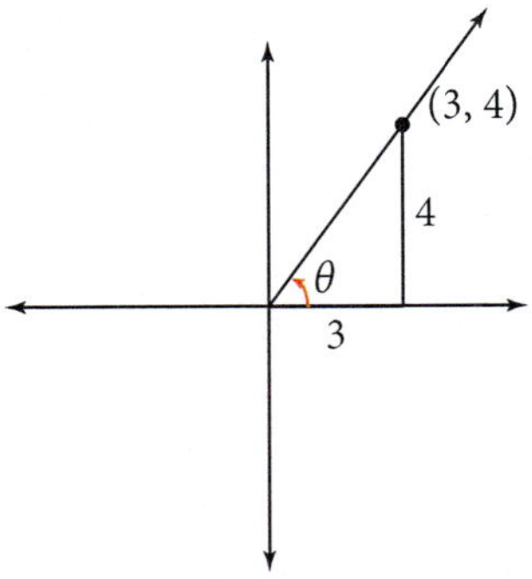

(−3, 4) still a 3, 4, 5 triangle

$\sin\theta=\frac{4}{5}$ $\cos\theta=\frac{-3}{5}$ $\tan\theta=\frac{4}{-3}=-\frac{4}{3}$

$\arcsin\left(\frac{4}{5}\right)=53.13°$ $\arccos\left(\frac{-3}{5}\right)=126.87°$ $\arctan\left(\frac{4}{-3}\right)=-53.13°$

reference angle is 53.13°

$\theta=180°-53.13°=126.87°$ $90-53.13°=36.87°$ $\theta=90°+36.87°=126.87°$

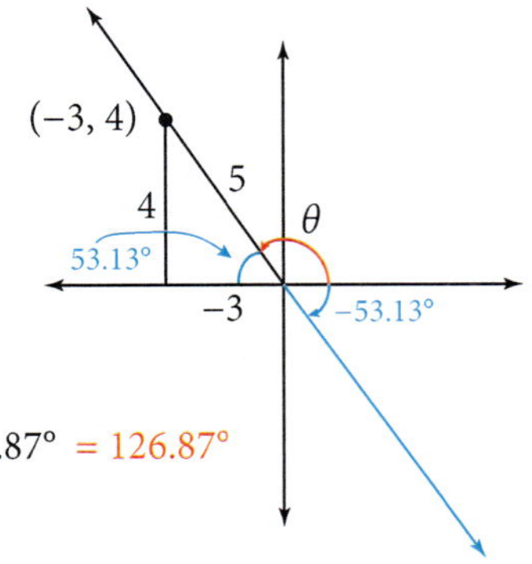

(−3, −4) still a 3, 4, 5 triangle

$\sin\theta=\frac{-4}{5}$ $\cos\theta=\frac{-3}{5}$ $\tan\theta=\frac{-4}{-3}=\frac{4}{3}$

reference angle is 53.13°

$\arcsin\left(\frac{-4}{5}\right)=-53.13$ $\arccos\left(\frac{-3}{5}\right)=126.87°$ $\arctan\left(\frac{4}{3}\right)=53.13°$

$\theta=180°+53.13°=133.13°$

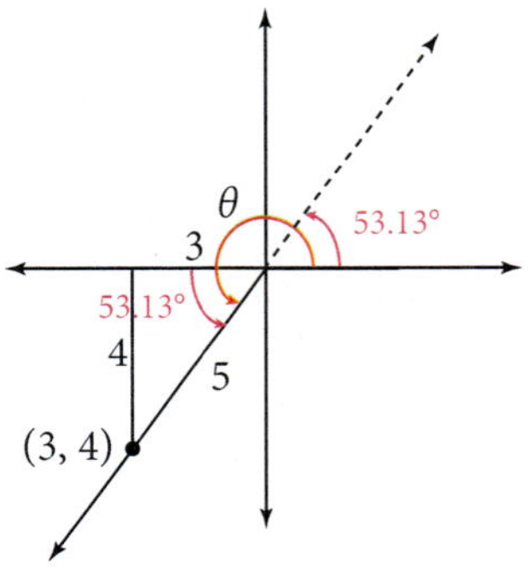

(3, −4) still a 3, 4, 5 triangle

$\sin\theta=\frac{-4}{5}$ $\cos\theta=\frac{3}{5}$ $\tan\theta=\frac{-4}{3}$

reference angle is 53.13°

$\arcsin\left(\frac{-4}{5}\right)=-53.13$ $\arccos\left(\frac{3}{5}\right)=53.13°$ $\arctan\left(\frac{-4}{3}\right)=-53.13°$

$\theta=360°-53.13°=306.87°$ $\theta=270°+36.87°=306.87°$

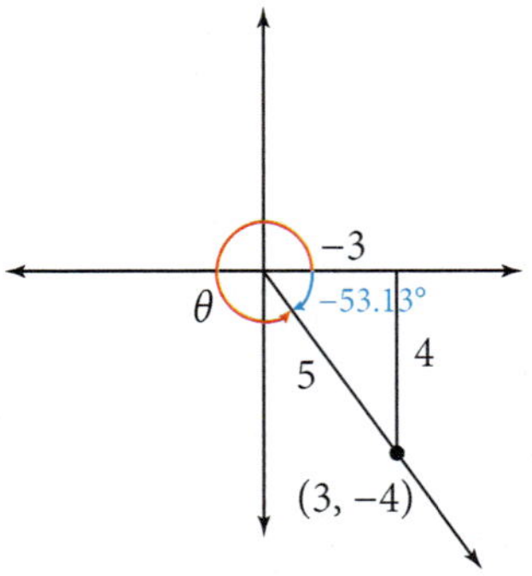

Facilitation Notes

This version of the problem string lists short notes for important teacher moves during the string. After you've done the string yourself and studied the relationships involved, you might make similar notes for the things you want a reminder of or deem important.

(3, 4)	*Let's keep reviewing right triangle trig with a twist.* *If we rotate a ray anchored at the origin along the positive x-axis until it intersects w/ this point, and create this angle, call it θ (theta), what can you tell me about the angle?* *Find sinθ, cosθ, tanθ and θ, in any order.* *Which did you find first and why?* *The hypotenuse? How did you find it? Pythagorean Theorem or triple.* *Say ratios as ratios (3/5 as 3 to 5, not three-fifths).* *How did you find θ? Same no matter which arc function? I wonder if that will always be true...*
(−3, 4)	*Repeat.* *Why are the results of different arc functions not the same?* *Reference angle.* *Define a negative angle as it comes up.* *Why does a calculator do that? Let's hang on to that for now.*
(−3, −4)	*Repeat.* *Look at all problems. What patterns are you finding? Note equivalence of absolute values of trig values. Note sign differences. Note connection between triangle side lengths and coordinates.* *Start anchor chart using coordinates.*
(3, −4)	*Based on patterns, predict first. Then find all.* *Continue anchor chart with sign of trig values in different quadrants.*

Sample Anchor Chart

Your anchor chart could look like this at the end of the problem string:

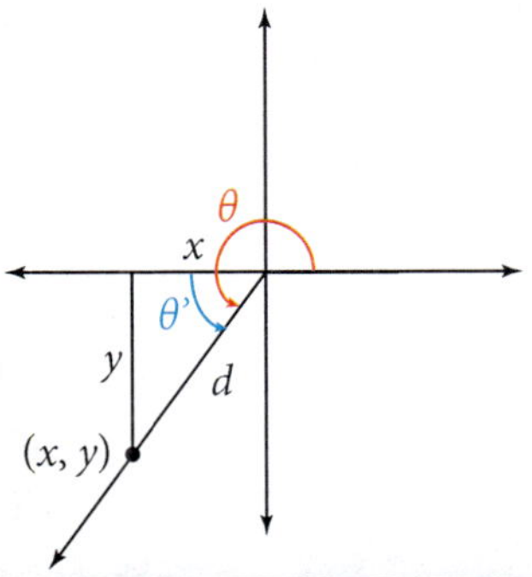

$$\cos\theta = \frac{x}{d}$$

$$\sin\theta = \frac{y}{d}$$

$$\tan\theta = \frac{y}{x}$$

cos θ − sin θ + tan θ −	cos θ + sin θ + tan θ +
cos θ − sin θ − tan θ +	cos θ + sin θ − tan θ −

PROBLEM STRINGS LESSON

7.3 Quadrant Angles

At a Glance	
Find the sine, cosine, and tangent of: 0° 90° 180° 270°	**Objectives** The goal of this problem string is to get students to use reasoning about right triangles to find the sine, cosine, and tangent of the quadrant angles and foreshadow the periodicity of the sine, cosine, and tangent functions. **Placement** After students have begun to generalize about the three trigonometry relationships in angles in standard form in the coordinate grid, they sometimes struggle to make sense of the special quadrant angles. This short string helps students reason using triangles that are just smaller or larger than the quadrant angles together with the right triangle trigonometry definitions to reason about the trigonometry values of the quadrant angles. For this string, students need prior experience with right triangle trigonometry and angles in standard position in the coordinate grid. You can use this problem string to follow up textbook Lesson 7.2 Extending Trigonometry or to introduce Lesson 7.3 Defining the Circular Functions. **Guiding the Problem String** In this short problem string of just four problems, ask students to reason about the sine, cosine, and tangent using angles that are just smaller or just bigger than the quadrant angles to make sense of the values connected to the right triangle trigonometry they are used to. Help connect the reasoning with the anchor chart from problem string 7.2. Encourage students to reason about the generalizations on the anchor chart rather than rote-memorization. You might refer to the work you did with rational functions to help students make sense of the fact that division by zero is undefined. Look specifically at values close to zero for the parent $y = \frac{1}{x}$. As the string progresses, encourage students to predict first and then reason about their predictions in relation to the actual values, if needed. **About the Mathematics** The quadrant angles as measured in degrees are important anchor points in the circular functions and as measured in radians for the real number functions. These values of −1 and 1 provide the extreme values of the sine and cosine and the value of 0 is the midline of those same functions.

Important Questions

Use the following as you plan how to elicit and model student strategies.

- *If 90° doesn't make a reference triangle, is there a point close to the y-axis we can use to make a triangle? In quadrant I? Quadrant II? Can this help us reason about* sin 90°?
- *Looking back at all of the sine, cosine, and tangent values, what is the largest value we just found? Why is* 1 *the largest value?*
- *What is the smallest value we just found? Why is* −1 *the smallest value?*

- *What is happening in between the −1 and 1, as you open up the angle from quadrant angle to the next quadrant angle, in the sine function? How do you know? How are you using ratios to help you?*
- *What happens in between the −1 and 1 in the cosine function, as you open up the angles from quadrant angle to the next quadrant angle? How do you know? How are you using ratios to help you?*
- *What happens in between the −1 and 1 in the tangent function, as you open up the angles from quadrant angle to the next quadrant angle? How do you know? How are you using ratios to help you?*
- *How often will sine repeat itself?*
- *How often will cosine repeat itself?*
- *How often will tangent repeat itself?*

How would you summarize some of the things that came up in this string today?

- *You can reason about the trigonometric values of the quadrant angles by creating pseudo triangles in standard position.*
- *It might help to reason about angles slightly smaller or larger than the given quadrant angle.*
- *We think that the trigonometric values repeat themselves. We think that sine and cosine repeat every 360°. We are not sure about the tangent but we are thinking it repeats too.*
- *Division by zero is undefined. Zero divided by anything (except 0) is zero.*

(continued)

Sample Final Display

Your display could look like this at the end of the problem string:

0°

$$\cos\theta = \frac{x}{d} = \frac{1}{1} = 1$$

$$\sin\theta = \frac{y}{d} = \frac{0}{1} = 0$$

$$\tan\theta = \frac{y}{x} = \frac{0}{1} = 0$$

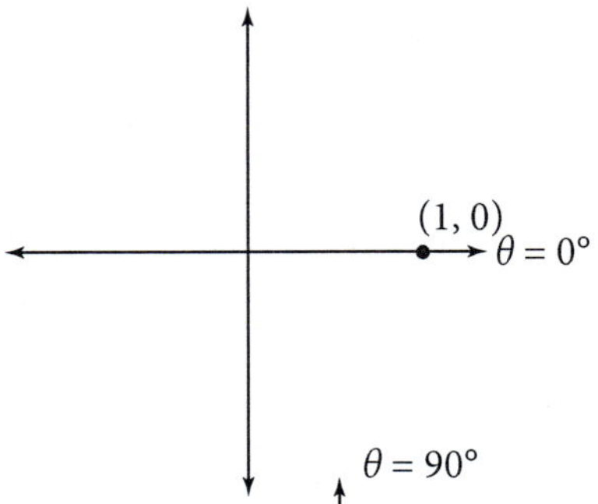

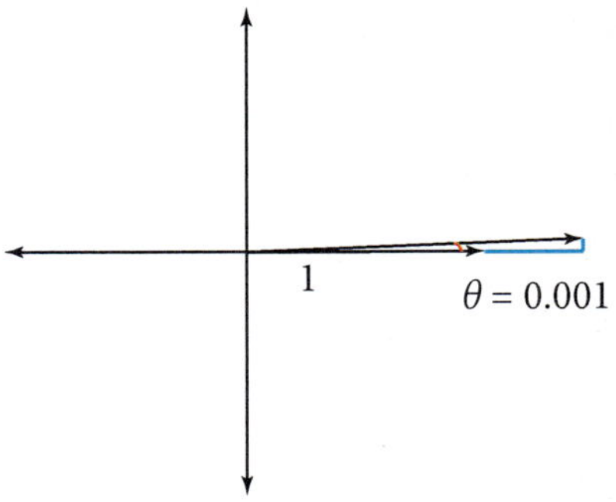

90°

$$\cos\theta = \frac{x}{d} = \frac{0}{1} = 0$$

$$\sin\theta = \frac{y}{d} = \frac{1}{1} = 1$$

$$\tan\theta = \frac{y}{x} = \frac{1}{0} \text{ undefined}$$

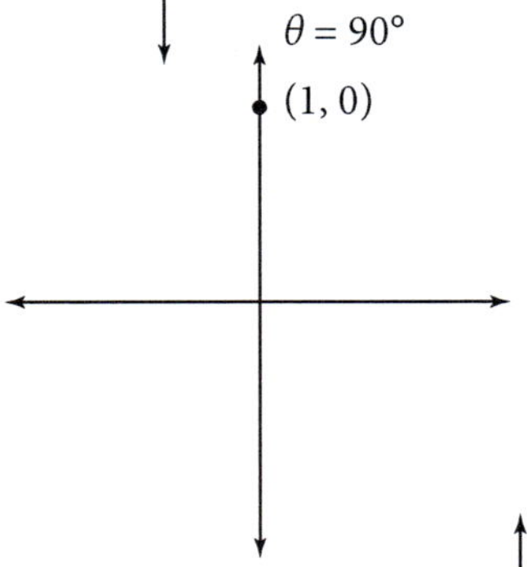

180°

$$\cos\theta = \frac{x}{d} = \frac{-1}{1} = -1$$

$$\sin\theta = \frac{y}{d} = \frac{0}{1} = 0$$

$$\tan\theta = \frac{y}{x} = \frac{0}{-1} = 0$$

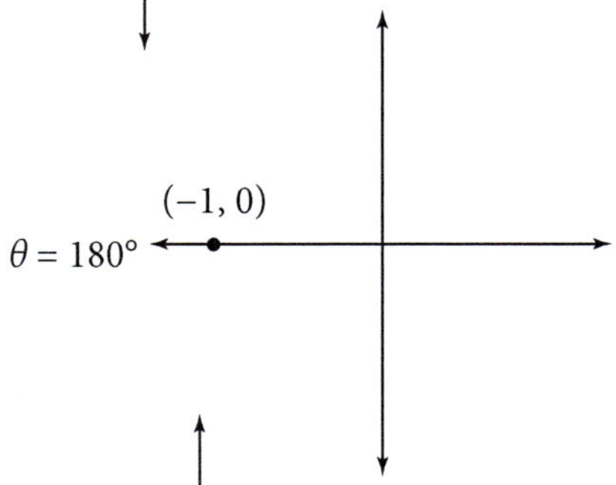

270°

$$\cos\theta = \frac{x}{d} = \frac{0}{1} = 0$$

$$\sin\theta = \frac{y}{d} = \frac{-1}{1} = -1$$

$$\tan\theta = \frac{y}{x} = \frac{-1}{0} \text{ undefined}$$

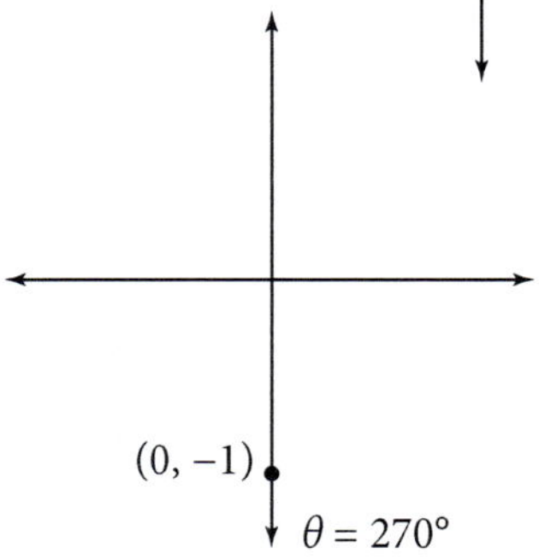

Facilitation Notes

This version of the problem string lists short notes for important teacher moves during the string. After you've done the string yourself and studied the relationships involved, you might make similar notes for the things you want a reminder of or deem important.

0°	*Let's find the cosine, sine, and tangent.* *The first angle is 0 degrees. Where is that? How can you make sense of the "triangle" for 0°?* *Look at a triangle that measures very small, like 0.001°. What are the lengths?* *What is 0 divided by anything? Why?* *Find the cosine, sine, and tangent for 0°*
90°	*Now let's look at a familiar angle, 90°. Where is the "triangle" for 90°?* *How will the horizontal side length of 0 affect the trig values? Which ones will it affect? Why?* *Find the cosine, sine, and tangent of 90°.* *What is anything divided by 0? Why?*
180°	*Where is 180°? Where is the "triangle" for 180°?* *How will the vertical side length of 0 affect the trig values? Which ones will it affect? Why?* *Find the cosine, sine, and tangent of 180°.*
270°	*Anyone want to guess the next problem? Yes, 270°!* *Before you do any figuring, look at the other angles and trig values. What do you predict? Why?* *Where is the "triangle" for 270°?* *Find the cosine, sine, and tangent for 270°.* *What will happen if we keep rotating the ray?* *What will happen when the ray gets to 360°, then 450°, 540°, 630° and so on? Why? Interesting...*

PROBLEM STRINGS
LESSON

7.4 Radian Measure

At a Glance

π

$\frac{\pi}{2}$

$\frac{3\pi}{2}$

$\frac{\pi}{4}$

$\frac{2\pi}{3}$

$\frac{7\pi}{4}$

$\frac{11\pi}{6}$

$\frac{5\pi}{3}$

$\frac{5\pi}{6}$

$-\frac{3\pi}{4}$*

*optional problem

Objectives

The goal of this problem string is to build students' facility with radian measure. The order of the problems encourages students to establish and then use landmark angles to reason about other angles in standard position.

Placement

This problem string is meant to help students connect their experience with angles in standard position measured in degrees to those same angles measured in radians. The string assumes that students have been introduced to the relationship between radians and degrees.

You can use this problem string after you have introduced the relationship between radian measure and degrees in textbook Lesson 7.4 Radian Measure.

Guiding the Problem String

The first three problems establish the important quadrant angles in radian measure. For the first 2–3 problems, model both the work of any students who are using the relationship between radians and degrees, and also students who are reasoning using radian benchmarks. Gradually shift to only modeling students who are reasoning given benchmark radian and degree measures.

For the angles that are only one unit fraction (like $\frac{\pi}{3}$, $\frac{\pi}{4}$, or $\frac{\pi}{6}$) away from π or 2π, look for and model an "over" strategy. For example, to reason about the location of $\frac{11\pi}{6}$, one could consider an angle that is just over it and back up, $\frac{12\pi}{6} - \frac{\pi}{6} = \frac{11\pi}{6}$.

You might also look to see if students are noticing the sort of "reciprocal looking" relationship between 30° and $\frac{\pi}{6}$, and between 60° and $\frac{\pi}{3}$. If no one is noticing, you might wonder about it aloud.

About the Mathematics

One radian is the measure of the angle that cuts off an arc whose length is equal to the radius of the circle. There are 2π radians in a circle. In this problem string, we assume that the angles are in standard position (the initial side is along the x-axis and the vertex is at the origin).

Convention often has teachers naming radian angles in ways that are less helpful to understand the size of the angle, such as naming $\frac{11\pi}{6}$ as "eleven π sixths." You might consider also naming it "eleven-sixths π" or "eleven-sixths of π."

The operator meaning of fractions, that $\frac{11\pi}{6}$ can be thought of as $\frac{11}{6}$ of π, is an important, but often not explicitly named relationship. Using unit fractions and multiplicative language such as thinking about $\frac{11}{6}$ of π as 11 one-sixths of π (or 11 times one-sixths of π) can help students think multiplicatively about the fractions involved. Using both of these meanings and being explicit about that use can help students reason with more sophistication about the fraction relationships.

Important Questions

Use the following as you plan how to elicit and model student strategies.

- *What are the four quadrant angles in radian measure?*
- *How can you use those benchmarks to help you find other angles in radian measure?*
- *How can you use your understanding of fractions to help you situate an angle in between two benchmarks?*
- *How can you use an angle that is just too big to help you reason about the location of an angle?*
- *How are you reasoning about 30°, 60°, $\frac{\pi}{3}$, and $\frac{\pi}{6}$?*
- *What are good benchmark radian measures?*

How would you summarize some of the things that came up in this string today?

- *Radians measure angles just like degrees, but instead of 0° to 360°, radians go from 0 radians to 2π radians.*
- *You can leave off the word "radian" or you can leave it on for clarity. You should always use the degree symbol when measuring in degrees.*
- *There are 2π radians in one circle.*
- *The quadrant angles are 0, $\frac{\pi}{2}$, π, $\frac{3\pi}{2}$, and 2π.*
- *We can reason about the location of angles in radian measure by comparing them to benchmarks we know.*
- *We can use an angle that is a bit too big to reason about an angle by subtracting from the bigger angle.*
- *Math teachers might say the radian measures in a way that is potentially confusing, but I can use what I know about fractions to reason about the meaning of angles like $\frac{11\pi}{6}$ as eleven-sixths of π.*
- *I can reason about the relationship between 30°, 60°, $\frac{\pi}{6}$, and $\frac{\pi}{3}$. One way is that 30° is smaller than 60° and $\frac{1}{6}$ of π is smaller than $\frac{1}{3}$ of π.*

(continued)

Sample Final Display

Your display could look like this at the end of the problem string:

π radians $\pi = \frac{\pi}{180}(n)$ half circle, straight line, π radians = 180°
$180° = n$

$\frac{\pi}{2}$ $\frac{\pi}{2} = \frac{\pi}{180}(n)$ $\frac{1}{2}$ of π, $\frac{1}{4}$ of a circle
can be referred to as π halves or half of π
$90° = n$

$\frac{3\pi}{2}$ $\frac{3}{2}$ of π or 1.5 π so 1.5 (180°) = 270°, $\frac{3}{4}$ of a circle
can be referred to as 3π halves or 3 halves of π

$\frac{\pi}{4}$ $\frac{1}{4}$ of π so $\frac{1}{4}$ of 180° = 45°, $\frac{1}{8}$ of a circle
can be referred to as π fourths or one-fourth of π

$\frac{2\pi}{3}$ $\frac{2}{3}$ of π, $\frac{1}{3}$ of π (180°) is 60° so $\frac{2}{3}$ is 120°, between $\frac{\pi}{2}$ and π
can be referred to as 2π thirds or two-thirds π
reference angle is $\frac{\pi}{3}$

$\frac{7\pi}{4}$ $\frac{7}{4}$ of π, $\frac{4}{4}$ π + $\frac{3}{4}$ π (180° + 135° = 315°) or $\frac{8}{4}$ π − $\frac{1}{4}$ π (360° − 45° = 315°)
between $\frac{6}{4}$ π and $\frac{8}{4}$ π
can be referred to as 7π fourths of seven-fourths π
reference angle is $\frac{\pi}{4}$

$\frac{11\pi}{6}$ $\frac{11}{6}$ of π, $\frac{6}{6}$ π + $\frac{5}{6}$ π (180° + 150° = 330°) or $\frac{12}{6}$ π − $\frac{1}{6}$ π (360° − 30° = 330°)
between $\frac{10}{6}$ π and $\frac{12}{6}$ π
can be referred to as 11π sixths or 11-sixths π
reference angle is $\frac{\pi}{6}$

$\frac{5\pi}{3}$ $\frac{5}{3}$ of π, $\frac{3}{3}$ π + $\frac{2}{3}$ π (180° + 120° = 300°) or $\frac{6}{3}$ π − $\frac{1}{3}$ π (360° − 60° = 300°)
between $\frac{4}{3}$ π and $\frac{6}{3}$ π
can be referred to as 5π thirds or five-thirds π
reference angle is $\frac{\pi}{3}$

$\frac{5\pi}{6}$ $\frac{5}{6}$ of π, $\frac{6}{6}$ π − $\frac{1}{6}$ π (180° − 30° = 150°)
between $\frac{4}{6}$ π and $\frac{6}{6}$ π
can be referred to as 5π sixths or five-sixths π
reference angle is $\frac{\pi}{6}$

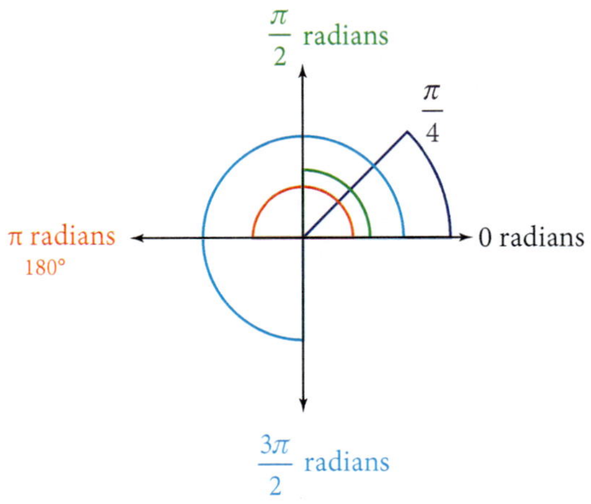

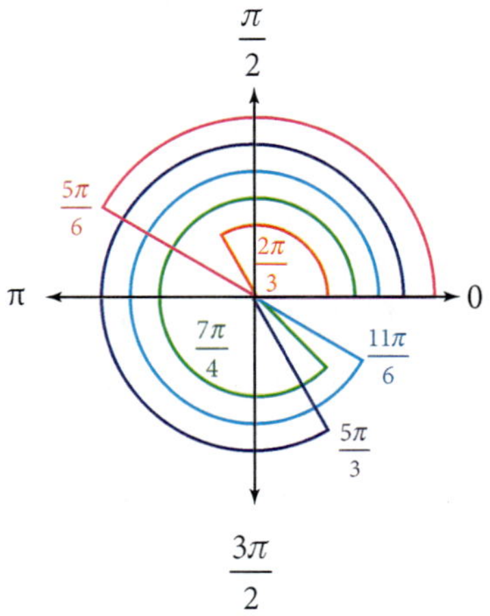

Facilitation Notes

This version of the problem string lists short notes for important teacher moves during the string. After you've done the string yourself and studied the relationships involved, you might make similar notes for the things you want a reminder of or deem important.

π radians	*We'll start with an angle in standard position. If it opens up to π radians, where is that?* *Find and model a degree conversion strategy.* *Find and model a strategy based on knowing that there are 2π radians in a circle, so half of the circle must be π.* *Label the angle in standard position. Draw in an arc as you motion opening up the angle.*
$\frac{\pi}{2}$	*Where is an angle that measures π halves, or half of π?* *Math teachers call this π halves, but it's just half of π.* *Share degree conversion strategy. Share using π or 2π.* *Label the angle. Draw in an arc as you motion opening up the angle.*
$\frac{3\pi}{2}$	*Repeat.* *Called 3π halves; it's 1.5 of π. Where's 1 π and another half of π?*
$\frac{\pi}{4}$	*Repeat.* *Called π fourths; it's ¼ of π. Where's 1 π, cut into fourths?* *Where's half of π, cut in half? Where's 2π cut in half, half, half, or in eighths?*
$\frac{2\pi}{3}$	*Start a new circle where the above are labeled (without the arcs drawn).* *Called 2π thirds; it's two-thirds of π. Where's one-third of π? Where's 2 of those one-thirds?* *How do two-thirds of π relate to three-thirds of π?* *So, you can use a friendly measure that's smaller, one-third π, or that's too big three-thirds π, to help?* *We call that one-third π the reference angle.*
$\frac{7\pi}{4}$	*Called 7π fourths; it's seven-fourths of π. What are some friendly fourths that could help?* *Where's one-fourth of π? Where's 7 of those one-fourths? 4 of them and 3 more?* *How do eight-fourths of π relate to seven-fourths of π? Or six-fourths π?* *So, you can use a friendly measure that's smaller, one-fourth π, or that's too big eight-fourths π, to help?* *We call that one-fourth π the reference angle.*
$\frac{11\pi}{6}$	*Called 11π sixths; it's eleven-sixths of π. What are some friendly sixths that could help?* *Where's one-sixth of π? Where's 11 of those one-sixths? 6 of them and 5 more?* *How do twelve-sixths of π relate to eleven-sixths of π? Or nine-sixths? Or ten-sixths?* *So, you can use a friendly measure that's smaller, one-sixth π, or that's too big 12-sixths π, to help?* *What is the reference angle?*
$\frac{5\pi}{3}$	*What do you think this is called? What does it mean? What are some friendly thirds that could help?* *Where's one-third of π? Where's 5 of those one-thirds? 3 of them and 2 more?* *How do six-thirds of π relate to five-thirds of π? Or four-thirds?* *How are you using friendly angle measures to help that are too small? That are too big?* *What is the reference angle?*
$\frac{5\pi}{6}$	*What do you think this is called? What does it mean? What are some friendly sixths that could help?* *What is the reference angle?* *How is this related to the previous? Double? Half? How do you know?* *How does one-sixth π relate to one-third π? 60° to 30°? How can you keep this relationship straight?* *What are good benchmark radian measures?*

PROBLEM STRINGS
LESSON

7.5 Sinusoid Viewing Windows

At a Glance

$y = \sin x$*

$y = \sin x + 9$

$y = 50\sin 2x$

$y = 0.01\sin 0.5x$

$y = \sin 0.1 - 4$

$y = \sin(x - 1.5\pi)$

Follow-up problem string:

$y = \cos x$*

$y = 2\cos x - 10$

$y = \cos\left(\frac{x - 5\pi}{10}\right)$

$y = 0.1\cos(0.2x)$

$y = 75\cos 4x$

$y = \cos(x + 1.5\pi)$

*optional problem

Objectives

The goal of this problem string is to strengthen students' understand of amplitude, period, and midline in the equations and graphs of sinusoid functions. The vehicle to create or strengthen these relationships is finding viewing windows to see the important aspects of the graph.

Placement

This problem string can be use to introduce or strengthen students' understanding and facility with sinusoids, specifically their amplitude, period, and midline. Students should have some prior experience with right triangle trigonometry, radian measure, and the notion of period functions.

You can use this problem string in textbook Lesson 7.5 Graphing Trigonometric Functions.

Guiding the Problem String

Each of these problems are designed so that in a Desmos home window of approximately $-10 \le x \le 10$, $-7 \le y \le 7$, they do not appear or they only partially appear. Therefore, when first graphed, a potentially confusing or deceiving graph appears. As students work out that confusion by finding an appropriate viewing window, learning occurs as they settle that disequilibrium. Because Desmos resizes the graph depending on the browser window size, you may need to set your browser window so that when you display the functions in this string, you get similar results to those pictured in the Sample Interactions. Ideally, $y = \sin x + 9$ is just off the screen in the home window.

If students are beginning to learn about sinusoids, start with the optional first function, the parent function. Ask questions that help students make connections between the equations and the graphs, naming the different attributes as they come up in the discussion. Continue the string by graphing the functions and having students suggest different things to try until you see the functions. Once students find a graph, nudge them to consider the period of the function. Use the zoom and swiping features at first, but gradually press students to predict and have you change the window manually. Ask students for justification. Help students generalize the attributes of the periodic motion.

If students are already familiar with sinusoids, ask students to predict first, pressing for justification before showing the first graph. Continue to press for reasoning as you change the window manually, choosing values based on student suggestions. Do not use the zoom or swiping features. Students must reason using what they know or are learning about. If students suggest incorrect moves, make them and let them learn from the missteps.

There are several viewing window problem strings in chapter 4 and Lesson 6.5 Polynomial Viewing Windows that you can use to help you plan your lesson.

About the Mathematics

The standard transformations have special names for sinusoids. The amplitude of a sinusoid is the absolute value of the vertical scale factor, or $|b|$. For $y = \sin x$ and $y = \cos x$, the amplitude is 1 unit. The horizontal translation value is called the phase shift. The vertical translation value, k, gives you the equation of the midline, or average value. The midline for the basic sine and cosine functions are $y = 0$, the x-axis. The horizontal dilation factor, a, must be multiplied by 2π to determine the period. This is because the period of either $y = \sin x$ and $y = \cos x$ is 2π.

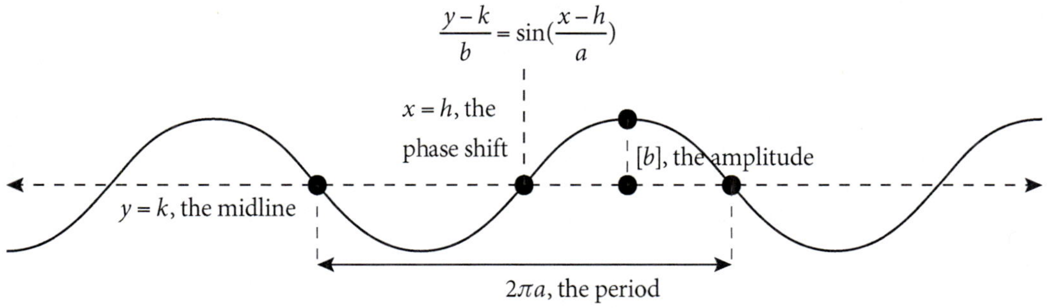

Sample Interactions

Use the following as you plan how to elicit and model student strategies. This is not meant as a script, but as a view into the relationships involved and the intent of the problem string.

Teacher: *The first problem in today's string is to look at the parent sinusoid function,* $y = \sin x$. *Here is a graph in the home window. What do you see?*
Since it's only oscillating between -1 *and* 1*, let's change the y's in the window. What would be a good range for y?*

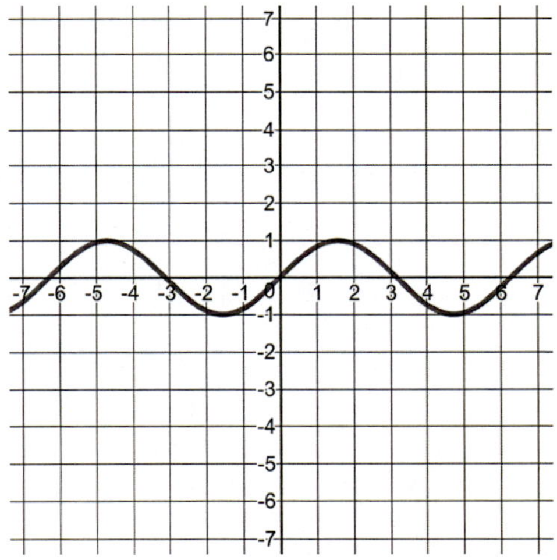

(continued)

Teacher: *What else do you see? You can see that it's repeating, over and over? We call one cycle the "period" and we describe the function as "periodic."*

What do you think would happen if we zoom out? Let's do it. Does that support that this function is periodic? I'll zoom back in now.

If we start here at the x = 0, where does one cycle end and then start over? That is its period, the smallest distance between values of x before the cycle begins to repeat. About how far does that seem? About 6? And where is the half-way point? About 3?

Hmmm. Those values sound familiar. A little more than 3. We've been studying the unit circle. Anyone wondering if π might be involved?

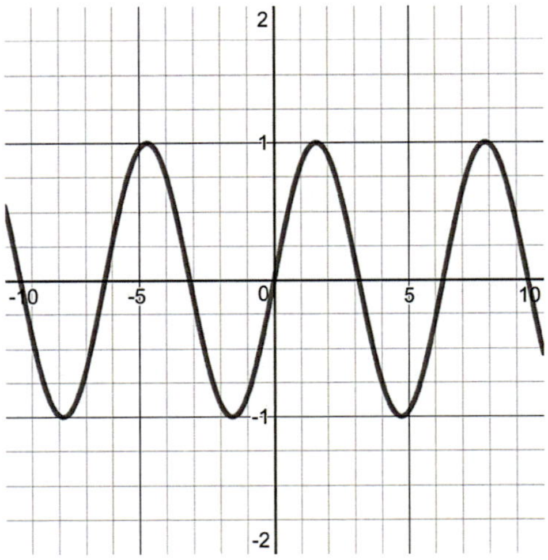

Teacher: *I'm going to change the scale so that the tick marks are at multiples of π. So, for this parent function, what is the period? Yes, in this case, it's 2π.*

Let's change the window so we just see one period. How could we do that? What should we change?

The window is opened just to the left and right to show one period clearly.

What is the vertical change from top to bottom?

Where is the midline, the middle of the vertical change?

We call the distance from the midline to the top or bottom the amplitude. It's related to the vertical dilation factor, which for this parent function is 1.

The teacher draws a smaller version of the graph with the attributes labeled for reference throughout the rest of the string. This could become an anchor chart for future use.

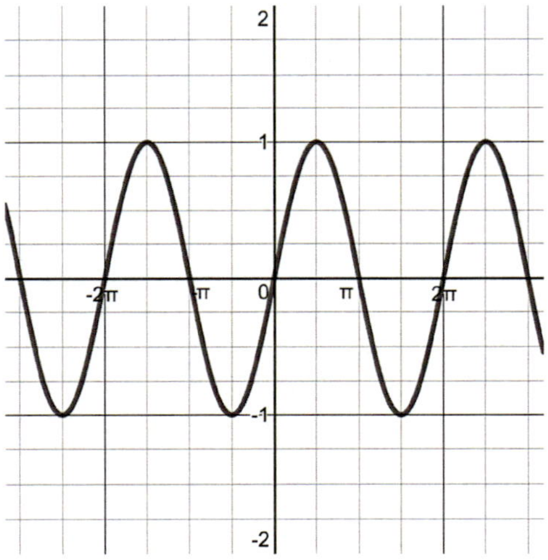

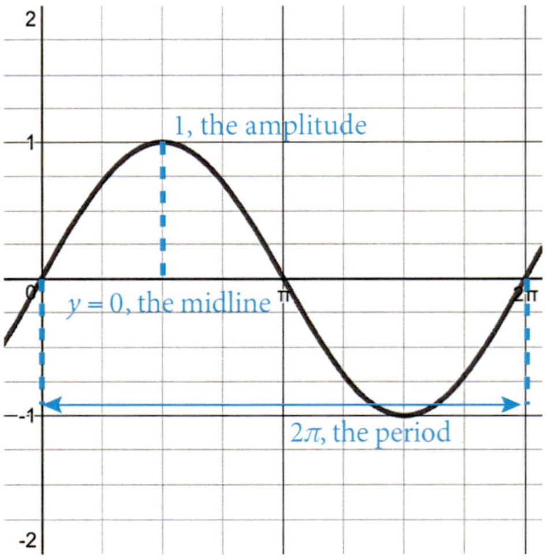

The teacher changes the window back to the home viewing window.

Teacher: *Here is the second problem in our string today,* $y = \sin x + 9$*. Any ideas what it would look like? Predict the graph by telling your partner. What do you see? Nothing? Where should we look? There it is, up 9! What about the rest of it? Did anything else change?*

So we could use the same x-values for the window as the parent function to show just one period? What is one period of the parent function again? What would be good y-values so we have a window without too much extra space, but still shows the x-axis?

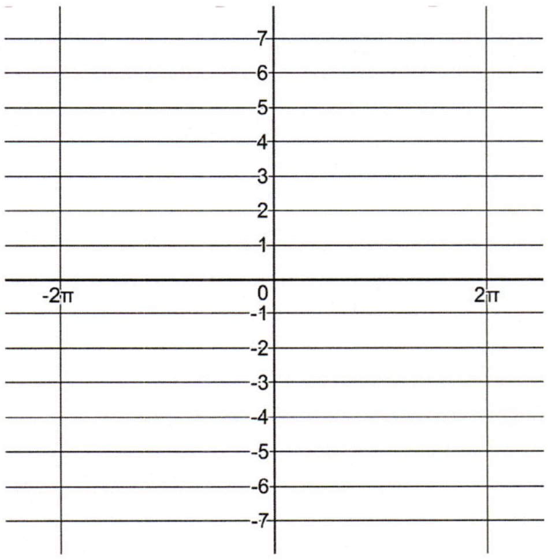

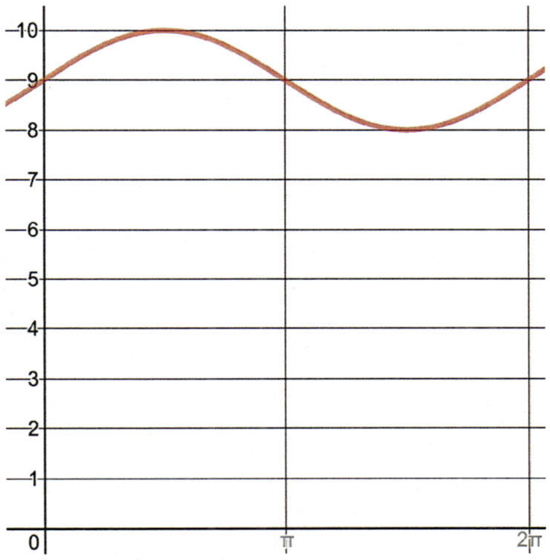

The teacher changes the window back to the home viewing window.

Teacher: *Here is the third problem in our string today,* $y = 50\sin 2x$

Predict what it would look like using your knowledge of transformations.

What do you see? A bunch of vertical lines? What happened to our periodic function? Ahhh, you think we just need to open up the window to see more? How can we rewrite the function to see the vertical and horizontal scale factors?

$$\frac{y}{50} = \sin\left(\frac{x}{\frac{1}{2}}\right)$$

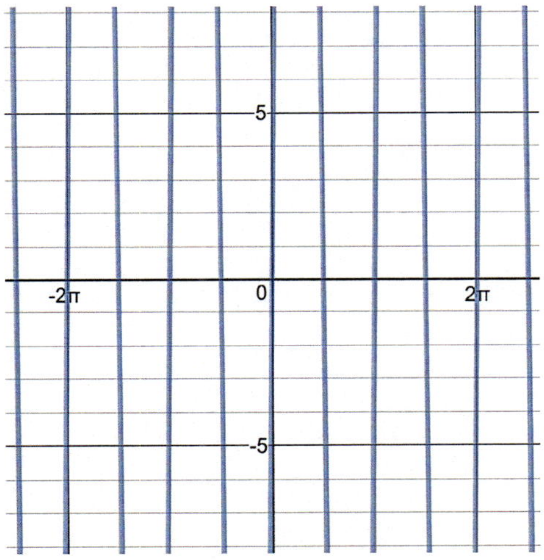

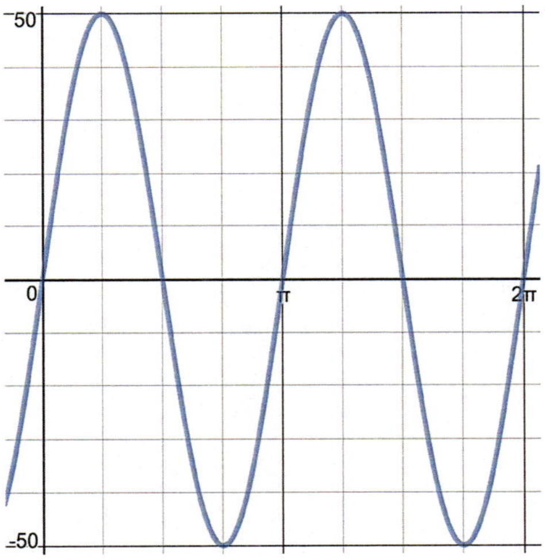

(continued)

Teacher: *Here is the fourth problem in our string today,* $y = 0.01 \sin 0.5x$
Predict what it would look like using your knowledge of transformations. Let's graph it in the home window. What do you see? A horizontal line? Weird. What's going on? How can we rewrite the function to see the vertical and horizontal scale factors?

$$\frac{y}{0.01} = \sin\left(\frac{x}{2}\right)$$

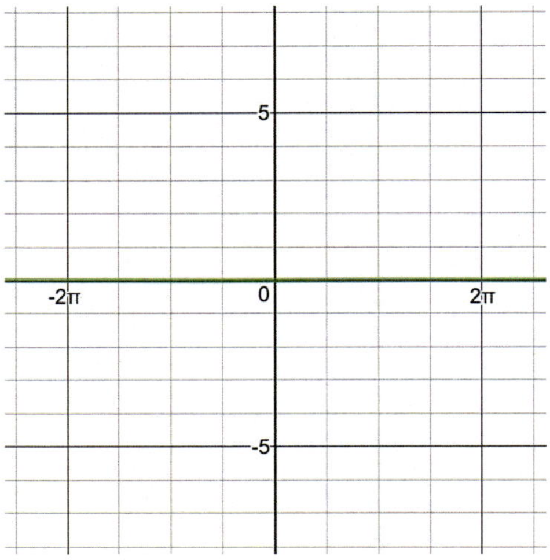

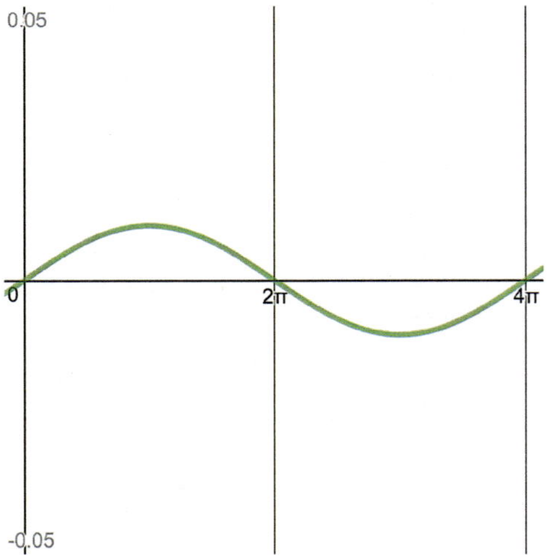

Teacher: *Here is the fifth problem in our string today,* $y = \sin 0.1 - 4$
Predict what it would look like using your knowledge of transformations. How can we rewrite the function to see the horizontal scale factor?

$$y = \sin\left(\frac{x}{10}\right) - 4$$

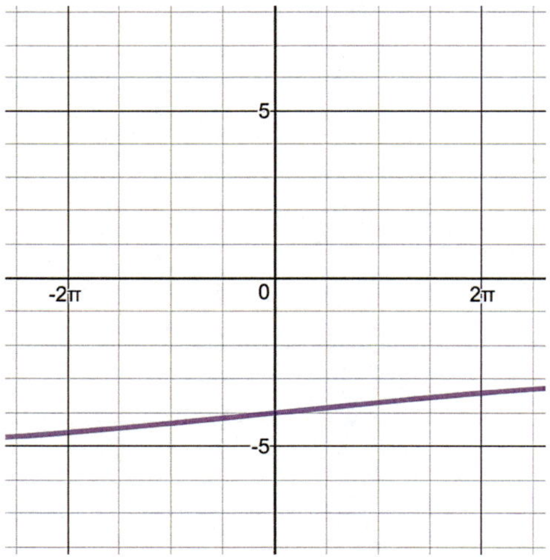

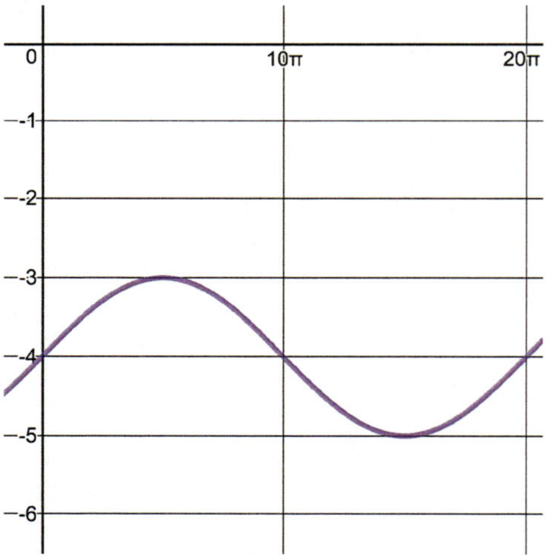

Teacher: *Here is the sixth problem in our string today,* $y = \sin(x - 1.5\pi)$
Predict what it would look like using your knowledge of transformations. What's the only transformation happening? What if the phase shift would have been left 1.5π*?*

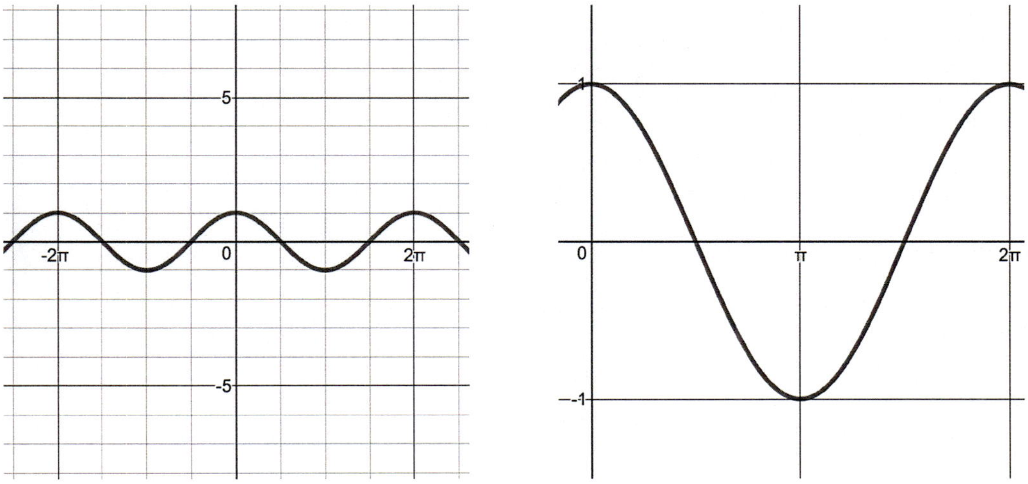

Teacher: *How would you summarize some of the things that came up in this string today?*

Elicit the following:

- *The graph* $y = \sin x$ *is periodic. For this parent function, the period is* 2π*, the amplitude is* 1*, and the midline is the x-axis*
- *We can transform sinusoids just like other functions. The transformation have terms specific to periodic functions, like phase shift, amplitude, period, and midline.*
- *Period refers to the smallest distance between values of x before the cycle begins to repeat.*

Sample Final Display

Your display could look like this at the end of the problem string:

$y = \sin x$

$y = \sin x + 9$

vertical translation up 9

$y = 50 \sin 2x$ $\quad \frac{y}{50} = \sin \frac{x}{\frac{1}{2}}$

amplitude 50
(vertical dilation by factor 50)
period is π
(horizonal dilation by factor ½)

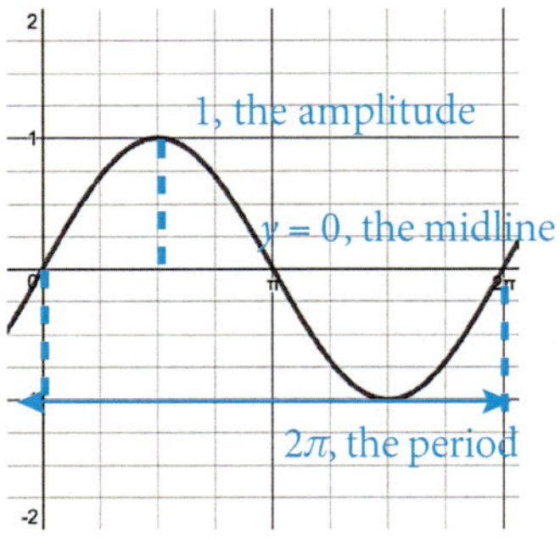

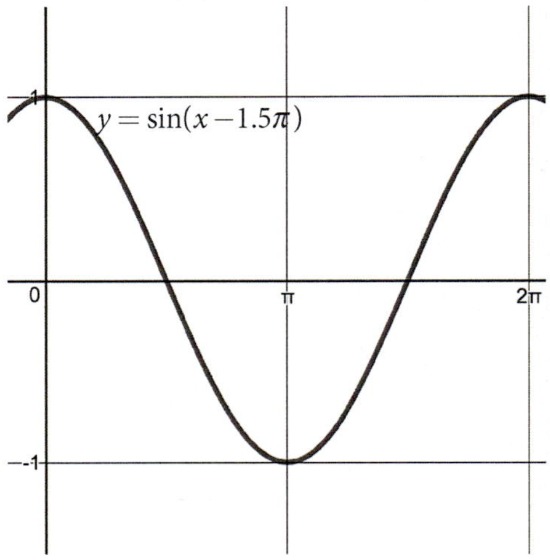

$y = 0.01 \sin 0.5x$ $\quad \frac{y}{0.01} = \sin \frac{x}{2}$

amplitude 0.01 (vertical dilation by factor 0.01)
period is 4π (horizonal dilation by factor 2)

$y = \sin 0.1 - 4$ $\quad y = \sin \frac{x}{10} - 4$

vertical translation down 4
period is 20π (horizontal dilation by factor 10)

$y = \sin(x - 1.5\pi)$

phase shift to the right 1.5π

Facilitation Notes

This version of the problem string lists short notes for important teacher moves during the string. After you've done the string yourself and studied the relationships involved, you might make similar notes for the things you want a reminder of or deem important.

$y = \sin x$	*Here is a function, we call it "sine of x". Any ideas what it would look like? Home window. What do you see? Connect transformations to period, amplitude, midline. Find an appropriate window for one period.*
$y = \sin x + 9$	*Use your knowledge of transformations: Predict behavior? Home window. Where is it? What changed? What didn't change? Find an appropriate window for one period.*
$y = 50 \sin 2x$	*Predict first. Show graph in home window. What is happening? Find an appropriate window for one period.*
$y = 0.01 \sin 0.5x$	*Repeat. As student describe, use precise mathematical vocabulary.*
$y = \sin 0.1 - 4$	*Repeat.*
$y = \sin(x - 1.5\pi)$	*Repeat. Phase shift. Compare to parent. Foreshadow y = cos x.*

PROBLEM STRINGS LESSON

7.6 A Bouncing Spring Follow Up

At a Glance

$$h(t) = 0.652 + 0.076\cos\left[\frac{2\pi(t-0.43)}{0.8064}\right]$$

$$h(t) = 0.8 + 0.076\cos\left[\frac{2\pi(t-0.43)}{0.8064}\right]$$

$$h(t) = 0.5 + 0.076\cos\left[\frac{2\pi(t-0.43)}{0.8064}\right]$$

$$h(t) = 0.652 + 0.1\cos\left[\frac{2\pi(t-0.43)}{0.8064}\right]$$

$$h(t) = 0.652 + 0.05\cos\left[\frac{2\pi(t-0.43)}{0.8064}\right]$$

$$h(t) = 0.652 + 0.076\cos\left[\frac{2\pi(t-0.23)}{0.8064}\right]$$

$$h(t) = 0.652 + 0.076\cos\left[\frac{2\pi(t-0.63)}{0.8064}\right]$$

$$h(t) = 0.652 + 0.076\cos\left[\frac{2\pi(t-0.43)}{0.5}\right]$$

$$h(t) = 0.652 + 0.076\cos\left[\frac{2\pi(t-0.43)}{0.9}\right]$$

Objectives

The goal of this problem string to solidify the learning about sinusoidal functions that took place in the "A Bounding Spring" Investigation in which students collected or simulated spring bouncing data (height of the bottom of a stretched and released spring as it bounces over time) and wrote a sinusoid function to model the data.

Placement

This string could come right after or the day after students have used their collected or simulated data to write a sinusoid function for the height of the bottom of a stretched and released spring over time.

You could deliver this string any time after the "A Bouncing Spring" Investigation in textbook Lesson 7.6 Modeling with Trigonometric Equations.

Guiding the Problem String

The first problem is based on the data from the answers in the textbook Teacher Edition. You could change the numbers to match the data your class collected for (elapsed time, height) of a stretched and released spring. You could enter all of the equations before you start the string and just make them appear one at a time or you could cut and paste each time, just changing the one value that changes.

The purpose of the first problem is to provide an anchor to the investigation that students can refer to throughout the string. Use the display grapher to graph the functions. Throughout the string, ask students to predict what they will see and decide what changed in the spring experience to create each new graph. The string is designed as partner problems that compare to the first problem. Keep the first problem's graph displayed during the entire string, adding each graph of a set one at a time. After you have compared both graphs of a set to the first problem, turn off the two graphs and move on to the next problem. Repeat. The Sample Final Display shows the first problem's graph with the last set of two graphs.

Take a little time on the second problem to predict and discuss similarities and differences. Push on answers for justification about the vertical shift. The third problem should go quickly. The fourth and fifth problems deal with the vertical stretch that affects the amplitude. The sixth and seventh problems give functions with phase shifts, which really just means that the spring hit the maximum height at a different time. The last two problems change the period of the function, which means that you are dealing with different springs, one less and one more springy.

About the Mathematics

In reality in this experiment, the amount of motion gradually decreases over time as the mass springs up and down, and eventually the mass returns to rest. However, if the initial motion is small, then the decrease in the motion occurs more slowly and can be ignored in the first few seconds.

One reason to use a cosine function, instead of a sine function, to fit such data is that it is often easier to identify the maximum value as the start value of each cycle.

(continued)

Important Questions

Use the following as you plan how to elicit and model student strategies.

- *What is a in* $\frac{y-k}{b}=\cos\left(\frac{x-h}{a}\right)$*? Where does it show up in the graph? What does it mean in the function?*
- *What is b in* $\frac{y-k}{b}=\cos\left(\frac{x-h}{a}\right)$*? Where does it show up in the graph? What does it mean in the function?*
- *What is h in* $\frac{y-k}{b}=\cos\left(\frac{x-h}{a}\right)$*? Where does it show up in the graph? What does it mean in the function?*
- *What is k in* $\frac{y-k}{b}=\cos\left(\frac{x-h}{a}\right)$*? Where does it show up in the graph? What does it mean in the function?*

How would you summarize some of the things that came up in this string today?

- *In the sinusoid* $\frac{y-k}{b}=\cos\left(\frac{x-h}{a}\right)$*, a is the horizontal dilation factor and* $2\pi a$ *is the period. In this scenario, the different a values represent different springiness of different springs.*
- *Here b is the vertical dilation factor and corresponds to the amplitude. In this scenario, the different b values represent different heights to which the mass at the end of the spring was dragged.*
- *Here h is the horizontal shift and is called a phase shift. In this scenario, the different h values represent the fact that the spring was released at a different time in reference to the starting time because the graph reaches its maximum height at a different time.*
- *And k is the vertical shift which shifts the midline. In this scenario, the different k values represent the different lengths of different springs, some longer and starting closer to the motion detector and vice versa.*

Sample Final Display

Your display could look like this at the end of the problem string:

$$h(t) = 0.652 + 0.076\cos\frac{2\pi(t-0.43)}{0.8064}$$

$$h(t) = 0.8 + 0.076\cos\frac{2\pi(t-0.43)}{0.8064}$$

$$h(t) = 0.5 + 0.076\cos\frac{2\pi(t-0.43)}{0.8064}$$

different springs (different lengths of the spring), vertical shift

$$h(t) = 0.652 + 0.1\cos\frac{2\pi(t-0.43)}{0.8064}$$

$$h(t) = 0.652 + 0.05\cos\frac{2\pi(t-0.43)}{0.8064}$$

stretch the spring more or less, amplitude change

$$h(t) = 0.652 + 0.076\cos\frac{2\pi(t+0.23)}{0.8064}$$

$$h(t) = 0.652 + 0.076\cos\frac{2\pi(t+0.63)}{0.8064}$$

time the spring reached the max height, phase shift

$$h(t) = 0.652 + 0.076\cos\frac{2\pi(t-0.43)}{0.5}$$

$$h(t) = 0.652 + 0.076\cos\frac{2\pi(t-0.43)}{0.9}$$

different springs (more or less stretchy) period changes

$$h(t) = 0.652 + 0.076\cos\frac{2\pi(t-0.43)}{0.8064}$$

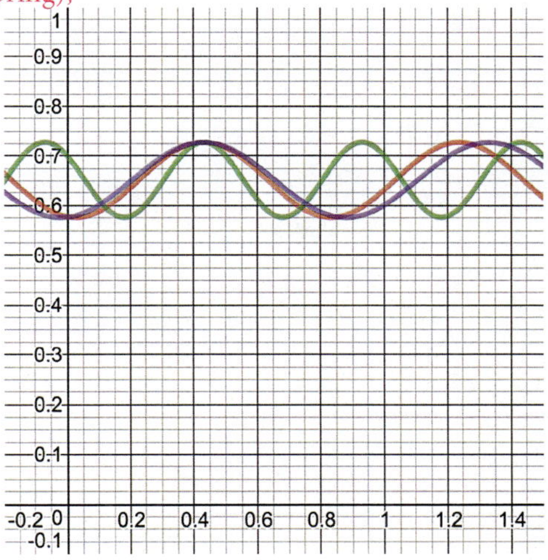

$$h(t) = 0.652 + 0.076\cos\frac{2\pi(t-0.43)}{0.5}$$

$$h(t) = 0.652 + 0.076\cos\frac{2\pi(t-0.43)}{0.9}$$

(continued)

Facilitation Notes

This version of the problem string lists short notes for important teacher moves during the string. After you've done the string yourself and studied the relationships involved, you might make similar notes for the things you want a reminder of or deem important.

$h(t)=0.652+0.076\cos\dfrac{2\pi(t-0.43)}{0.8064}$	*Remember the spring data? What does this represent?* *Predict the graph. Display on grapher in good window.* *Ask 1 meaning of the transformations.*
$h(t)=0.8+0.076\cos\dfrac{2\pi(t-0.43)}{0.8064}$	*What changed in the function? What stayed the same?* *What changed in the scenario? How?* *Predict the graph. Compare predictions to display graph.*
$h(t)=0.5+0.076\cos\dfrac{2\pi(t-0.43)}{0.8064}$	*Repeat.* *What changed here and in problem 2?* *Vertical shift, shifts midline—shorter or longer springs, (closer or farther from detector at rest). Turn off #2, 3*
$h(t)=0.652+0.1\cos\dfrac{2\pi(t-0.43)}{0.8064}$	*What changed in the function from the first problem?* *What changed in the scenario? How?* *Predict the graph. Compare predictions to display graph.*
$h(t)=0.652+0.05\cos\dfrac{2\pi(t-0.43)}{0.8064}$	*Repeat.* *What changed here and in problem 4?* *Vertical dilation, amplitude—how close to detector you stretched the spring. Turn off #4, 5.*
$h(t)=0.652+0.076\cos\dfrac{2\pi(t-0.23)}{0.8064}$	*What changed in the function from the first problem?* *What changed in the scenario? How?* *Predict the graph. Compare predictions to display graph.*
$h(t)=0.652+0.076\cos\dfrac{2\pi(t-0.63)}{0.8064}$	*Repeat.* *What changed here and in problem 6?* *Horizontal shift, phase shift—time at which spring reaches maximum distance. Turn off #6, 7.*
$h(t)=0.652+0.076\cos\dfrac{2\pi(t-0.43)}{0.5}$	*What changed in the function from the first problem?* *What changed in the scenario? How?* *Predict the graph. Compare predictions to display graph.*
$h(t)=0.652+0.076\cos\dfrac{2\pi(t-0.43)}{0.9}$	*Repeat.* *What changed here and in problem 8?* *Horizontal dilation, period change—2πa is period, takes that long to complete 1 cycle.*

PROBLEM STRINGS
LESSON

7.7 Solving Trigonometric Equations

At a Glance

$\sin x = 0$

$\cos x = 0$

$\sin x = \frac{1}{2}$

$\cos x = \frac{1}{2}$

Objectives

The goal of this problem string is for students to use what they know to solve simple trigonometric equations that have either sine or cosine. By helping students make connections and use relationships from right triangle trigonometry, the unit circle, radian measure, and the sine and cosine functions, students strengthen their understanding and solve simple trigonometric equations preparing them up to solve more complicated problems.

Placement

This problem string can be used to introduce solving simple trigonometric equations. It can also be used to bring together everything students are learning about right triangle trigonometry, the unit circle, radian measure, and the sine and cosine functions.

You can use this problem string to follow up textbook Lesson 7.6 Modeling with Trigonometric Equations or to introduce Lesson 7.7 Pythagorean Identities.

Guiding the Problem String

This problem string consists of four simple, related trigonometric equations. The problems alternate between the sine function and the cosine function, potentially suggesting using solutions of one to help find solutions of the other.

Use student thinking about right triangle trigonometry to help find the exact answers, but press students to realize the periodic nature of the functions. Use the unit circle and the sine and cosine functions to help students consider more than one solution. Graph the functions on the display grapher and mark the solution points. If students use technology with graphs and arc functions to find approximate answers, press them for justification about the exact answers. If students only use paper and pencil, press them to use technology to find the approximations. Help students connect their mental ideas with the technology's ability to graph precisely and quickly. Exact answers are great when possible, but most trigonometric equations do not have exact answers, unless they are in textbook designed problems meant to be solved without technology.

About the Mathematics

By convention θ typically refers to angle measures in degrees, whereas the variable x refers to real number measures in radians. Thus we don't typically solve the equation $\sin\theta = \frac{1}{2}$ for more than one value—the value that represents the angle for which the ratio of the opposite to the hypotenuse is 1:2. We do solve $\sin x = \frac{1}{2}$ for multiple values, where x represents all of the intersection points of $y = \sin x$ and $y = \frac{1}{2}$ in a specified range.

(continued)

Sample Interactions

Use the following as you plan how to elicit and model student strategies. This is not meant as a script, but as a view into the relationships involved and the intent of the problem string.

Teacher: *Let's get warmed up today with a quick problem string. The first problem is* $\sin x = 0$. *What is x if the sine of x is zero? I'm interested in values between zero and 2π. Go ahead and solve it any way you want.* Students work and the teacher circulates, looking for students who used the graph of $y = \sin x$, the unit circle, right triangles, arcsine on a calculator, and taking note of any other strategies. **Teacher:** *What is x?* **Student:** 0. π. Here, the teacher did not hear the solution 2π, but will continue to listen for it. If it does not come up as students discuss their strategies, the teacher plans to ask questions to help students realize that 2π is also a solution.	$\sin x = 0, 0 \leq x \leq 2\pi$ $x = 0, \pi$
Teacher: *I saw you use a calculator. Please tell us about that.* **Student:** *I just put in* $\arcsin 0$ *and it was zero.* **Teacher:** *Okay, so you agree with everyone who said x is zero?*	$\arcsin 0 = 0$
Teacher: *I overheard you two talking about triangles. What about triangles?* **Student:** *We know that sine is opposite over hypotenuse. In this case opposite over hypotenuse has to be zero. So the opposite has to be zero.* **Student:** *We were talking about that little tiny triangle we looked at the other day, right on the axis. If you started at zero and just barely opened the angle.* **Teacher:** *You are referring to the angles in standard position, like this?*	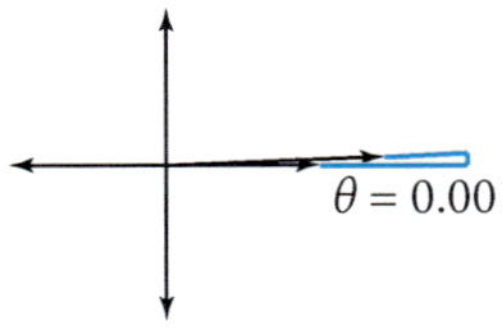
Student: *Yes, so that really short triangle, if the angle got smaller, until it was zero, then the "opposite" side of the triangle would be zero. Not really, but it helps me think about it.* **Student:** *I remember that. That's what helped us create the anchor chart where we talked about thinking about the angles as opening up. Zero is right there and 180° or π is when you open it up to the other side.* **Student:** *And we used a radius of 1. So since we're talking about sine being zero, put the points (1, 0) and (−1, 0).* **Student:** *That's where we got the π answer. Sine of x is zero both when the angle is at zero and when it opens up to π.*	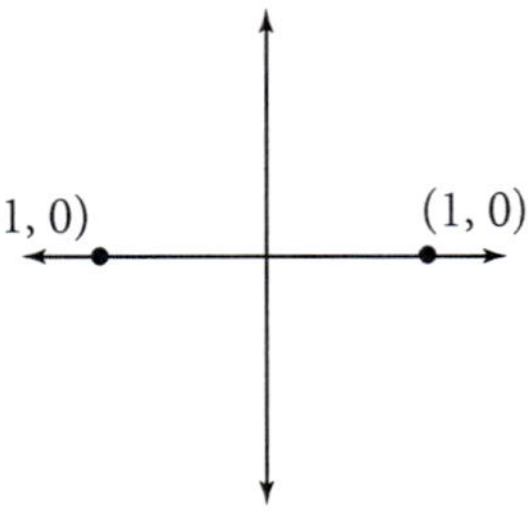
Teacher: *Does everyone agree that the answer is both zero and π? Why did the calculator only return zero?* **Student:** *It only gives one answer. We have to understand that there are more and we can use what we know to find them.*	

Teacher: *I saw you drawing the function for sine. Tell us about that.*

Student: *We've been graphing sine and cosine, so we thought about where the function* $y = \sin x$ *is zero.*

Teacher: *I'm going to put the graph on the display grapher.*

So, here it is. What are you looking for?

Student: *Where the function is zero, the y-values. And it's there at zero and π like we said. But now that we're looking at the graph, we can see that it's also zero at 2π. We hadn't drawn our graph out far enough.*

Teacher: *What does everyone think? Do you think we should include 2π in our solution set?*

Student: *Yes. And it should've been when we were thinking about opening angles. You would just keep going until you opened the angle all the way to 2π.*

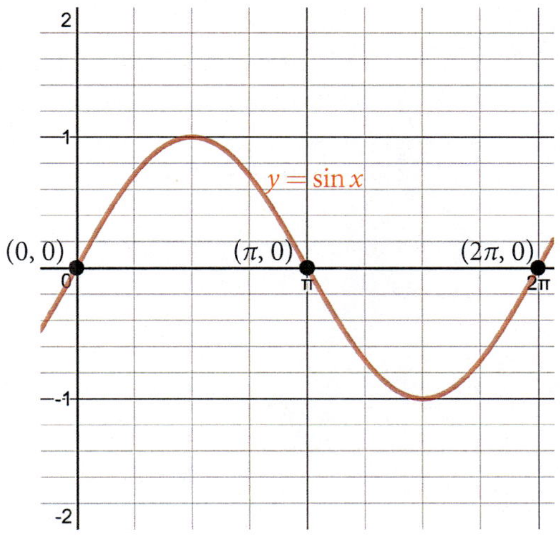

$x = 0, \pi, 2\pi$

Teacher: *But then, shouldn't we keep going?*

Student: *Right! It should also be 3π and 4π and ...*

Student: *But, in the beginning you said only between zero and 2π.*

Teacher: *Right. So, does everyone agree that if there were no restrictions, there would be an infinite number of solutions?*

Student: *Just think about the graph of* $y = \sin x$. *It just keeps going and going and going.*

Student: *It's like going around the unit circle, again and again and again.*

Teacher: *What about to the left?*

Student: *Lots of solutions back there too. Thanks for giving us the restriction!*

$\sin x = 0 \quad 0 \le x \le 2\pi$

$x = 0, \pi, 2\pi$

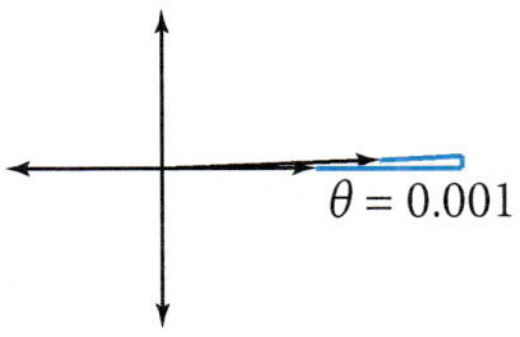

$\arcsin 0 = 0$

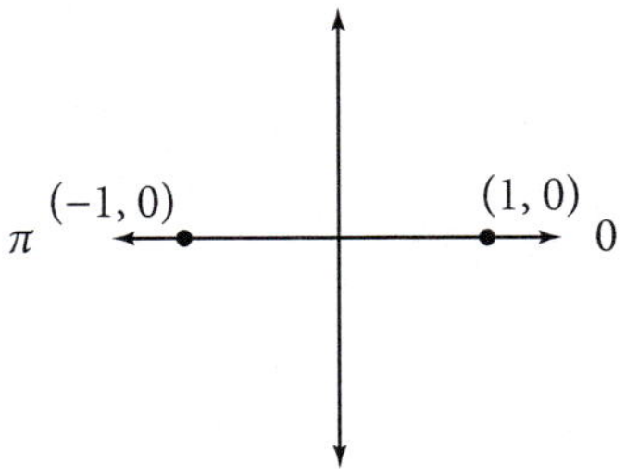

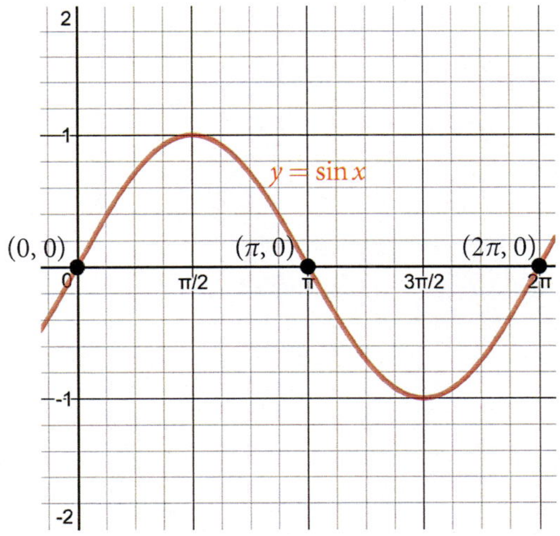

(continued)

Teacher: *The next problem in our string is* $\cos x = 0$. *For what values of x is that true? See if you can confirm your answer using a different way of thinking. We have several ways of thinking about our first problem on the board. Go!*

The students work while the teacher circulates, asking questions and pressing students for justification. Then the teacher asks students to share their solutions and strategies, while the teacher models on the board.

$\sin x = 0 \quad 0 \le x \le 2\pi$

$x = 0, \pi, 2\pi$

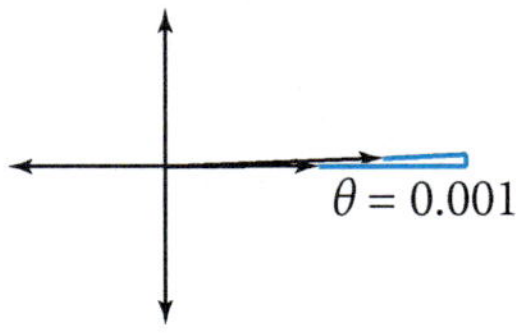

$\arcsin 0 = 0$

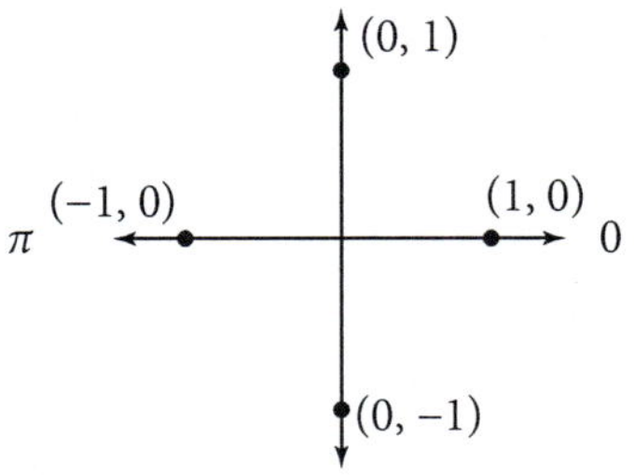

$\cos x = 0 \quad 0 \le x \le 2\pi$

$x = \frac{\pi}{2}, \frac{3\pi}{2}$

$\arccos 0 \approx 1.57$

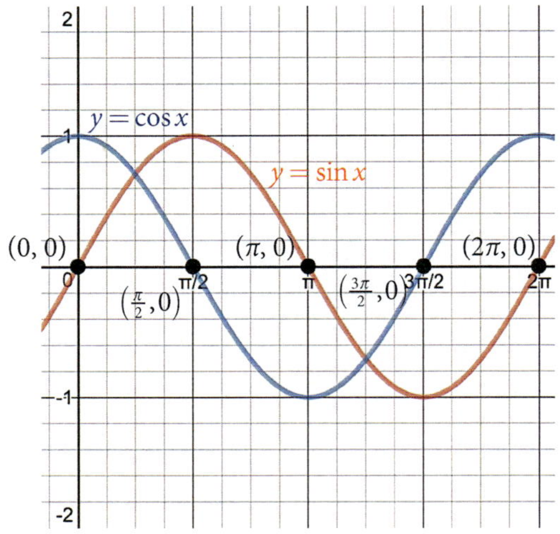

Teacher: *How do the solutions to* $\sin x = 0$ *and* $\cos x = 0$ *compare? What do you notice?*

Student: *It's like they're translated to the right.*

Student: *Yeah, by half of π.*

Teacher: *Are the graphs of the parent functions of sine and cosine always like that?*

Student: *Yes, they are just the same but cosine is half of π further to the right.*

Teacher: *That seems like a helpful thing.*

Student: *You mean that if we know the solution to sine, then we might be able to use that to help us find the solution to the same equation with cosine?*

Teacher: *That's an interesting idea.*

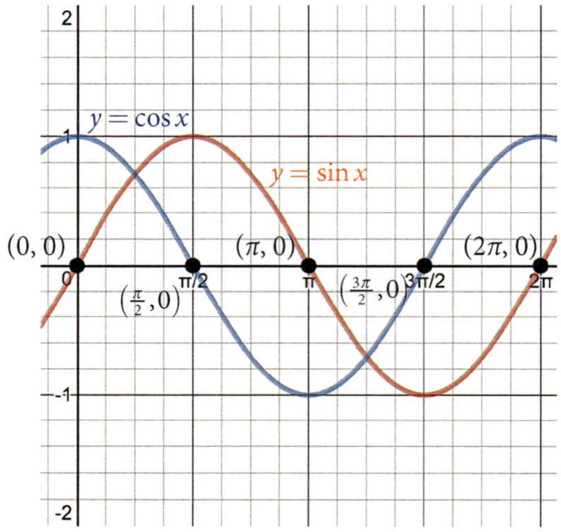

Teacher: *The next problem in our string is to solve* $\sin x = \frac{1}{2}$. *What is x when the sine of x is one-half?*

Students work while the teacher circulates, looking for students who are using right triangles, the unit circle, the function $y = \sin x$ and noting other strategies.

$\sin x = \frac{1}{2}$

Teacher: *I noticed you were using triangles. Please tell us about that thinking.* **Student:** *I know that sine is the opposite to the hypotenuse. So I drew a triangle with 1 as the opposite and 2 as the hypotenuse. I used the Pythagorean Theorem and recognized that it's a 30°, 60°, 90°. So, the small angle is across from the small side, so the x is 30°.* **Teacher:** *Did you need to figure out the other side, with the square root of 3?* **Student:** *Not really, except I didn't realize that it was a 30°, 60°, 90° until I did.*	$\sin x = \frac{1}{2} \quad 0 \leq x \leq 2\pi$ 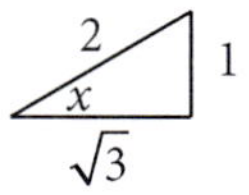
Teacher: *So x is 30°? What is that in radians? I ask that, because by convention, when we use θ, we usually mean degrees and when we use x, we usually mean a real number, a radian measure.* **Student:** *Thirty degrees is the same as* $\frac{\pi}{6}$.	$x = \frac{\pi}{6}$
Teacher: *Are we done? Is there only one possibility for x? I saw you two putting a triangle on the unit circle. Tell us about that.* **Student:** *We kind of thought about the triangle too but we looked at opening up an angle so that the opposite side would be one-half and the hypotenuse would be 1. We know we are dealing with a 30°, 60°, 90°, so we know the values are either one-half or* $\frac{\sqrt{3}}{2}$. *We set it up so that the y-coordinate of that point is one-half. Since that's shorter than the other side,* $\frac{\sqrt{3}}{2}$, *the angle has to be* $\frac{\pi}{6}$.	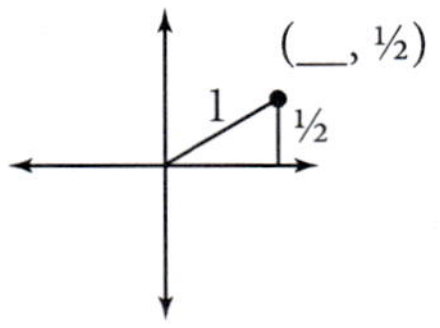
Teacher: *Can someone else join in? What are these two thinking about?* **Student:** *I did the same thing. I know that there are two places in the first quadrant on the unit circle that have one-half. One angle has one-half as the opposite, or the y-value. The other angle has one-half as the adjacent, the x-value. So, since this problem has one-half as the y-value, then the angle is* $\frac{\pi}{6}$. **Student:** *I get how there are the two angles,* $\frac{\pi}{6}$ *and* $\frac{\pi}{3}$ *and that they both deal with one-half. But how do you know which one it is here?* **Student:** *See how the one-half is the y-value because it's sine? In that triangle, it's opposite the angle, right? One-half is the smaller side, it's smaller than* $\frac{\sqrt{3}}{2}$ *so it has to be opposite the smallest angle. And* $\frac{\pi}{6}$, *one-sixth of π is smaller than* $\frac{\pi}{3}$, *one-third of π.* **Student:** *Ah, it's like you know it's one of those two and you just have to figure out if it's the small one or the big one. Got it.*	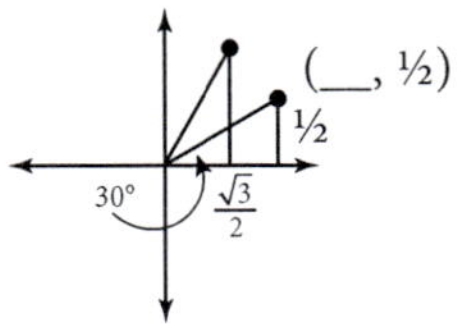

(continued)

Teacher: *Are we done?*

Student: *No, I thought about the unit circle. The sine is positive in both the first and second quadrants. So x can also be the angle where the opposite is one-half in the second quadrant.*

Student: *That's 150° which is the same as, uh, π minus one-sixth of π is five-sixths of π,* $\frac{5\pi}{6}$.

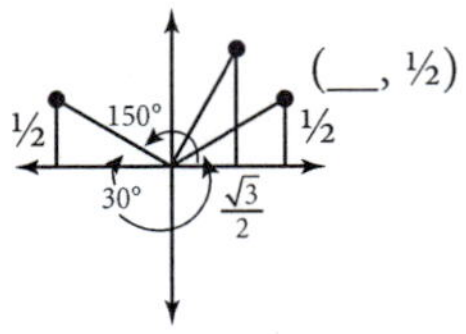

Teacher: *Did anyone use the graph of sine this time?*

Student: *I figured out the* $\frac{\pi}{6}$ *first and then I thought about the graph to make sure I had them all. I could tell that* $y = \sin x$ *was at one-half two times before π, but the graph didn't give me exact answers, just decimal approximations.*

The teacher graphs the line $y = 0.5$ and writes in the point coordinates.

$$\sin x = \frac{1}{2} \qquad 0 \leq x \leq 2\pi$$

$$x = \frac{\pi}{6}, \frac{5\pi}{6}$$

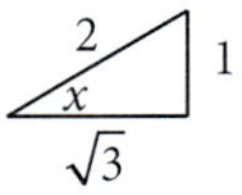

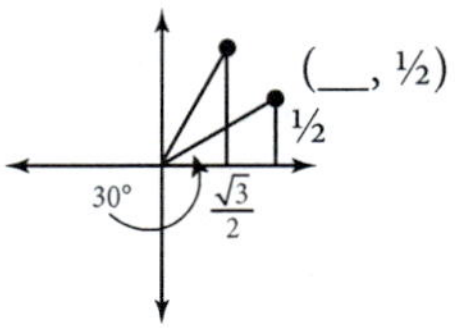

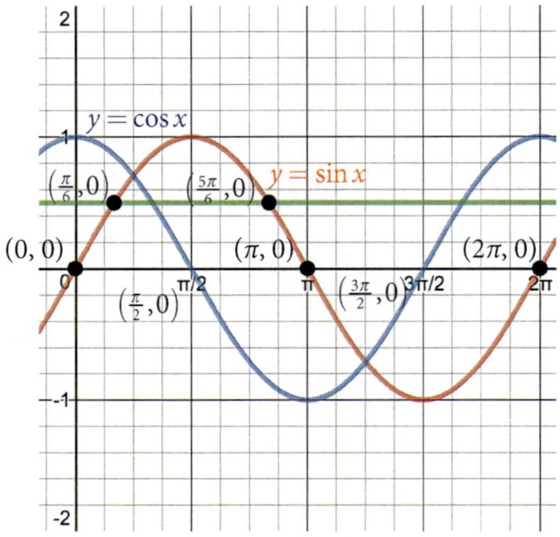

The teacher facilitates a similar conversation about the last problem, $\cos x = \frac{1}{2}$. The teacher looks for students who are beginning to use the previous equation to help and elicits those relationships.

$$\cos x = \frac{1}{2}$$

$$\cos x = \frac{1}{2} \qquad 0 \leq x \leq 2\pi$$

$$x = \frac{\pi}{3}, \frac{5\pi}{3}$$

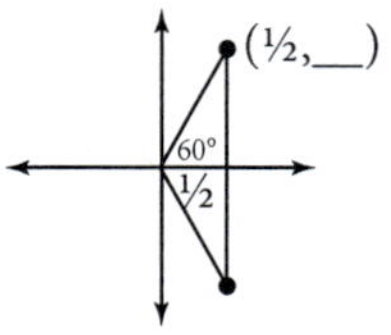

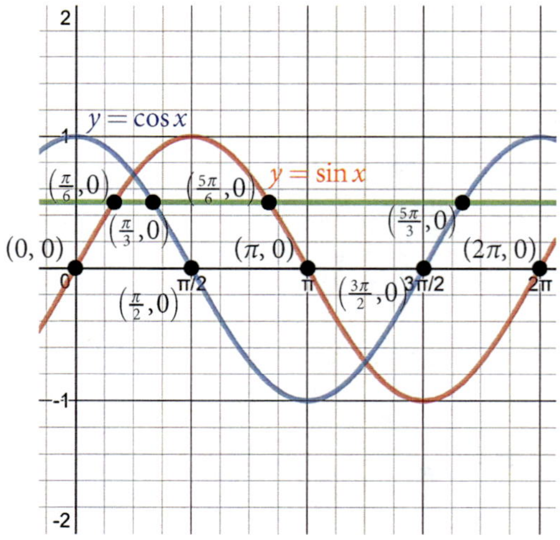

Teacher: *How would you summarize some of the things that came up in this string today?*

Elicit the following:

- *You can use special right triangles to help you find the exact solutions to some trigonometric equations.*
- *If you use right triangles, you have to consider the periodic nature of the trigonometric functions and consider all of the possible answers within the specified domain.*
- *The sine and cosine functions are translations of each other. This can help find solutions to equations.*

Sample Final Display

Your display could look like this at the end of the problem string:

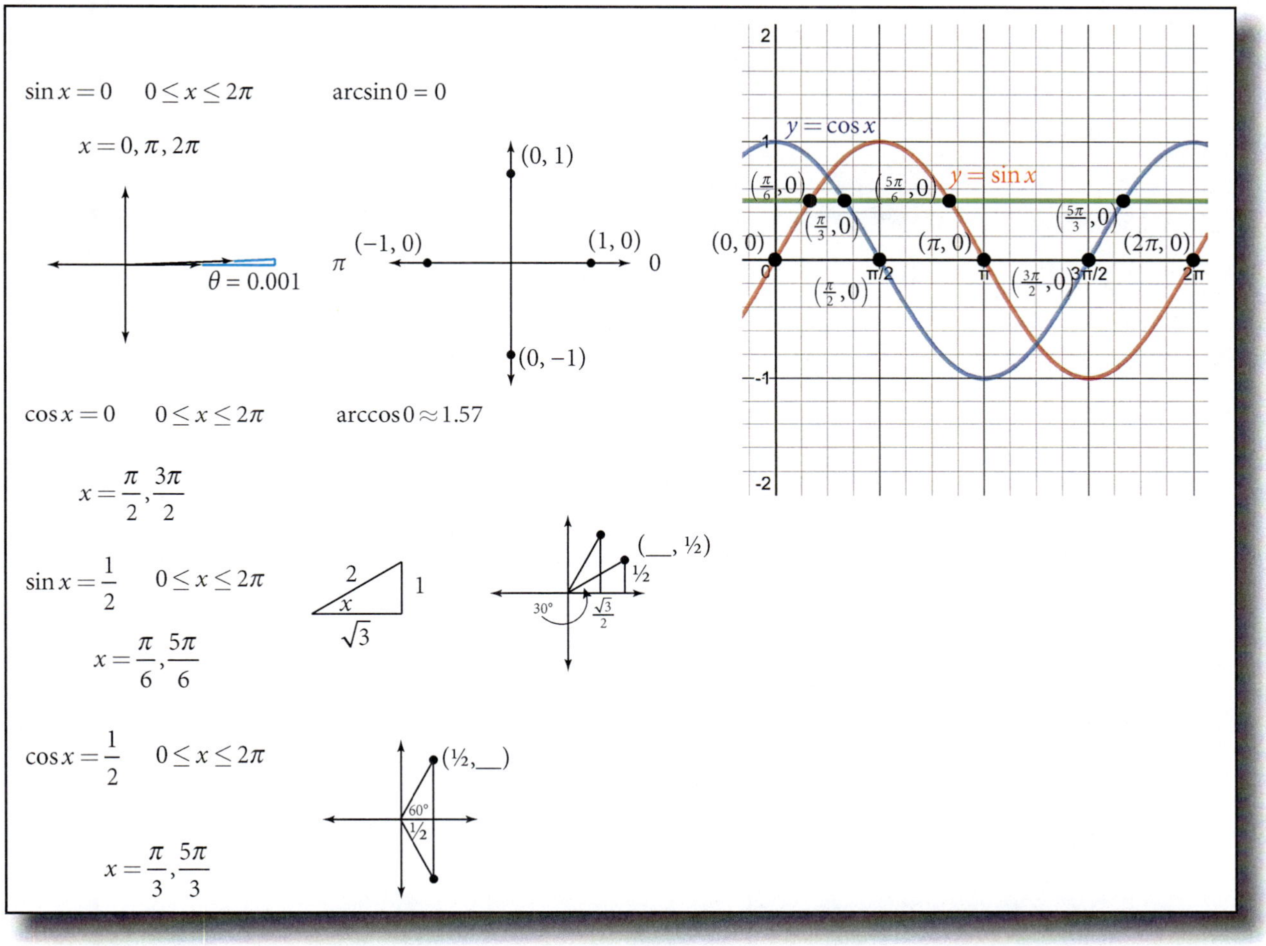

(continued)

Facilitation Notes

This version of the problem string lists short notes for important teacher moves during the string. After you've done the string yourself and studied the relationships involved, you might make similar notes for the things you want a reminder of or deem important.

$\sin x = 0$	$0 \le x \le 2\pi$	What is x if the sine of x is 0? Model thinking about right triangles, unit circle, arcsine, graph of y = sin x. Make sure you get both solutions!
$\cos x = 0$	$0 \le x \le 2\pi$	Repeat. Quicker. Add y = cos x and solutions to class graph. How are y = sin x and y = cos x related? Could that help?
$\sin x = \frac{1}{2}$	$0 \le x \le 2\pi$	Repeat. Do you know anything about one-half and special angles that could help? Add y = 0.5 and solutions to class graph. How are y = sin x and y = cos x related? Could that help?
$\cos x = \frac{1}{2}$	$0 \le x \le 2\pi$	Repeat. Could the previous problem help? How? Add solutions to class graph.

PROBLEM STRINGS
LESSON

8.1 Probability

At a Glance

What's the probability, with two 6-sided dice or one 12-sided die, of rolling a total of:

exactly 6?

exactly 12?

exactly 2?

exactly 4?

What else has probability of $\frac{3}{36}$?

Objectives

The goal of this short problem string is to get students to figure several probabilities to compare them in a situation that might seem the same, rolling one 12-sided die versus rolling two 6-sided dice.

Placement

This problem string can be a quick reminder for students of theoretical probability of events that have equally likely outcomes.

You can use this problem string to introduce textbook Lesson 8.1 Randomness and Probability.

Guiding the Problem String

This problem string is a set of problems based on the same context, comparing the probabilities of several outcomes of two seemingly, but not, equivalent situations: rolling a 12-sided die versus rolling two 6-sided dice. Use the first problem to remind students, if needed, how to find the probability of an event and to model a way of organizing the possible outcomes of rolling two 6-sided dice using a grid. The second and third problems should go quickly because there is only one possible way to get either a 12 or 2 in both scenarios. The fourth problem should only take a little longer. Press students to justify how they know they have them all. The last problem turns the question around asking students to identify a target amount that also has a probability of $3/36$. You can end the string nudging students towards notions of the expected value of each scenario by asking which scenario they think would win over time.

You can use a scenario of a role playing game, where a great axe and a great sword inflict damage by rolling a 12-sided die or two 6-sided dice respectfully. Which weapon would you choose if your goal is to inflict the most damage most of the time?

About the Mathematics

The events (outcomes of each roll) in these scenarios are each equally likely because the dice are fair and therefore each face has the same chance of landing up. The probability being found in this string is theoretical which is found by counting the number of ways a desired event can happen and comparing this number to the total number of equally likely possible outcomes.

(continued)

Sample Interactions

Use the following as you plan how to elicit and model student strategies. This is not meant as a script, but as a view into the relationships involved and the intent of the problem string.

Teacher: *Does anyone like to play board games? How about role playing games? Our problem string today has some questions about a role playing game. In this game, if you choose to use a great sword, you roll two 6-sided dice to determine the damage done, but if you choose to use a great axe, you roll one 12-sided die. I'm wondering if the game makers are just wanting you to buy extra dice. I wonder if the probabilities are all the same for these weapons because you're basically rolling the same thing, right? The first problem is to find the probability of getting exactly a score of 6 either rolling two 6-sided dice or one 12-sided die. Do you think the probabilities are the same?*

Brief think time.

What's the probability?

rolling a score of:	one 12-sided die	two 6-sided dice
exactly 6?		

Teacher: *Before we get going too far, someone remind us all please. What is probability?*

Student: *It's like the chance that something will happen.*

Student: *You need to know the number of times something can happen and the total.*

Teacher: *The total?*

Student: *Yeah, like the total number of things.*

Teacher: *Let's talk about the 12-sided die. How many different ways can you get exactly a 6?*

Student: *Just one way, rolling a 6.*

Teacher: *How many possible outcomes are there when you roll that 12-sided die?*

Student: *Twelve.*

Teacher: *Here is a definition of the probability of an event. In this case, rolling a 6 can occur just one time out of the total 12 equally likely outcomes possible.*

$$P(E) = \frac{\textit{number of different ways an event can occur}}{\textit{total number of equally likely outcomes possible}}$$

What's the probability?

rolling a score of:	one 12-sided die
exactly 6?	$\frac{1}{12}$

Teacher: *What about the probability of getting exactly 6 when rolling two 6-sided dice?*

The teacher circulates while students work, modeling the total outcomes with a grid like the following.

Teacher: *Tell us about your thinking.*

Student: *We were counting all of the ways to add to six, but we weren't sure we had them all. You came over and started writing down the ones we had already found.*

Teacher: *How was that helpful?*

Student: *It helped us think about how to organize it all.*

Teacher: *So, I'll project a grid like the one we were working on. Someone tell us what you see.*

Student: *Ahhh, yes, I've seen something like that before. One die is down the side and one across.*

Student: *So, what are the stars?*

Student: *Possible totals. The top left where the one and one meet is the sum of 2. The top right is one and six, so seven.*

red die (across), **blue die** (down)

blue die \ red die	1	2	3	4	5	6
1	*	*	*	*	*	*
2	*	*	*	*	*	*
3	*	*	*	*	*	*
4	*	*	*	*	*	*
5	*	*	*	*	*	*
6	*	*	*	*	*	*

Student: *Right, so now I can find all of the 6s. So, one and five....*

As the student lists the combinations and directs the teacher to them, the teacher circles them.

Teacher: *So how many possible ways to get 6 and how many total possibilities?*

$$P(E) = \frac{\textit{number of different ways an event can occur}}{\textit{total number of equally likely outcomes possible}}$$

What's the probability?

rolling a score of:	one 12-sided die	with two 6-sided dice
exactly 6?	$\frac{1}{12}$	$\frac{5}{36}$

red die (across), **blue die** (down)

blue die \ red die	1	2	3	4	5	6
1	*	*	*	*	*	*
2	*	*	*	*	*	*
3	*	*	*	*	*	*
4	*	*	*	*	*	*
5	*	*	*	*	*	*
6	*	*	*	*	*	*

(continued)

Student: *So, if you want a 6, the probability is better if you roll two 6-sided dice.*

Teacher: *Do you think that is true for every possibility? Let's check out another one. The next problem today is to find the probability of getting exactly 12. Same two options, 12-sided die or two 6-sided dice.*

This and the next problem, getting exactly 2, go quickly, with students establishing that there is only one possible way to get 12 or to get 2, $1/12$ for one 12-sided die or $1/36$ for two 6-sided dice.

$$P(E) = \frac{\textit{number of different ways an event can occur}}{\textit{total number of equally likely outcomes possible}}$$

What's the probability?

rolling a score of:	one 12-sided die	with two 6-sided dice
exactly 6?	$\frac{1}{12}$	$\frac{5}{36}$
exactly 12?	$\frac{1}{12}$	$\frac{1}{36}$
exactly 2?	$\frac{1}{12}$	$\frac{1}{36}$

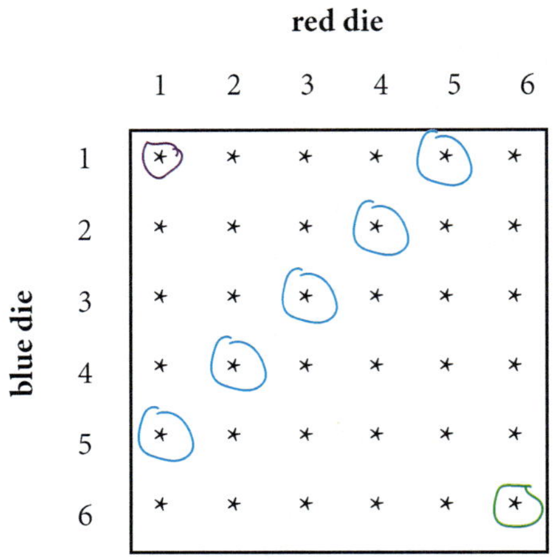

Teacher: *What do you think about these results so far?*

Student: *It's much more likely that you'll get the high 12 if you roll the one 12-sided die.*

Student: *But you're equally likely to roll a really low 1.*

Student: *With the two 6-sided dice, you're not very likely, one out of 36, to get a 12 or to get a 2.*

Teacher: *What do you think about the next problem, getting exactly 4? Go!*

After brief work time, the teacher elicits thinking from students.

$$P(E) = \frac{\textit{number of different ways an event can occur}}{\textit{total number of equally likely outcomes possible}}$$

What's the probability?

rolling a score of:	one 12-sided die	with two 6-sided dice
exactly 6?	$\frac{1}{12}$	$\frac{5}{36}$
exactly 12?	$\frac{1}{12}$	$\frac{1}{36}$
exactly 2?	$\frac{1}{12}$	$\frac{1}{36}$
exactly 4?	$\frac{1}{12}$	$\frac{3}{36} = \frac{1}{12}$

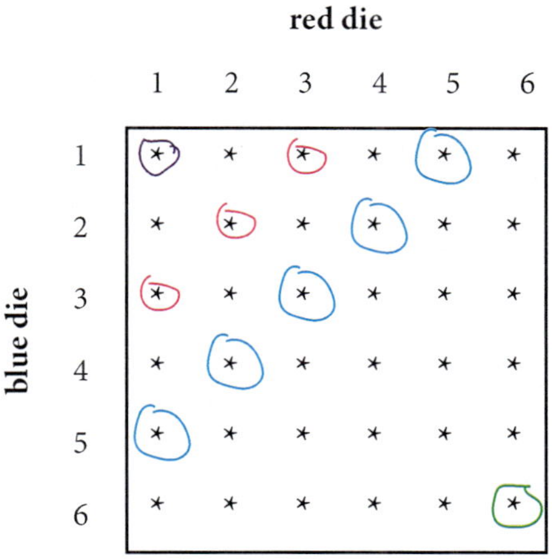

Teacher: *The last problem of the string is to find other values that have the probability of $^3/_{36} = {}^1/_{12}$.*

Student: *You mean with either the 12-sided or the two 6-sided?*

Teacher: *Yes, in both scenarios, find combinations that have that probability.*

What's the probability?

rolling a score of:	one 12-sided die	with two 6-sided dice
exactly 6?	$\frac{1}{12}$	$\frac{5}{36}$
exactly 12?	$\frac{1}{12}$	$\frac{1}{36}$
exactly 2?	$\frac{1}{12}$	$\frac{1}{36}$
exactly 4?	$\frac{1}{12}$	$\frac{3}{36} = \frac{1}{12}$
What's another combination for:	$\frac{1}{12}$	$\frac{3}{36} = \frac{1}{12}$

(continued)

The teacher ask students to share and records their ideas.

Teacher: *How do you know you have them all?*

$$P(E) = \frac{\textit{number of different ways an event can occur}}{\textit{total number of equally likely outcomes possible}}$$

What's the probability?

rolling a score of:	one 12-sided die	with two 6-sided dice
exactly 6?	$\frac{1}{12}$	$\frac{5}{36}$
exactly 12?	$\frac{1}{12}$	$\frac{1}{36}$
exactly 2?	$\frac{1}{12}$	$\frac{1}{36}$
exactly 4?	$\frac{1}{12}$	$\frac{3}{36} = \frac{1}{12}$
What's another combination for:	$\frac{1}{12}$ all of them!	$\frac{3}{36} = \frac{1}{12}$ score of exactly 10

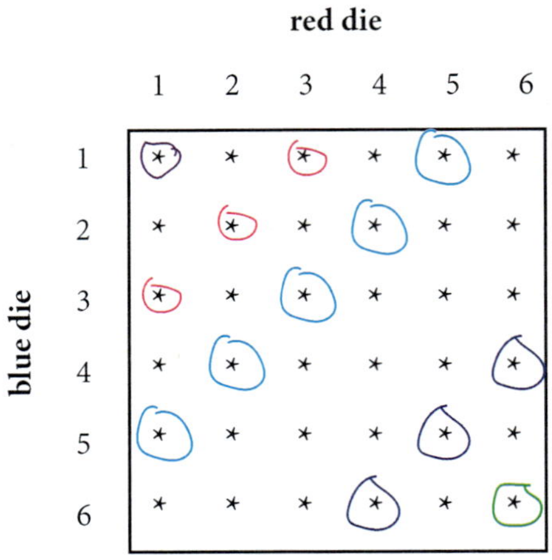

The teacher revisits students initial predictions about the two scenarios.

Teacher: *So, great sword or great axe?*

Teacher: *How would you summarize some of the things that came up in this string today?*

Elicit the following:

- *The probability of an event is the number of different ways an event can occur divided by the total number of equally likely outcomes possible.*
- *Organizing the total possible outcomes can be helpful.*

Sample Final Display

Your display could look like this at the end of the problem string:

$$P(E) = \frac{\textit{number of different ways an event can occur}}{\textit{total number of equally likely outcomes possible}}$$

What's the probability?

rolling a score of:	one 12-sided die	with two 6-sided dice
exactly 6?	$\frac{1}{12}$	$\frac{5}{36}$
exactly 12?	$\frac{1}{12}$	$\frac{1}{36}$
exactly 2?	$\frac{1}{12}$	$\frac{1}{36}$
exactly 4?	$\frac{1}{12}$	$\frac{3}{36} = \frac{1}{12}$
What's another combination for:	$\frac{1}{12}$ all of them!	$\frac{3}{36} = \frac{1}{12}$ score of exactly 10

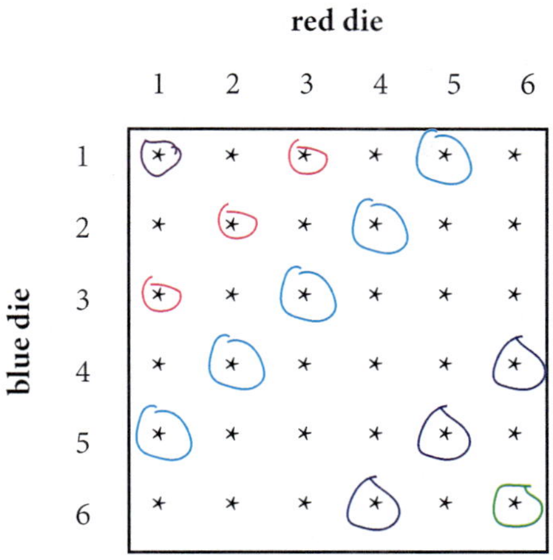

Facilitation Notes

This version of the problem string lists short notes for important teacher moves during the string. After you've done the string yourself and studied the relationships involved, you might make similar notes for the things you want a reminder of or deem important.

What's the probability of rolling a score of the following with a 12-sided die or two 6-sided dice?

exactly 6?	*Surely these are the same? Find the probability for a 12-side die and two 6-sided dice.* *What is probability when events are equally likely?* *Model organizing the total outcomes with a grid.*
exactly 12?	*Find both probabilities. How do the probabilities compare?* *Quick.*
exactly 2?	*Repeat.* *Quick.*
exactly 4?	*Repeat.* *Press for justification that they found them all.*
What's another combination for $\frac{3}{36} = \frac{1}{12}$	*Reverse it. What are combinations that have the probability of 3/36?* *Which weapon do you think will win over time?*

PROBLEM STRINGS
LESSON

8.2 Multiplication Rules of Probability

At a Glance

Which has the greater probability?

- 90% then 10% or 10% then 90%
- ¼ then ½ or ½ then ¼
- 84% then 75% or 75% then 84%
- 24% then 5% or 5% then 24%
- 20% then 32% or 30% then 22%

(contexts for each question follow)

Objectives

The goal of this problem string is to give students lots of experience with the multiplication rule of probability having students compare different scenarios and use numerical reasoning to find the probabilities. The string encourages students to let the numbers influence the multiplication strategy, strengthening their numeracy and probability at the same time.

Placement

This problem string is meant to follow an introduction to the multiplication rule of probability. The string will help students find the probabilities of compound events with multiple outcomes joined by the word "and."

You can use this problem string with textbook Lesson 8.2 Multiplication Rules of Probability.

Guiding the Problem String

This problem string consists of five scenarios meant to given students experience with compound events where students multiply the probability of simple events to get the probability of compound events joined by the word "and." The first four problems are designed to have equal probabilities for two scenarios that might sound like they do not. This equivalence can cause intrigue and nudge students to realize that they can use the commutative property of multiplication to make finding the probabilities easier. Each problem must be delivered in context or the necessary disequilibrium to prompt intrigue may not occur. For the first problem, model student thinking with a tree diagram. As students notice the equivalence, wonder aloud if that will always be true. Make sure you write the second problem using fractions as this might delay the realization just a bit and also give students a different way of reasoning about the equivalence, with fraction multiplication. After students have solved problem three, ask students if anyone considered which problem to solve first. Model finding three-fourths of 84. Ask the fourth problem specifically wondering which relationship might be easier to use since the answers will be equivalent. The last problem is the only one in the string where the probabilities in question are not equivalent. By this point, students should be reasoning about the numbers involved. Encourage them to think about how they could use the numbers to find the probabilities based on relationships they know.

About the Mathematics

If n_1, n_2, n_3, and so on, represent events, then the probability that this sequence of events will occur can be found by multiplying the conditional probabilities of the events.

$$P(n_1 \text{ and } n_2 \text{ and } n_3 \text{ and} \ldots) = P(n_1) \cdot P(n_2 | n_1) \cdot P\big(n_3 | (n_1 \text{ } and \text{ } n_2)\big) \cdots$$

Since you are multiplying the conditional probabilities, you can consider using the commutative property if it makes the multiplication easier. For example, many people find 75% of 44% easier than 44% of 75%, or especially easier than $0.44 \cdot 0.75$.

Sample Interactions

Use the following as you plan how to elicit and model student strategies. This is not meant as a script, but as a view into the relationships involved and the intent of the problem string.

Teacher: *Okay, we've been working with this multiplication rule of probability. Let's do a quick problem string to give you some practice. The first problem involves two locations. One is a dry, desert place where the chance of rain is 10%. If it rains there, the chance of it being a thunderstorm is 90%. Does anyone know of a place like that? Here's the other place. It's a coastal, rainy city where the chance of rain is 90%. If it rains there, the chance of a thunderstorm is only 10%. So, I'm wondering which place has the higher probability of a thunderstorm?*
Does everyone understand the question? Someone repeat it for me? What's your task?

Student: *Which place has the higher probability of a thunderstorm?*

Teacher: *Before you start calculating, predict with your partner. What's your guess? Then figure it out.*

Students work and the teacher circulates, looking for students who are reasoning about the multiplication, especially finding 10%.

Which has the higher probability?

Of a thunderstorm:
Desert area—10% chance of rain, if it rains 90% chance of thunderstorm
Coastal city—90% chance of rain, if it rains 10% chance of thunderstorm

Teacher: *Some of you are smiling. What's up?*

Student: *We found that they have the same probability.*

Teacher: *One of you tell us about the desert*

Student: *We know that it's a 10% chance that the desert gets rain and then if it does, 90% that there's a thunderstorm. That's 90% of the 10%, 0.09 or 9%.*

Teacher: *It sounds like you thought about 90% of 10%, even though the problem has 10% listed first, then the 90%. What's up?*

Student: *Well, the chance of thunderstorms is only 90% of the 10% chance of rain.*

90% of 10% is 9%

Teacher: *I saw you two drawing a tree diagram. Tell us about that.*

Student: *We drew the tree so that it had 10% rain on top and 90% no rain on the bottom. Then we only drew from the rain the 90% chance of thunderstorm. Yes, like that. So we knew it was the 10% times the 90%.*

Teacher: *It looks like you thought of the multiplication differently?*

Student: *Yeah, they thought about 90% of 10% and we multiplied 0.1 times 0.9.*

Teacher: *Do both work?*

Students: *Yes.*

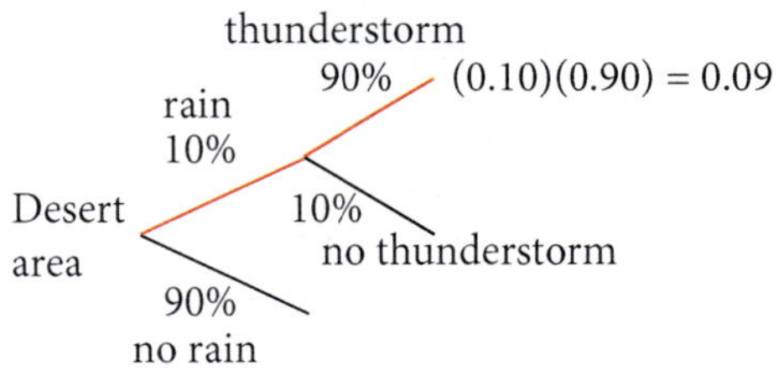

(continued)

Teacher: *What about the coastal city?*

Student: *The coastal city has 90% chance of rain and if it does, then 10% chance of a thunderstorm, so that's 10% of the 90%. That's also 9%.*

Student: *And we drew the tree diagram.*

The teacher draws the tree diagram as the student explains.

Teacher: *It looks like we have multiplication happening with the coastal city in two different orders as well.*

Which has the higher probability?

Of a thunderstorm:
Desert area—10% chance of rain, if it rains 90% chance of thunderstorm
Coastal city—90% chance of rain, if it rains 10% chance of thunderstorm

9% 90% of 10% is 9% 10% of 90% is 9%

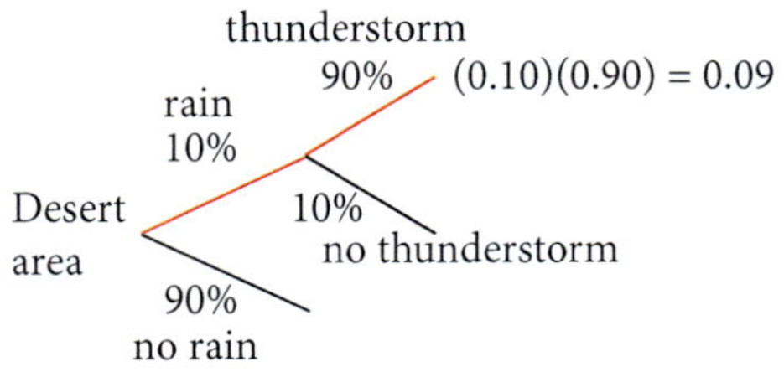

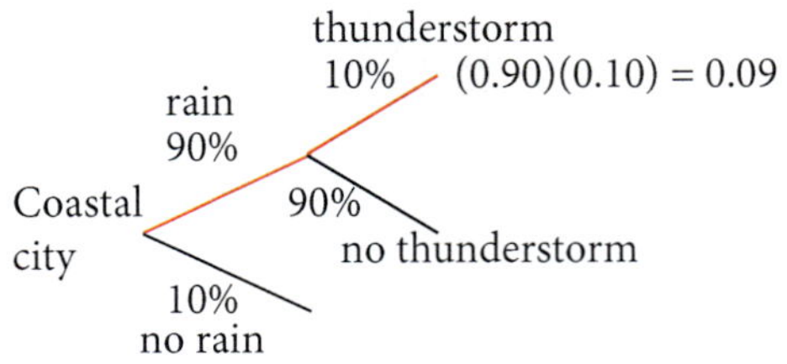

Teacher: *That's interesting. Many of you were convinced that the coastal city has such a high chance of rain that it would also have the higher chance of thunderstorms. But actually it doesn't. Interesting. I heard you two wondering about something. What was that?*

Student: *We were saying that both problems ended up with 0.9 times 0.1 or 0.1 times 0.9. That's why the answers are the same.*

Teacher: *That's the commutative property of multiplication at work.*

$(0.10)(0.90) = (0.90)(0.10)$

Teacher: *Our next problem today deals with those electronic claw games, where you pay to operate a claw to try to snag an item in the bin. We've got two bins full of plush animal toys. In bin #1, one-fourth of the toys are dogs, and one-half of those toy dogs have blue eyes. In bin #2, half of the plush toys have blue eyes, and one-fourth of those blue eyed toys are dogs. Which claw game bin has the higher probability of getting a blue-eyed dog toy?*

Of grabbing a blue-eyed dog:
Electronic Claw Game #1—¼ are dog plush toys, if you got a dog ½ have blue eyes
Electronic Claw Game #2—½ of the plush toys have blue eyes, if you got a blue-eyed toy ¼ are dogs.

<table>
<tr><td>
As the teacher circulates, some students are working at finding the probabilities and some students are smirking. The teacher challenges those who think they know that the probabilities are the same to prove it.
Teacher: So, if I want a blue-eyed plush toy, which claw game should I play?
Student: They are the same.
Teacher: Are you sure? How do you know?
Student: It's the same numbers just flipped.
Student: You're multiplying the one-half and the one-fourth, just in different orders.
Teacher: Does everyone agree with them? Does the product of one-half times one-fourth equal one-fourth times one-half?
</td><td>Is $\frac{1}{4} \times \frac{1}{2} = \frac{1}{2} \times \frac{1}{4}$? Yes!</td></tr>
<tr><td>
Teacher: You guys were doing some interesting work. Tell us how you figured those.
Student: To find a fourth of a half, we knew that a half of a half is a fourth. Then a half of that fourth is an eighth.
Student: And for the other one, one-half of a fourth is just one-eighth.
Teacher: It sounds like you guys were not following some rule of fraction multiplication, but just thinking of the relationships. Nice work.
</td><td>$\frac{1}{4} \times \frac{1}{2} = \frac{1}{8}$ ($\frac{1}{2}$ of $\frac{1}{2}$ is $\frac{1}{4}$, so $\frac{1}{2}$ of $\frac{1}{4}$ is $\frac{1}{8}$)
$\frac{1}{2} \times \frac{1}{4} = \frac{1}{8}$</td></tr>
<tr><td colspan="2">
Teacher: Do you think it's just these two examples or can you figure out probabilities like these in either order of multiplication? I wonder why you would even care. So, the next problem of our string is about schools closing for bad weather, either for too cold of temperatures or for too much snow. We've got one school in a cold state with an 84% chance of having temperatures below 0°F. If the temperature is below 0°F, there is a 75% chance they will close school. A different school is in a snowy town that has a 75% chance of snow. If it snows, there's an 84% chance they will close school. Tell me what you're thinking about these schools?

Of a school closing:
Cold state—84% chance of temperatures below 0°F, if it's below 0°F 75% chance of schools closing
Snowy town—75% chance of snow, if it snows 84% chance of school closing
</td></tr>
<tr><td colspan="2">
Student: You're not going to trick us this time. The probabilities are the same!
Teacher: I can't get anything past you, can I? Fine, but the question is, what is the probability that school will close?
As students work, the teacher finds students who found the probability of school closing by multiplying 0.84 by 0.75, without considering using the commutative property.
</td></tr>
<tr><td>
Teacher: What is the probability?
Student: We multiplied 0.84 times 0.75 in our calculators and got 0.63 so 63%.
Teacher: Which place were you thinking about?
Student: We thought about the cold state. The first probability is 84% so we multiplied that by the next one, 75%.
</td><td>63%

$0.84 \cdot 0.75 = 0.63$</td></tr>
</table>

(continued)

<table>
<tr><td>

Teacher: *Did anyone think about finding the probability of snowy town first? Or in other words using the commutative property?*

Student: *No, why would we do that?*

Student: *Oh! I know. Because you can think about 75% of 84. One-quarter of 84 is 21, then times 3 is 63.*

Teacher: *So you can think about 75% as three-quarters and think about three quarters as three one-quarters? Nice. So, it might make sense to consider the numbers first, before you choose a multiplication strategy.*

</td><td>

$0.75 \cdot 84 = \frac{1}{4} \cdot 84 \cdot 3 = 21 \cdot 3 = 63$

</td></tr>
<tr><td colspan="2">

Teacher: *New scenario. What if there is a windy town, like Emerald City, where there is a 24% chance of high winds. If there are high winds, there's a 5% chance of a tornado. Scary! Another place called Town of Oz has only a 5% chance of high winds, but if there are high winds, there is a 24% chance of a tornado. Which of these do you think you might want to figure out first?*

Student: *Or only.*

Teacher: *What do you mean?*

Student: *They're the same, so we only need to figure out one.*

Teacher: *Choose wisely. Let's see what relationships you can use.*

Of a tornado:
Emerald City—24% chance of high winds, if there are high winds, 5% chance of a tornado
Town of Oz—5% chance of high winds, if high winds, 24% chance of a tornado

</td></tr>
<tr><td>

As students work, the teacher finds students who found the probability using relationships they know.

Teacher: *What is the probability?*

Student: 1.2%

Teacher: *I overheard what you were thinking about. Please explain.*

Student: *We thought about finding twenty-four 0.05s, like twenty-four nickels. That's just twenty nickels and ...*

Teacher: *And twenty nickels is?*

Student: $1.00 *and then four more nickels is 20 cents, so 24 nickels is* $1.20. *So scale down to* 2.4 *and* 0.12 *and then to* 0.24 *and* 0.012. *That's* 1.2%.

</td><td>

using 24 nickels:

1	20	4	24	2.4	0.24
0.05	1.00	0.20	1.20	0.12	0.012

</td></tr>
<tr><td>

Teacher: *Did anyone think about finding five 24s?*

Student: *We did. We know that ten 24s is 240, so five is half of that, so 120. But it's 5% and 24%, so then you need four place-value shifts, 1.2%.*

Teacher: *So, you can use your calculators, but it's pretty cool when you use your brain first.*

</td><td>

$10 \cdot 24 = 240$

$5 \cdot 24 = 120$

$0.5 \cdot 0.24 = 0.012$

</td></tr>
</table>

Teacher: *The last problem of today is about free-throws. Sometimes in basketball a coach gets to assign who will shoot free-throws, like after a technical foul. If the coach needs to decide between these two players, who do you think the coach should pick?*

Making the free throw:
Player #54—20% chance of making the first shot, if makes the first shot 32% chance of making the second shot
Player #36—30% chance of making the first shot, if makes the first shot 22% chance of making the second shot

Student: *Drat, this time they're not the same.*

Teacher: *Are you sure? The numbers look close to me. I wonder if you can use what you know to reason about these probabilities?*

As students share, the teacher models their thinking.

Making the free throw:
Player #54—20% chance of making the first shot, if makes the first shot 32% chance of making the second shot
Player #36—30% chance of making the first shot, if makes the first shot 22% chance of making the second shot

Player #36 has 6.6%

20% · 32%   10% of 32% is 3.2%, then double that to get 20% of 32% is 6.4%

2 · 32% is 64%, then scale down to 6.4%

30% · 22%   10% · 22% is 2.2%, then triple that to get 30% of 22% is 6.6%

3 · 22% is 66%, then scale down to 6.6%

Teacher: *Nicely reasoned through everyone! It is good to use what we know and let the numbers suggest a strategy. Which of the strategies do you like the most? Which do you wish your brain would go toward the next time you encounter numbers like these?*

Teacher: *How would you summarize some of the things that came up in this string today?*

Elicit the following:

- *If you are finding the probability of a sequence of events, you can multiply the conditional probabilities.*
- *When you multiply, you can look for relationships that might make the multiplication easier.*
- *You can use the commutative property to make the multiplication easier.*
- *Those basketball players had terrible free-throw percentages. They should practice more!*

(continued)

Sample Final Display

Your display could look like this at the end of the problem string:

Which has the higher probability?

Of a thunderstorm:
Desert area—10% chance of rain, if it rains 90% chance of thunderstorm
Coastal city—90% chance of rain, if it rains 10% chance of thunderstorm

9% 90% of 10% is 9% 10% of 90% is 9%

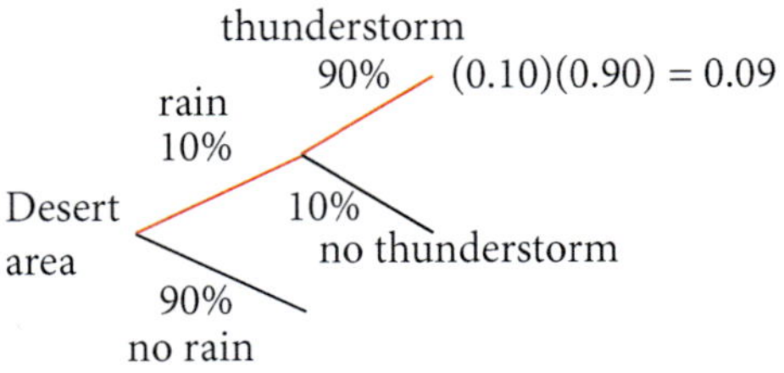

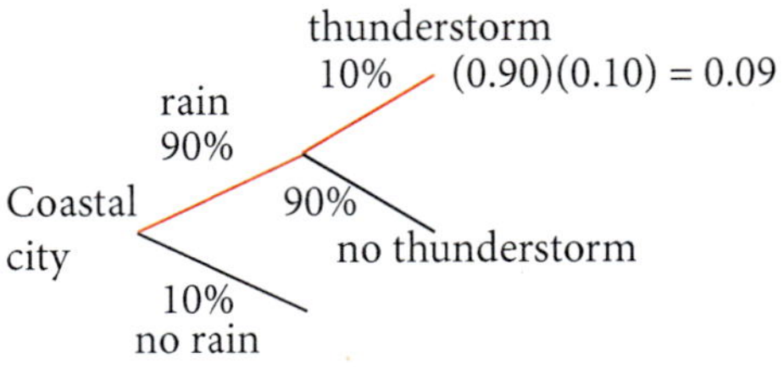

Of grabbing a blue-eyed dog:
Electronic Claw Game #1—$\frac{1}{4}$ are dogs, if you got a dog $\frac{1}{2}$ have blue eyes
Electronic Claw Game #2—$\frac{1}{2}$ of the plush toys have blue eyes, if you got a blue-eyed toy $\frac{1}{4}$ are dogs.

$\frac{1}{8}$ Is $\frac{1}{4} \times \frac{1}{2} = \frac{1}{2} \times \frac{1}{4}$? Yes! $\frac{1}{4} \times \frac{1}{2} = \frac{1}{8}$ ($\frac{1}{2}$ of $\frac{1}{2}$ is $\frac{1}{4}$, so $\frac{1}{2}$ of $\frac{1}{4}$ is $\frac{1}{8}$)
$\frac{1}{2} \times \frac{1}{4} = \frac{1}{8}$

Of a school closing:
Cold state—84% chance of temperatures below 0°F, if it's below 0°F 75% chance of school's closing
Snowy town—75% chance of snow, if it snows 84% chance of school closing

63% $0.84 \cdot 0.75 = 0.63$ $0.75 \cdot 84 = \frac{1}{4} \cdot 84 \cdot 3 = 21 \cdot 3 = 63$

Of tornados:
Emerald City—24% chance of high winds, if there are high winds 5% chance of tornado
Town of Oz—5% chance of high winds, if high winds 24% chance of tornados

1.2%

using 24 nickels:

1	20	4	24	2.4	0.24
0.05	1.00	0.20	1.20	0.12	0.012

5 24s: $10 \cdot 24 = 240$
$5 \cdot 24 = 120$
$0.5 \cdot 0.24 = 0.012$

Making the free throw:
Player #54—20% chance of making the first shot, if makes the first shot 32% chance of making the second shot
Player #36—30% chance of making the first shot, if makes the first shot 22% chance of making the second shot

Player #36 has 6.6%

20% · 32% 10% of 32% is 3.2%, then double that to get 20% of 32% is 6.4%

2 · 32% is 64%, then scale down to 6.4%

30% · 22% 10% · 22% is 2.2%, then triple that to get 30% of 22% is 6.6%

3 · 22% is 66%, then scale down to 6.6%

Facilitation Notes

This version of the problem string lists short notes for important teacher moves during the string. After you've done the string yourself and studied the relationships involved, you might make similar notes for the things you want a reminder of or deem important.

Which has the higher probability?	
Of a thunderstorm: Desert area—10% chance of rain, if it rains 90% chance of thunderstorm Coastal city—90% chance of rain, if it rains 10% chance of thunderstorm	*Whoa! Where would you want to live?* *Predict w/ partner.* *Find probabilities.* *Same? Interesting.*
Of grabbing a blue-eyed dog: Electronic Claw Game #1—¼ are dogs, if you got a dog ½ have blue eyes Electronic Claw Game #2—½ of the plush toys have blue eyes, if you got a blue-eyed toy ¼ are dogs	*Read as fractions.* *Find probabilities.* *Still the same? Prove it.* *I wonder if you'd ever want to use that? Hmm...*
Of a school closing: Snowy town—75% chance of snow, if it snows 84% chance of school closing Cold state—84% chance of temperatures below freezing, if it's below freezing 75% chance of school's closing	*Same again? Fine, what is either probability?* *Anyone find the cold state first? Why would you?*
Of tornado: Emerald City—24% chance of high winds, if there are high winds 5% chance of tornado Town of Oz—5% chance of high winds, if high winds 24% chance of tornado	*Which of these would you rather choose to find?* *Let's see what relationships you use.* *Why? How?* *Model 24 5s and 5 24s and place-value shifts.*
Making the free throw: Player #54—20% chance of making the first shot, if makes the first shot 32% chance of making the second shot Player #36—30% chance of making the first shot, if makes the first shot 22% chance of making the second shot	*Some times the coach can choose who shoots.* *Who should the coach choose?* *Model several strategies.* *Which do you wish your brain would go to next time?* *These guys should practice!*

PROBLEM STRINGS LESSON

8.3 Addition Rules of Probability

At a Glance

Probability That People Like Abby's Social Media Posts:

- P(Grandmother)
- P(friend)
- P(Mom)
- P(brother)
- P(Mom and Grandmother)
- P(Mom or Grandmother)
- P(friend and brother)
- P(friend or brother)
- P(brother or Mom)

Objectives

The goal of this problem string is to immerse students in a scenario where they figure out when to multiply probabilities and when to add probabilities.

Placement

This problem string deals with both "and" scenarios that use the multiplication rules of probability, and "or" scenarios that use the addition rules of probability. The string can support the work you have done to teach these relationships by having students compute both, while at the same time computing by reasoning about the numerical relationships.

You can use this problem string to support the work in textbook Lesson 8.3 Addition Rules of Probability.

Guiding the Problem String

This string of problems is based on one scenario of a teen counting the number of reactions she has received on 40 social media posts and then using those percentages to predict the probability of reactions on future posts by the same people. The first four problems are the simple probabilities of four people reacting. The next four problems are in helper-problem sets, where the first problem's answer can be used to solve the second problem. The last problem is given without a helper. All of the computations are accessible, so encourage students to think before they grab a calculator. If needed to keep students in the context, ask students to predict how an answer will relate to a previous problem.

About the Mathematics

The general addition probability rule is that if n_1 and n_2 represent event 1 and event 2, then the probability that at least one of the events will occur can be found by adding the probabilities of the events and subtracting the probability that both will occur.

$$P(n_1 \text{ or } n_2) = P(n_1) + P(n_2) - P(n_1 \text{ and } n_2)$$

Sample Interactions

Use the following as you plan how to elicit and model student strategies. This is not meant as a script, but as a view into the relationships involved and the intent of the problem string.

Teacher: *In today's problem string we meet Abby, a teenager who posts on social media. One day she was bored and she went back through her last 40 posts and noted who reacted to the posts in some way. She found that her grandmother reacted to 38 posts. If we take this data to be predictive, what is the probability that Grandma will react to a future post?* Students work briefly and the teacher circulates, encouraging students to reason using relationships to find the probability.	<u>"Reacting" on 40 Posts:</u> Grandmother: 38
Teacher: *What's the experimental probability of Grandma reacting?* **Student:** 95% **Teacher:** *Did everyone get 95%? Those of you who used a relationship, what were you thinking?* **Student:** *I thought about 38 out of 40, scaled down to 19 to 20, and then scaled up to 95 to 100.* **Teacher:** *Does anyone have any questions about this?* **Student:** *Why did you scale down to 20?* **Student:** *Because I was trying to get too 100 for 100%. I knew I could get there from 20. I used 20 times 5 to help me figure 19 times 5.*	$P(\text{Grandmother}) = 95\%$ $P(G) = \frac{38}{40} = \frac{19}{20} = \frac{95}{100}$ (÷ 2, then × 5, on numerator and denominator)
Teacher: *Anyone else think you had a nice use of relationships, nice enough that it wasn't worth pulling out your calculator?* **Student:** *I also started with 38 to 40, but I scaled down to 10. That's 9.5 to 10. Then scale that up to 95 to 100.* **Student:** *How do you know 38 divided by 4?* **Student:** *It's just like 36 divided by 4 and 2 divided by 4.*	$\frac{38}{40} = \frac{9.5}{10} = \frac{95}{100}$ (÷ 4, then × 10, on numerator and denominator) $\frac{38}{4} = \frac{36}{4} + \frac{2}{4} = 9.5$
Teacher: *So Grandmother is on social media a lot! Abby also noted that her friend reacted to her posts 30 out of the 40 times. If we take that as predictive, what's the probability that Abby's friend will react to a post?*	<u>"Reacting" on 40 Posts:</u> Grandmother: 38 friend: 30
Student: *Thirty out of 40, that's just three-fourths.* **Student:** *Seventy-five percent.* **Teacher:** *Everyone agree? Nice relationships. No calculator needed.*	$P(\text{friend}) = 75\%$

(continued)

The teacher gives the next two problems, the number of post reactions by Abby's mother and then brother, one at a time. The teacher encourages students to reason using relationships, and asks a couple of students to share their strategies as the teacher models the thinking.

<u>"Reacting" on 40 Posts:</u>

Grandmother:	38
friend:	30
Mom:	4
brother:	8

(Mom to brother: × 2)

$P(\text{Mom}) = 10\%$

× 2

$P(\text{brother}) = 20\%$

$P(M) = \dfrac{4}{40} = \dfrac{1}{10}$ (÷ 4, ÷ 4) $\qquad \dfrac{4}{40} = \dfrac{10}{100}$ (× 2.5, × 2.5)

$P(b) = \dfrac{8}{40} = \dfrac{4}{20} = \dfrac{20}{100}$ (÷ 2, ÷ 2; × 5, × 5) $\qquad \dfrac{8}{40} = \dfrac{1}{5} = \dfrac{20}{100}$ (÷ 8, ÷ 8; × 20, × 20) $\qquad \dfrac{8}{40} = \dfrac{2}{10}$ (÷ 4, ÷ 4)

Teacher: *Here's the next problem. What is the probability that Mom and Grandmother both react to a post?* The teacher circulates while students briefly work.	$P(\text{Mom and Grandmother})$
Teacher: *What is the probability that Mom and Grandma react?* **Student:** 9.5% *because* 10% *times* 95% *is* 9.5%. **Teacher:** *Why did you multiply?* **Student:** *It's Mom and Grandmother. That's the probability of Mom reacting and Grandma reacting. That's* 10% *of* 95%. **Teacher:** *Is everyone clear on that? No questions?*	$P(\text{Mom and Grandmother}) = 9.5\%$ $10\% \cdot 95\% = 9.5\%$
Teacher: *The next problem is to find the probability that Mom or Grandma react. Same question? Different question?* **Student:** *This is different. The "and" question was finding the probability that Grandmother reacts given that Mom has reacted. But this is the probability that either of them reacts. It's an "or" question.* **Teacher:** *Do you think the probability that either Mom or Grandma react is higher or lower than the probability that Mom and Grandma react?* **Student:** *The "or" should be higher because it can be either Mom or Grandma. So it counts if just one of them reacts.* **Teacher:** *Go ahead and find the probability. Use relationships.*	$P(\text{Mom or Grandmother})$

<table>
<tr><td>

Teacher: *What is the probability of Mom or Grandma reacting?*

Student: *It's* 95.5%.

Student: *I think it's 105%.*

Teacher: *Why?*

Student: *It's 105% because it's 10% and 95% so that's 105%.*

Student: *I agree that you have to think about adding their probabilities but then you have to subtract the overlap. You can't have 105% probability, right? That made me take another look and realize that I needed to subtract the overlap. So,* 10% + 95% *minus* 10% *times* 95%.

Teacher: *What do you think? Why? When do you have to worry about overlap? Turn to your partner. Also brainstorm efficient ways to compute.*

Students turn and talk while the teacher listens in.

Teacher: *Why subtract the overlap?*

Student: *They're not mutually exclusive. Mom can react. Grandma can react. So there is overlap.*

Teacher: *So in this case, the events, or reactions, are not mutually exclusive so you have to subtract the overlap.*

</td><td>

$P(\text{Mom or Grandmother}) = 95.5\%$

$10\% + 95\% - (10\% \cdot 95\%)$

</td></tr>
<tr><td>

Teacher: *Let's talk about how you computed. I heard you say you already knew part of it?*

Student: *Yeah, we had already figured the* 10% *times* 95%.

Teacher: *How did you use that?*

Student: *We figured out* 10% *plus* 95% *by thinking about money. It's like* 10 *cents and* 95 *cents then subtract the* 9.5 *cents, that's* 95.5%.

Teacher: *I'll model your thinking.*

</td><td>

$\frac{1}{10} + \frac{95}{100} - 9.5\% = \frac{10}{100} + \frac{95}{100} - \frac{9.5}{100} = \frac{95.5}{100} = 95.5\%$

</td></tr>
<tr><td colspan="2">

The teacher repeats with the next questions, the probability of the friend and brother reacting and then the probability of the friend or brother. In the discussion, the teacher asks someone to share who used the "and" answer to help with the "or" scenario. Because the order switches from "friend and brother" to "brother and friend" the teacher asks students to reason about whether order matters, in an "and" scenario and in an "or" scenario.

$P(\text{friend and brother}) = 15\%$ $\quad 75\% \cdot 20\% = (3 \cdot 25\%) \cdot 20\% = 3 \cdot (25\% \cdot 20\%) = 3 \cdot 5\% = 15\%$

$P(\text{brother or friend}) = 80\%$ $\quad 20\% + 75\% - (20\% \cdot 75\%)$ $\quad 75\% + 20\% - (75\% \cdot 20\%)$

$\frac{20}{100} + \frac{75}{100} - \frac{15}{100} = \frac{80}{100} = 80\%$ $\quad \frac{75}{100} + \frac{20}{100} - \frac{15}{100} = \frac{80}{100} = 80\%$

</td></tr>
</table>

(continued)

The teacher intentionally sets up the helper-problem sequence and tries to nudge students to think about the relationships. **Teacher:** *You might have noticed that these last four problems were in the form helper-problem, where the first problem in the set helps solve the second one. For this last one, there is no helper, you're on your own. Maybe you could come up with your own helper?* *I've noticed that some of you are still using your calculator every time. Well, you might really need it for this problem. I don't know if you even want to try to reason about these values. The last problem today is to find the probability that the brother or Mom reacts.* Students work briefly.	*P*(brother or Mom)
Teacher: *What is the probability that the brother or Mom reacts?* **Student:** *You tried to make us think that we needed to use our calculators, but this one was easy to just think about. It's just 20% + 10% minus 20% times 10%. That's just 30% minus 2% so 28%. Wow. Sometimes you really can just think.*	*P*(brother or Mom) = 28% $20\% + 10\% - (20\% \cdot 10\%)$ $30\% - 2\% = 28\%$

Teacher: *How would you summarize some of the things that came up in this string today?*

Elicit the following:

- *"And" questions are about multiplying probabilities because you are finding the probability of something given the probability of the other thing. A percentage of a percentage can be found by using relationships and multiplying.*
- *"Or" questions are about adding the probabilities. If the events are not mutually exclusive, then you have to subtract the overlap.*

Sample Final Display

Your display could look like this at the end of the problem string:

"Reacting" on 40 Posts:

Grandmother: 38
friend: 30
Mom: 4) × 2
brother: 8

$P(\text{Grandmother}) = 95\%$ $\quad P(G) = \frac{38}{40} = \frac{19}{20} = \frac{95}{100}$ (÷ 2, × 5) $\quad \frac{38}{40} = \frac{9.5}{10} = \frac{95}{100}$ (÷ 4, × 10) $\quad \frac{38}{4} = \frac{36}{4} + \frac{2}{4} = 9.5$

$P(\text{friend}) = 75\%$ $\quad P(f) = \frac{30}{40} = \frac{3}{4}$ (÷ 10)

$P(\text{Mom}) = 10\%$ $\quad P(M) = \frac{4}{40} = \frac{1}{10}$ (÷ 4) $\quad \frac{4}{40} = \frac{10}{100}$ (× 2.5)

× 2

$P(\text{brother}) = 20\%$ $\quad P(b) = \frac{8}{40} = \frac{4}{20} = \frac{20}{100}$ (÷ 2, × 5) $\quad \frac{8}{40} = \frac{1}{5} = \frac{20}{100}$ (÷ 8, × 20) $\quad \frac{8}{40} = \frac{2}{10}$ (÷ 4)

$P(\text{Mom and Grandmother}) = 9.5\%$ $\quad 10\% \cdot 95\% = 9.5\%$

$P(\text{Mom or Grandmother}) = 95.5\%$ $\quad 10\% + 95\% - (10\% \cdot 95\%)$

$$\frac{1}{10} + \frac{95}{100} - 9.5\% = \frac{10}{100} + \frac{95}{100} - \frac{9.5}{100} = \frac{95.5}{100} = 95.5\%$$

$P(\text{friend and brother}) = 15\%$ $\quad 75\% \cdot 20\% = (3 \cdot 25\%) \cdot 20\% = 3 \cdot (25\% \cdot 20\%) = 3 \cdot 5\% = 15\%$

$P(\text{brother or friend}) = 80\%$

$20\% + 75\% - (20\% \cdot 75\%)$ $\quad \frac{20}{100} + \frac{75}{100} - \frac{15}{100} = \frac{80}{100} = 80\%$

$75\% + 20\% - (75\% \cdot 20\%)$ $\quad \frac{75}{100} + \frac{20}{100} - \frac{15}{100} = \frac{80}{100} = 80\%$

$P(\text{brother or Mom}) = 28\%$ $\quad 20\% + 10\% - (20\% \cdot 10\%)$

$30\% - 2\% = 28\%$

(continued)

Facilitation Notes

This version of the problem string lists short notes for important teacher moves during the string. After you've done the string yourself and studied the relationships involved, you might make similar notes for the things you want a reminder of or deem important.

"Reacting" on 40 Posts: Grandmother: 38 friend: 30 Mom: 4 brother: 8	
P(Grandmother) P(friend) P(Mom) P(brother)	*If we think of this data as predictive, what is the probability that of Grandmother reacting to a future post?* *Share scaling up and down strategies.* *Repeat for next three problems.* *These four are quick.*
P(Mom and Grandmother)	*Find the probability of both reacting.* *Why multiply?*
P(Mom or Grandmother)	*Find the probability one or the other reacting.* *Why add? Why subtract the overlap?* *Did anyone use the previous problem? How?* *What's mutually exclusive?*
P(friend and brother) P(friend or brother)	*Repeat for the next two problems, one at a time.* *Encourage students to use relationships by having students share.* *How did you use the first to help with the second?* *Does order matter? Why?*
P(brother or Mom)	*Prior problems were in helper-problem sets.* *No helper this time.* *Encourage reasoning.*

PROBLEM STRINGS
LESSON

8.4 Bivariant Independence

At a Glance

For the survey results below, find the interval likely to contain the true proportion:

	Side Item: Salad	Side Item: Fries
Soda	12	8
Diet Soda	10	20

Diet soda and salad, $n = 50$

Diet soda and salad, $n = 100$

Diet soda and salad, $n = 1{,}000$

Diet soda and salad, $n = 10{,}000$

Objectives

The goal of this short problem string is for students to understand that to tighten the interval that contains the true proportion for bivariant data, they need a large sample size.

Placement

This problem string can support your work with bivariant data. Students should have already found marginal frequencies and been introduced to the margin of error for sample proportions.

You can use this problem string to support the work of textbook Lesson 8.4 Bivariant Independence.

Guiding the Problem String

The problems in this string are based in the context of a fictitious survey given to teens about what soda and sides they order with a meal, regular soda or diet and salad or french fries. In the first problem, spend some time discussing the scenario, the data, and the formula for the margin of error for sample proportions, and the result for diet soda and salad combination. For the second problem, ask students to estimate what will happen if the survey had the exact same proportions but double the sample size. Then calculate and discuss results. Repeat for the third problem. For the fourth problem, you can either ask students to estimate what sample size they might try to get within a hundredth of 0.2 and try their suggestions, or give them the sample size of 10,000. Keep the problems in context—talk about sample size, intervals, and the 20% who chose diet sodas and salads.

Use technology to calculate each of the intervals. Copy and paste equations and just change the necessary value.

About the Mathematics

Given a sample proportion, $p = \frac{x}{n}$, the interval likely to contain the true proportion is:

$$p \pm 2\sqrt{\frac{p(1-p)}{n}}$$

Important Questions

Use the following as you plan how to elicit and model student strategies.

- *What does this table mean? What do the numbers in the table mean?*
- *What are the sample proportions?*
- *How confident are you that the sample proportions are close to the true proportion?*
- *What is the formula to find the interval likely to contain the true proportion? What are all of the variables? What does the plus and minus operations mean? How can we use this to figure out an interval?*
- *How wide is the interval? How tight is the interval? What does the interval tell you? What doesn't the interval tell you?*
- *How does increasing the sample size affect the interval? Does that make sense? Why?*

(continued)

How would you summarize some of the things that came up in this string today?

- *You can find the sample proportion by dividing each group of responses by the total number of responses.*
- *The sample proportion is in an interval that contains the true proportion. That interval tightens as the sample size increases.*

Sample Final Display

Your display could look like this at the end of the problem string:

	Side Item	
	Salad	Fries
Soda	12	8
Diet Soda	10	20

	Side Item	
	Salad	Fries
Soda	0.24	0.16
Diet Soda	0.20	0.40

Margin of Error:

Diet soda and salad, $n = 50$ $0.08 <$ true proportion < 0.31

Diet soda and salad, $n = 100$ $0.12 <$ true proportion < 0.28

Diet soda and salad, $n = 1{,}000$ $0.17 <$ true proportion < 0.23

Diet soda and salad, $n = 10{,}000$ $0.19 <$ true proportion < 0.21

$$.20 + 2\sqrt{\frac{.2(1-.2)}{50}} = 0.31313708499$$

$$.2 - 2\sqrt{\frac{.2(1-.2)}{50}} = 0.0868629150102$$

$$.2 + 2\sqrt{\frac{.2(1-.2)}{100}} = 0.28$$

$$.2 - 2\sqrt{\frac{.2(1-.2)}{100}} = 0.12$$

$$.2 + 2\sqrt{\frac{.2(1-.2)}{1000}} = 0.225298221281$$

$$.2 - 2\sqrt{\frac{.2(1-.2)}{1000}} = 0.174701778719$$

$$.2 + 2\sqrt{\frac{.2(1-.2)}{10000}} = 0.208$$

$$.2 - 2\sqrt{\frac{.2(1-.2)}{10000}} = 0.192$$

Facilitation Notes

This version of the problem string lists short notes for important teacher moves during the string. After you've done the string yourself and studied the relationships involved, you might make similar notes for the things you want a reminder of or deem important.

Diet soda and salad, $n = 50$	*Intro the scenario, write table. Find proportions by doubling (12/50 = 24/100).* *What do these values mean? What do they not mean?* *What is the interval likely to contain the true proportion?* *Find the interval using tech. What does this interval mean?*
Diet soda and salad, $n = 100$	*Predict—what will happen if the sample size doubles, asking 100 people? Why?* *What did we find? How does this interval compare?* *How confident do you feel about the sample proportion?*
Diet soda and salad, $n = 1{,}000$	*Repeat.*
Diet soda and salad, $n = 10{,}000$	*How big do you think the sample size needs to be to be within 0.01?* *Want to try 10,000?* *What does this interval mean?*

PROBLEM STRINGS LESSON

9.1 Experimental Design

At a Glance	
Which popcorn is best? The front door test The cafeteria probe The movie popcorn study Design a better popcorn study* *optional problem	**Objectives** The goal of this problem string is to help students learn to differentiate between three experimental designs and identify potential bias.
	Placement This problem string could be used to help students learn to differentiate between experimental, observational, and survey types of experimental design. You could use the string to introduce each design by having students describe each study in turn and then provide a name. Or you could use this problem string to support students' understanding after students have been introduced to the different types of studies. You can use this string to introduce or support the work in textbook Lesson 9.1 Experimental Design.
	Guiding the Problem String This problem string consists of three studies each based on finding the preferred popcorn for students. Each study has flaws in the design. The first problem is an observational study, the second is a survey, and the third problem is an experimental study. The last problem is an optional extension to design a study that would better determine the favorite popcorn. Give students each problem, one at a time, discussing the kind of study and what types of bias might be embedded. Ask students how they might change the study to minimize the bias, while keeping the study the same type.
	About the Mathematics See the summary points that follow for descriptions and traits of the three study types.

Important Questions

Use the following as you plan how to elicit and model student strategies.

- *What kind of experimental design is this study?*
- *What are some potential problems with the way the student has set up the study?*
- *How could you improve the study, while keeping it the same kind of study?*
- *Which kind of study do you think is best to decide what brand of popcorn the students prefer?*

How would you summarize some of the things that came up in this string today?

- *Experimental studies must have the treatments (assignments) made randomly by the researcher. The way subjects are assigned may affect the data.*
- *Observational studies have the time and location selected randomly by the researcher. The subjects choose the treatment without interference of the researcher, who merely records their choice. In an observational study, a flaw could be that some subjects are inadvertently omitted from consideration.*
- *In survey studies, subjects are chosen at random from the population of interest. Subjects choose responses, like from a survey. Surveys can be biased by the selection of the participants, by the truthfulness of the subjects, and by how the questions are phrased.*
- *Researchers need to consider bias and take measures to minimize bias that could invalidate the study. Bias could happen in many ways like in the way that subjects are chosen and in the way that treatments are presented.*

Sample Final Display

Your display could look like this at the end of the problem string:

The Front Door Test	Observational	Bias:	Only students who have time to stop? Is the best looking popcorn the best tasting? Do students even know there are 2 different kinds?
The Cafeteria Probe	Survey	Bias:	Student not selected at random, maybe only hungry students participated? How was question worded?
The Movie Popcorn Study	Experimental Study	Bias:	What time of the day were the classes, before or after lunch (were some full and others hungry)? Does amount eaten equal preference?

Facilitation Notes

This version of the problem string lists short notes for important teacher moves during the string. After you've done the string yourself and studied the relationships involved, you might make similar notes for the things you want a reminder of or deem important.

The Front Door Test	Which popcorn is better? Three students—different studies. Bowl of each by front door. Watches to see which is finished first. What kind of study? Why? What kinds of potential bias? How could you tweak to minimize that bias (but stay observational)?
The Cafeteria Probe	Walk around cafeteria with bag of each. Asks which they prefer and notes. What kind of study? Why? What kinds of potential bias? How could you tweak to minimize that bias (but stay as a survey)?
The Movie Popcorn Study	Two periods of foreign language, each get a different kind. Records how much popcorn was eaten. What kind of study? Why? What kinds of potential bias? How could you tweak to minimize that bias (but stay experimental)?

(continued)

Questions to Display

The Front Door Test
Mark placed a bowl of each kind of popcorn next to the front door of the school as classes were letting out for the day. He watched to see which bowl was finished first.

The Cafeteria Probe
Destiny walked around the cafeteria at lunch time with a bag of each kind of popcorn asking students to participate. She recorded volunteers' stated preference after they had tried some of each kind of popcorn.

The Movie Popcorn Study
Kelly chose two different periods of a foreign languages class. On the day the classes were watching a cartoon in the foreign language, Kelly provided a different kind of popcorn to each class. She recorded how much popcorn was eaten.

PROBLEM STRINGS LESSON 9.2

Normal Distribution

At a Glance

For the data of $\mu = 497$ cm, $\sigma = 99$ cm for adult male giraffes:

- What range of heights fall in the middle 68 percent?
- How tall is an adult male giraffe in the 99.85th percentile?
- How tall is an adult male giraffe in the 2nd percentile?
- An adult male giraffe is 200 cm tall. What percentile?
- Find the probability that a random adult male giraffe is taller than 695 cm? Shorter than 695 cm?

Objectives

The goal of this problem string is for students to gain facility using the mean and standard deviation with the 68–95–99.7 rule of normally distributed data.

Placement

This problem string helps students work with the mean and standard deviation of data that is normally distributed to find benchmark percentiles of data values or data values for benchmark percentiles. Students should have prior experience with normal distributions, means, standard deviation, and the 68–95–99.7 rule.

You can use this string to support the work in Lesson 9.2 Normal Distributions after students have been introduced to the 68–95–99.7 rule to help students use the rule and understand how standard deviation relates to the mean of normally distributed data.

Guiding the Problem String

This problem string works with the assumption that adult male giraffe height is normally distributed with the given mean and standard deviation. Help students realize that each question's answer can be found using the 68–95–99.7 rule of normally distributed data. Engage students in the data, estimating the sizes of giraffes and helping students realize that the normal curve represents percentages of data that are normally distributed and that individual data points can be found if they are at one of the benchmark percentiles.

About the Mathematics

In a normal distribution approximately 68% of values will fall within one standard deviation of the mean, approximately 95% of values will fall within two standard deviations of the mean, and approximately 99.7% of values will fall within three standard deviations of the mean. These correspond to benchmark percentiles shown: 0.15, 2.5, 16, 50, 84, 97.5, 99.85.

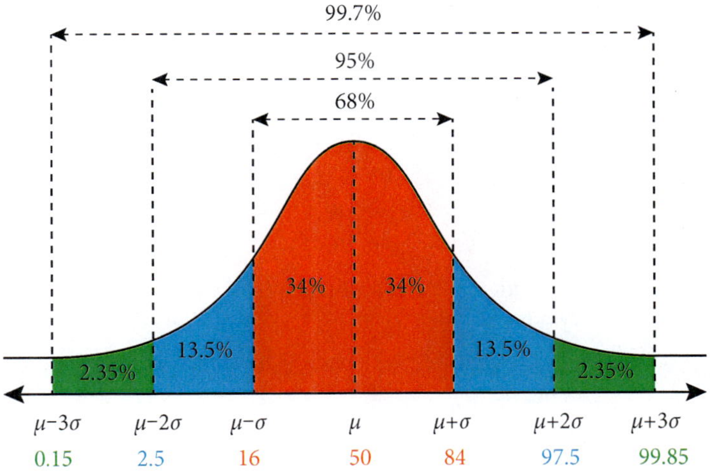

(continued)

Sample Interactions

Use the following as you plan how to elicit and model student strategies. This is not meant as a script, but as a view into the relationships involved and the intent of the problem string.

Teacher: *Help me sketch a normal curve up here, with the 68–95–99.7 benchmarks.*

Student: *It should look like a bell curve.*

Teacher: *How do I know where to put the 68–95–99.7 benchmarks?*

Student: *They are evenly spaced from the middle. Mark the mean in the middle. Then you need three relatively even spaces behind and in front of it, although the ends don't end.*

Teacher: *The ends keep going, they don't meet the axis?*

Student: *Right. Then go from the mean one standard deviation back and forward 34.*

Teacher: 34 *what?* 34 *data values?* 34 *percent?*

Student: *Percent.*

The teacher continues to ask questions until the students have helped the teacher fill in the percents.

Teacher: *So where is the 50th percentile?*

Student: *Right in the middle.*

Teacher: *What percentile would be at these benchmark dotted lines? How do you know?*

Student: *Go up* 34% *from* 50, *that's* 84.

The students continue to find the benchmark percentiles while the teacher fills them in.

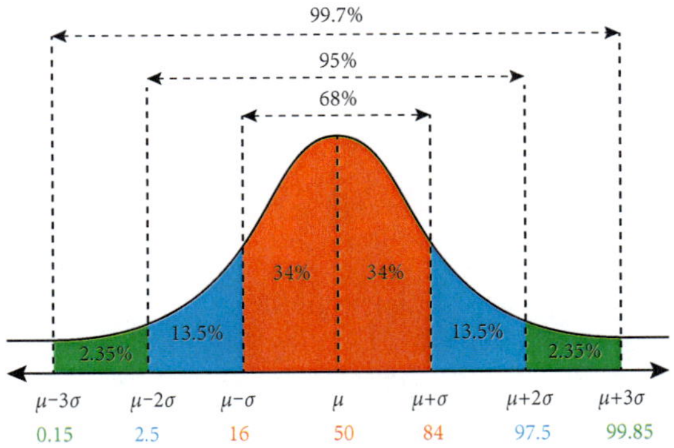

Teacher: *Suppose the heights of adult male giraffes is approximately normally distributed. What does that mean?*

Student: *That if we could measure all of the adult male giraffes the measurements would all fit in a normal curve.*

Student: *Most giraffes will be clustered around the mean.*

Student: *Only a few are really short or really tall.*

Student: *We could figure out the mean and standard deviation.*

Teacher: *So for our string today, I'll tell you that the mean is* 497 *cm and the standard deviation is* 99 *cm. Someone give me some ideas of what that means?*

Student: *The average height of an adult male giraffe is about* 5 *meters. And the standard deviation is about* 1 *meter.*

Teacher: *Those sound like helpful benchmarks. Where does the mean go on the normal curve?*

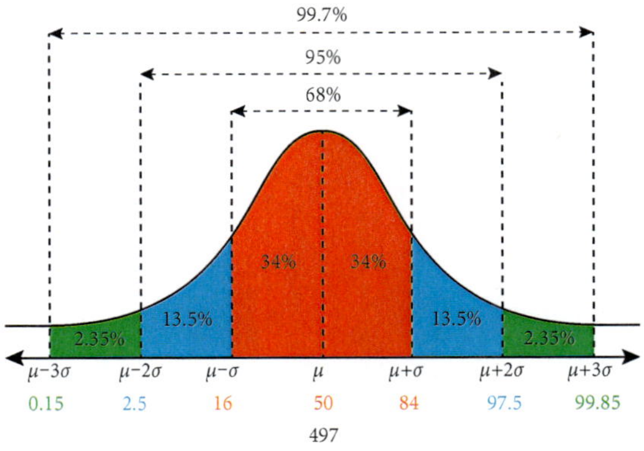

Teacher: *The first problem of our string today is what is the range of heights of adult male giraffes in the middle 68%? That's kind of a weird number, 68%. Sure wish I knew something about 68% and normal distribution. Turn and talk to your partner.*

Students turn and talk briefly while the teacher listens in.

Teacher: *What are you thinking?*

Student: *It's the middle? Isn't that just the two 34 sections? That's 68%.*

Student: *Yeah, but you asked what range of heights, right? So that has to be 99 on either side of 497.*

Teacher: *Why?*

Student: *You wanted to know heights of giraffes in the middle 68% so we need to find the heights. The 34 and 34 are 68, but we need to find what heights those correspond to.*

Teacher: *That makes sense. What is 99 on the right side of 497? And how do you know?*

Student: *It's 596 because 497 and 3 is 500. Then 500 and what's left 96 is just 596.*

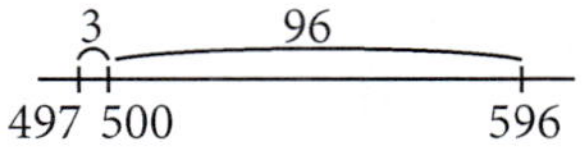

$497 + 99 = 500 + 96 = 596$

Student: *Or you could think about 99 is almost 100 so 100 and 496. Just give and take 1.*

Teacher: *Those are two fine ways to think about 99 to the right of 497.*

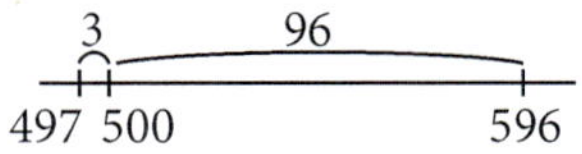

$497 + 99 = 500 + 96 = 596$

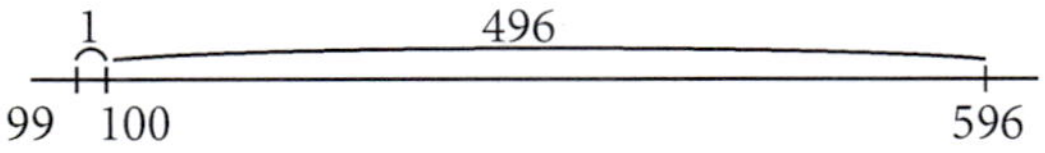

$99 + 497 = 100 + 496 = 596$

Teacher: *What about 99 to the left of 497?*

Student: *It's 398 because 497 minus 100 is 397, but I subtracted too much so it's just 398.*

$\mu - \sigma$
$= 497 - 99$

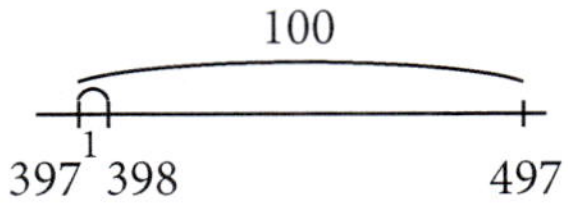

$497 - 99$
$= 497 - (100 - 1) = 497 - 100 + 1 = 398$

Student: *Or you can make an equivalent problem by adding one to each number so 498 minus 100 is 398.*

Teacher: *Nice using relationships.*

$\mu - \sigma$
$= 497 - 99$

$497 - 99 = 498 - 100 = 398$

(continued)

Teacher: *So the range of heights in the middle 68% is?*

Student: *398 to 596 cm.*

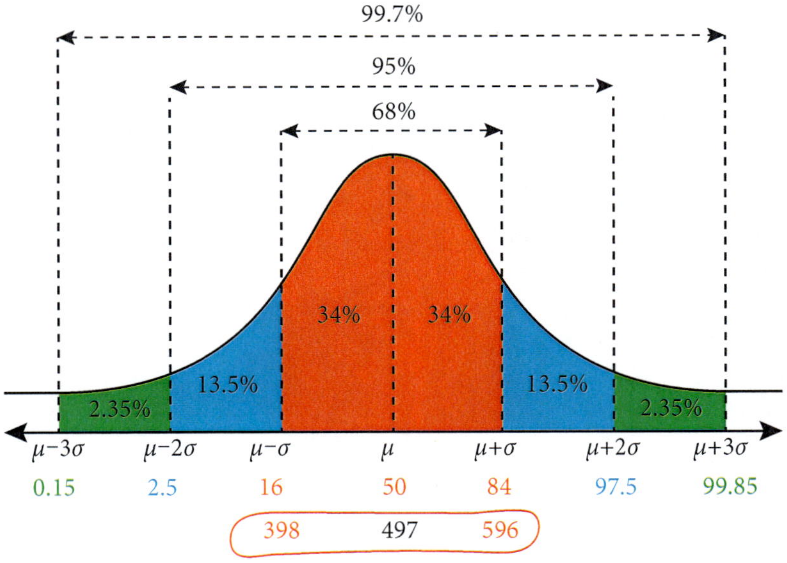

μ = 497 cm, σ = 99 cm, adult male giraffes, approximately normal

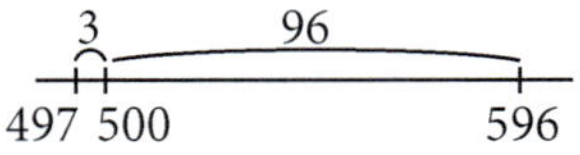

497 + 99 = 500 + 96 = 596

Range of heights in the middle 68 percent?

398–596 cm.

34 + 34 = 68

μ + σ
= 497 + 99

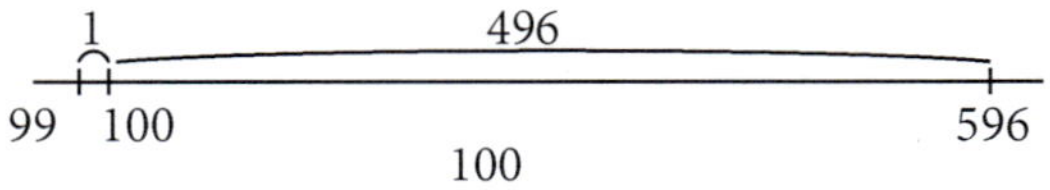

99 + 497 = 100 + 496 = 596

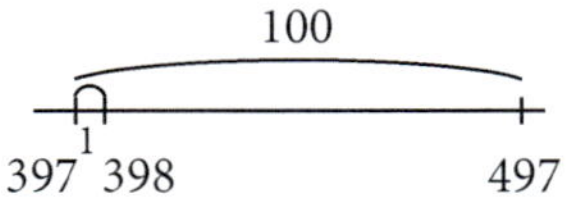

497 − 99
= 497 − (100 − 1) = 497 − 100 + 1 = 398

μ − σ
= 497 − 99

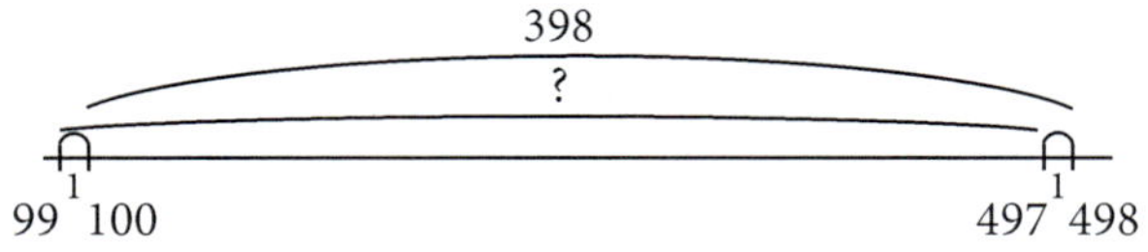

497 − 99 = 498 − 100 = 398

Teacher: *The next problem is to find how tall a giraffe is if we know it's in the 99.85th percentile?*

The teacher has students work together briefly and then asks students to share how they found the middle 99.7% and how they added three standard deviations of 99, perhaps by adding 300 and subtracting the extra 3 or by giving and taking 3 as shown. The teacher marks the 794 on the normal curve.

How tall if in the 99.85th percentile? + 3 standard deviations

794 cm.

μ + 3σ
= 497 + 3(99)

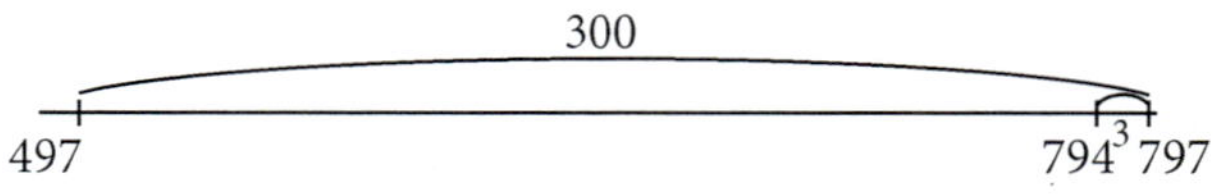

497 + 300 − 3 = 797 −3 = 794

497 + 297 = 500 + 294 = 794

Teacher: *The next problem is to find how tall an adult male giraffe is if he's in the 2nd percentile. Does that mean he's tall or short? And how do you know?*

Student: *Short! The last guy was in the 99.85% and he was almost 8 meters. This guy is on the opposite end.*

Student: *I wonder what is short for an adult giraffe?*

Teacher: *Go! How tall/short is he?*

After brief think time, the teacher leads a conversation about subtracting two standard deviations to find the "short" giraffe at nearly three meters. Students estimate three meters in their classroom if possible. The teacher records 299 on the normal curve.

How tall if in the 2nd percentile? – 2 standard deviations

299 cm. $\mu - 2\sigma$
$= 497 - 2(99)$

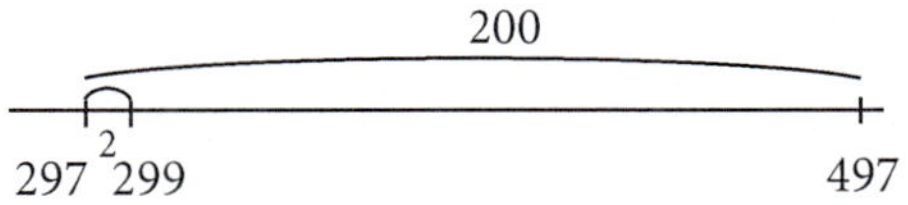

Teacher: *What about a giraffe that we know is 200 cm tall? What percentile? How do you know?*

The teacher leads a brief conversation about subtracting 200 from 497 and dividing by 99 or just using what is already on the normal curve to subtract 99 from 299 to get to the 0.15th percentile. The teacher records the 200 on the normal curve.

Teacher: *The last question of the day is to find the probability of a giraffe that is taller than 695 cm? And shorter than 695 cm?*

Student: *What does that mean?*

Teacher: *Good question. Thoughts?*

Student: *If we know what percentile it's at, that's the probability too. It's like asking what's the percent of giraffes that are taller and shorter than 695 cm.*

After brief think time, the teacher leads a conversation about finding 695 using what is already shown and how students might find it if it was the first question given.

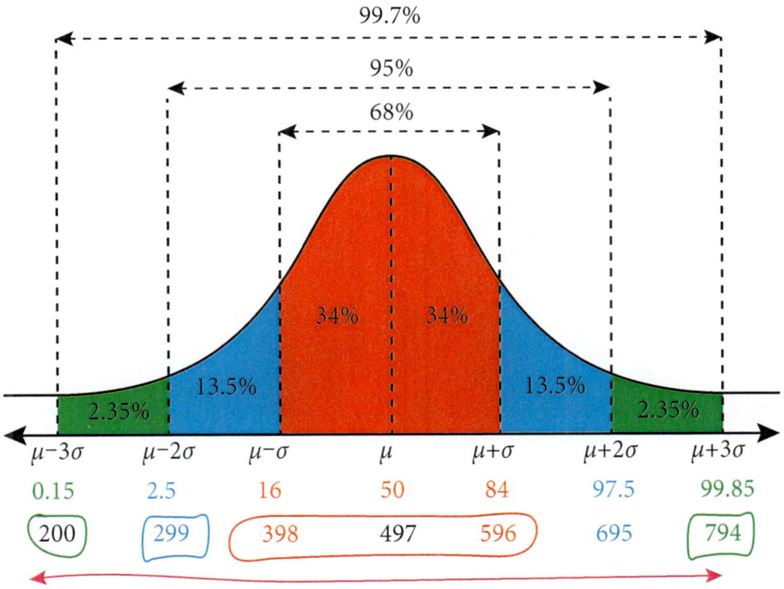

Probability of giraffe > 695 cm? Giraffe < 695 cm?

695 is at the 97.5th percentile so 2.5% of adult male giraffes are taller than 695 cm.but 97.5% are shorter.

(continued)

Teacher: *How would you summarize some of the things that came up in today's string?*

Elicit the following:

- *One, two, and three standard deviations away from the mean of normally distributed data are at the helpful benchmarks of* 34%, 13.5%, *and* 2.35%, *which correspond to the middle* 68%, *the* 95% *and the* 99.7% *of the data.*
- *We can use the* 68–95–99.7 *relationships of normally distributed data to find benchmark data values.*
- *Given data values that are close to benchmark percentiles, we can find approximate probabilities.*
- *The middle percent groupings are different from percentiles.*

Sample Final Display

Your display could look like this at the end of the problem string:

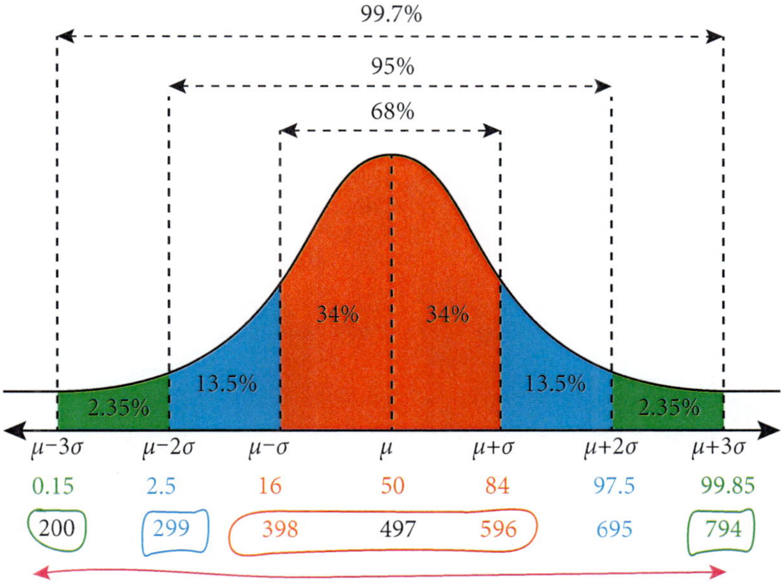

μ = 497 cm, σ = 99 cm, adult male giraffes, approximately normal

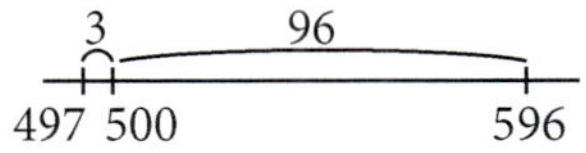

497 + 99 = 500 + 96 = 596

Range of heights in the middle 68 percent?

398–596 cm.

34 + 34 = 68

μ + σ
= 497 + 99

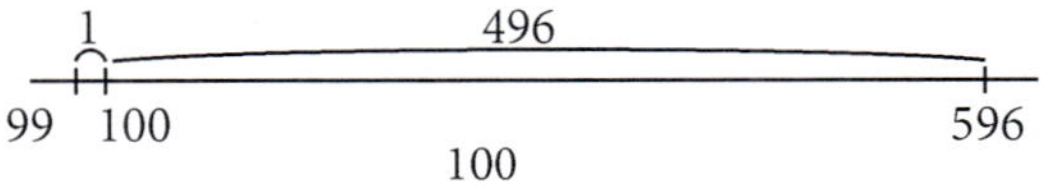

99 + 497 = 100 + 496 = 596

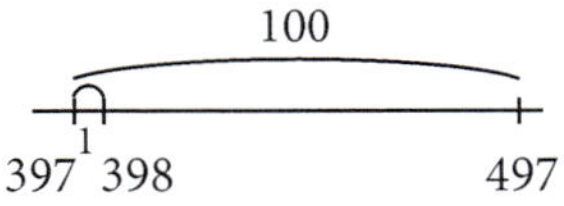

497 − 99
= 497 − (100 − 1) = 497 − 100 + 1 = 398

μ − σ
= 497 − 99

497 − 99 = 498 − 100 = 398

How tall if in the 99.85th percentile? + 3 standard deviations

794 cm.

μ + 3σ
= 497 + 3(99)

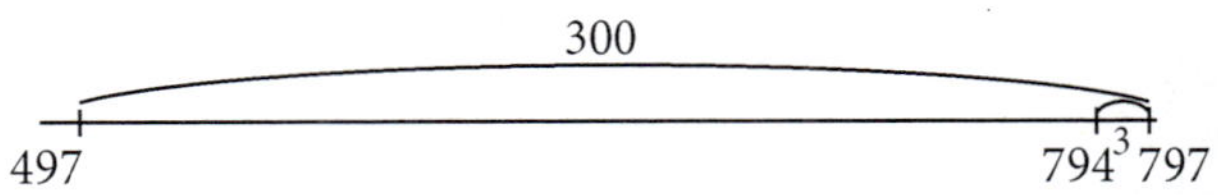

497 + 300 − 3 = 797 −3 = 794

497 + 297 = 500 + 294 = 794

How tall if in the 2nd percentile? − 2 standard deviations

299 cm.

μ − 2σ
= 497 − 2(99)

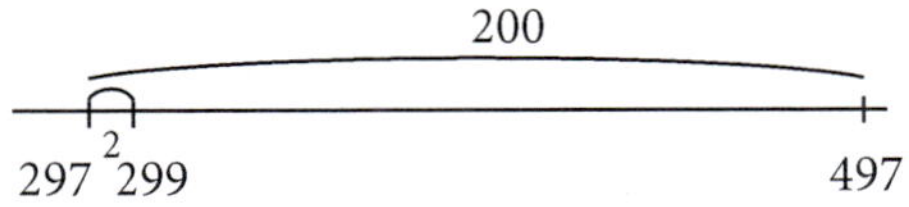

Giraffe is 200 cm tall. Percentile? 0.15th

Probability of giraffe > 695 cm? Giraffe < 695 cm?

695 is at the 97.5th percentile so 2.5% of adult male giraffes are taller than 695 cm.but 97.5% are shorter.

(continued)

Facilitation Notes

This version of the problem string lists short notes for important teacher moves during the string. After you've done the string yourself and studied the relationships involved, you might make similar notes for the things you want a reminder of or deem important.

$\mu = 497$ cm, $\sigma = 99$ cm, adult male giraffes, approximately normal

Range of heights in the middle 68 percent?	*Why 68?* *Sure wish I knew something about 68% and normal distribution...* *Mean in the middle, 34 percent on either side.* *497 + 99, Model over, give and take addition strategies.* *497 − 99, Model over, constant difference subtraction strategies.*
How tall if in the 99.85th percentile?	*Mean plus three standard deviations.* *497 + 3(99), Model over, give and take strategies.*
How tall if in the 2nd percentile?	*Mean minus two standard deviations.* *497 − 2(99), Model over strategy*
Giraffe is 200 cm tall. Percentile?	*200%? No, 200 cm.* *Sure wish I knew where 200 cm would be. What do we know?*
Probability of giraffe > 695 cm? Giraffe < 695 cm?	*What does this mean? Do we know where 695 cm is?* *How many giraffes are taller? So probability?* *Shorter?* *What did the mean and the standard deviation have to do with all of this?*

PROBLEM STRINGS
LESSON

9.3 Confidence Intervals

At a Glance

$n = 9, \overline{x} = 170, \sigma = 9$

$n = 100, \overline{x} = 170, \sigma = 9$

$n = 1{,}000, \overline{x} = 170, \sigma = 9$

$n = 100, \overline{x} = 170, \sigma = 8$

$n = 100, \overline{x} = 170, \sigma = 7$

$n = 100, \overline{x} = 170, \sigma = 1$

Objectives

The goal of this problem string is for students to realize the two variables that can tighten a confidence interval, either an increased sample size or a decreased standard deviation.

Placement

This problem string strengthens students' sense of the margin of error and it's relationship to the confidence interval. Prior to this string, students should have been introduced to these and all of the variables involved. Use this problem string to give students experience reasoning about margin of error and confidence intervals and the effect that changing variables have.

You can use the problem string to support the work in textbook Lesson 9.3 z-Values and Confidence Intervals.

Guiding the Problem String

This problem string consists of problems based on the context of normally distributed heights, where each problem changes one variable, first the sample size and then the standard deviation. None of the problems are difficult to compute, so do not spend time discussing computation strategy. Instead, the aim of the problem string is to help students make connections between the changing variables and the tightening confidence intervals.

About the Mathematics

The confidence interval $p\%$ for the population mean μ, a sample of size n, with sample mean $\overline{x}$, z the number of standard deviations from the mean within which $p\%$ of normally distributed data lie, and with standard deviation σ from a normally distributed population is:

$$\overline{x} - \frac{z\sigma}{\sqrt{n}} < \mu < \overline{x} + \frac{z\sigma}{\sqrt{n}}$$

The margin of error is $\frac{z\sigma}{\sqrt{n}}$.

(continued)

Sample Interactions

Use the following as you plan how to elicit and model student strategies. This is not meant as a script, but as a view into the relationships involved and the intent of the problem string.

Teacher: *We've been defining and discussing confidence intervals, the margin of error, and all of the variables involved. We've investigated predicting the true mean. Now, let's do a quick problem string to get to know something about confidence intervals a little more. For this entire string, we're going to consider a 95% confidence interval. Given that, what else do we know?* **Student:** *Well, 95% is one of the benchmark numbers for normally distributed data. It's two standard deviations, right?* **Student:** *Yes, 95% is two. So, don't we know z?* **Teacher:** *So, we know that 95% of normally distributed data falls within two standard deviations of the mean. And the z-score is two. For our first problem today, a student gathered the heights of nine fellow students and found a mean of 170 cm with a standard deviation of 9 cm. Do we know enough to figure out the margin of error? What variables do we know?* **Student:** *Yes, the mean is 170 cm, that's $\overline{x}$, the 9 students are n, and sigma is 9.* **Teacher:** *Great. What's the margin of error and the confidence interval?* Students work and the teacher circulates, looking for students who have graphed the points shown.	$\overline{x}-\frac{z\sigma}{\sqrt{n}}<\mu<\overline{x}+\frac{z\sigma}{\sqrt{n}}$ $\frac{z\sigma}{\sqrt{n}}$ 9 students, sample mean = 170, standard deviation = 9 $n=9, \overline{x}=170, \sigma=9$
Teacher: *What is the margin of error?* **Student:** *It's 6. It's z of 2 times a standard deviation of 9, divided by the square root of n, 9. That's 18 divided by 3, so 6.*	$\frac{z\sigma}{\sqrt{n}}=\frac{2\cdot 9}{\sqrt{9}}=\frac{18}{3}=6$

Teacher: *What's the confidence interval?* **Student:** *It's 164 to 176 because you take 170 plus and minus 6.* **Teacher:** *And what does this mean, that μ is between 164 and 176?* **Student:** *Well, the margin of error is 6 cm, so the population mean is between 164 and 176.* **Student:** *I thought we already knew the mean. You told us it was 170.* **Student:** *But that is the sample mean. We don't know the population mean for everyone's height, so now we're figuring out what the population mean is with this margin of error.* **Student:** *But doesn't the 95% mean something?* **Teacher:** *Yes, what does that mean?* **Student:** *We started by saying that we were going to be dealing with 95%. That's how we knew the z-score is 2. Right?* **Teacher:** *Yes, so we can say that we are 95% confident that the population mean falls within 164 and 176 cm. I wonder how we could tighten that interval?* **Student:** *Well, I noticed that they only measured 9 people. That's a really small n.* **Teacher:** *That is a really small sample size.*	$170-6<\mu<170+6$ $164<\mu<176$
Teacher: *The next problem is that the students decided to measure more people. They must have heard your complaint! They measured 100 people. And crazy, they got the same sample mean and standard deviation. Silently predict what should happen.* *What's the margin of error and confidence interval now?* Brief think time.	$n=100, \overline{x}=170, \sigma=9$
Teacher: *What did you find?* A student responds and the teacher records. **Student:** *Wow. That got a lot smaller.* **Teacher:** *What did?* **Student:** *Both the margin of error and the confidence interval.* **Teacher:** *Does that make sense? Convince us.* **Student:** *If you measure all of those people, then it makes sense that the population mean would be close to the sample mean.*	$\frac{z\sigma}{\sqrt{n}}=\frac{2\cdot 9}{\sqrt{100}}=\frac{18}{10}=1.8$ $170-1.8<\mu<170+1.8$ $168.2<\mu<171.8$

(continued)

Teacher: *What if they measured 1,000 people? Go!* The teacher repeats asking students to predict, figure, and model on the board. **Teacher:** *What's going on and why does it make sense?* **Student:** *As you increase the sample size, the confidence interval is getting smaller. This makes sense because the more people you measure, the closer the sample mean should be to the actual population mean.* **Teacher:** *Take a look at the fraction that represents the margin of error. Why does this make sense just looking at the fraction?* **Student:** *The sample size is in the denominator. As you increase the denominator, the whole fraction decreases. That means that the fraction you add and subtract from the sample mean is smaller so the confidence interval tightens.*	$n=1{,}000, \bar{x}=170, \sigma=9$ $\frac{z\sigma}{\sqrt{n}}=\frac{2\cdot 9}{\sqrt{1{,}000}}=\frac{18}{100}=0.18$ $170-0.18<\mu<170+0.18$ $169.82<\mu<170.18$
Teacher: *Let's change something else. What if those students are back to measuring 100 people, but they actually found that the standard deviation is 8, not 9. Predict first, what effect, if any, that will that have on the margin of error or the confidence interval.* **Student:** *Doesn't that mean that the measurements they found actually didn't vary as much?* **Student:** *Yeah, like the people weren't as far from the mean?* **Teacher:** *Well, at least their heights weren't, right? Okay, go ahead and find the margin of error and the confidence interval.* Brief work time.	$n=100, \bar{x}=170, \sigma=8$
Teacher: *Tell us what you got.* Students report out the margin of error and the confidence interval as the teacher models. **Teacher:** *Was your prediction correct?* **Student:** *Yes, the standard deviation is less than before so people's heights weren't varying as much so the margin of error was less than when the standard deviation was 9 and the sample size was 100.* **Teacher:** *Can anyone think of what could account for that? How would it be that the standard deviation was smaller?* **Student:** *Maybe at first they were measuring anyone and now they were measuring sports teams or something?* **Teacher:** *So, if they don't choose people randomly, the standard deviation might get smaller?*	$\frac{z\sigma}{\sqrt{n}}=\frac{2\cdot 8}{\sqrt{100}}=\frac{16}{10}=1.6$ $170-1.6<\mu<170+1.6$ $168.4<\mu<171.6$

The teacher repeats with the next question, asking students to predict, then find and discuss the results.	$n=100, \bar{x}=170, \sigma=7$ $\frac{z\sigma}{\sqrt{n}}=\frac{2\cdot 7}{\sqrt{100}}=\frac{14}{10}=1.4$ $170-1.4<\mu<170+1.4$ $168.6<\mu<171.4$
Teacher: *Let's stay with that sports team analogy. What if they measured lots of teams synchronized swimmers, where it would matter if people on the team were approximately the same height? What if everything for the next problem is the same, but the standard deviation is just 1 cm?* Student work briefly, and the teacher models the thinking. **Teacher:** *Tell us about changing the standard deviation and the affect it has on the margin of error and the confidence interval, both in context and just looking at the fraction.*	$n=100, \bar{x}=170, \sigma=1$ $\frac{z\sigma}{\sqrt{n}}=\frac{2\cdot 1}{\sqrt{100}}=\frac{2}{10}=0.2$ $170-0.2<\mu<170+0.2$ $169.8<\mu<170.2$
The teacher leads a conversation about the variables that affect the margin of error and the confidence interval. **Teacher:** *So, what's happening? What can make the confidence interval tighten?* **Student:** *There are two ways to affect the margin of error. You can either make the sample size bigger by measuring more people, or make the standard deviation smaller by measuring only people that are about the same height.* **Student:** *If you're really taking any kind of data, you really don't usually control the standard deviation. It is what it is. But you can always poll, measure, or study more people.* The teacher records students' generalizations.	$\frac{z\sigma}{\sqrt{n}}$ As n increases, the fraction decreases $\frac{z\sigma}{\sqrt{n}}$ As σ decreases, the fraction decreases

Teacher: *How would you summarize some of the things that came up in this string today?*

Elicit the following:

- *As the sample size increases, the margin of error decreases and the confidence interval tightens.*
- *As the standard deviation decreases, the margin of error decreases and the confidence interval tightens.*
- *The variable that a pollster has some control over is the sample size.*

(continued)

Sample Final Display

Your display could look like this at the end of the problem string:

$\frac{z\sigma}{\sqrt{n}}$ $\quad \bar{x}-\frac{z\sigma}{\sqrt{n}}<\mu<\bar{x}+\frac{z\sigma}{\sqrt{n}}$

9 students, sample mean = 170, standard deviation = 9

$n=9, \bar{x}=170, \sigma=9$ $\quad \frac{z\sigma}{\sqrt{n}}=\frac{2\cdot 9}{\sqrt{9}}=\frac{18}{3}=6$ $\quad 170-6<\mu<170+6$ $\quad 164<\mu<176$

$n=100, \bar{x}=170, \sigma=9$ $\quad \frac{z\sigma}{\sqrt{n}}=\frac{2\cdot 9}{\sqrt{100}}=\frac{18}{10}=1.8$ $\quad 170-1.8<\mu<170+1.8$ $\quad 168.2<\mu<171.8$

$\frac{z\sigma}{\sqrt{n}}$ As n increases, the fraction decreases

$n=1{,}000, \bar{x}=170, \sigma=9$ $\quad \frac{z\sigma}{\sqrt{n}}=\frac{2\cdot 9}{\sqrt{1{,}000}}=\frac{18}{100}=0.18$ $\quad 170-0.18<\mu<170+0.18$ $\quad 169.82<\mu<170.18$

$n=100, \bar{x}=170, \sigma=8$ $\quad \frac{z\sigma}{\sqrt{n}}=\frac{2\cdot 8}{\sqrt{100}}=\frac{16}{10}=1.6$ $\quad 170-1.6<\mu<170+1.6$ $\quad 168.4<\mu<171.6$

$n=100, \bar{x}=170, \sigma=7$ $\quad \frac{z\sigma}{\sqrt{n}}=\frac{2\cdot 7}{\sqrt{100}}=\frac{14}{10}=1.4$ $\quad 170-1.4<\mu<170+1.4$ $\quad 168.6<\mu<171.4$

$\frac{z\sigma}{\sqrt{n}}$ As σ decreases, the fraction decreases

$n=100, \bar{x}=170, \sigma=1$ $\quad \frac{z\sigma}{\sqrt{n}}=\frac{2\cdot 1}{\sqrt{100}}=\frac{2}{10}=0.2$ $\quad 170-0.2<\mu<170+0.2$ $\quad 169.8<\mu<170.2$

Facilitation Notes

This version of the problem string lists short notes for important teacher moves during the string. After you've done the string yourself and studied the relationships involved, you might make similar notes for the things you want a reminder of or deem important.

$n = 9, \bar{x} = 170, \sigma = 9$	*For the entire string, it's all 95% confidence interval.* *What else do we know? Yes, z is 2.* *Find the margin of error and the confidence interval.* *Discuss difference between sample mean and population mean if necessary,* *I wonder how we could tighten the interval?*
$n = 100, \bar{x} = 170, \sigma = 9$	*Find the margin of error and the confidence interval.* *How do the two problems compare? Why?*
$n = 1{,}000, \bar{x} = 170, \sigma = 9$	*Repeat.* *Generalize—what happens as the sample size increases?* *Why does that make sense in context? Looking at the fraction?*
$n = 100, \bar{x} = 170, \sigma = 8$	*Let's go back to a sample size of 100. Now the standard deviation is 8.* *Predict what effect that will have, if any?* *Was your prediction correct? Compare to 2nd problem.*
$n = 100, \bar{x} = 170, \sigma = 7$	*Repeat.* *How do the two problems compare? Why?*
$n = 100, \bar{x} = 170, \sigma = 1$	*Repeat.* *Generalize—what happens as the standard deviation decreases?* *Why does that make sense in context? Looking at the fraction?*

PROBLEM STRINGS
LESSON

9.4 Correlation versus Causation

At a Glance

Correlation or Causation?

- Smoking causes alcoholism.
- Smoking causes an increase in the risk of developing lung cancer.
- As levels of CO_2 have increased, obesity levels have increased. Do increased levels of CO_2 cause increased rates of obesity?
- As more lemons are imported from Mexico, the number of highway crashes in the USA decrease.
- Ice cream sales have decreased as sunscreen sales have decreased. Does that mean that ice cream causes sunburns?
- Students who are tutored tend to get worse grades than students who are not tutored.
- The number of times the pilot switches on the seat belt sign increases as the number of drinks spilled on an airplane increases.

Objectives

The goal of this problem string is to help students define and compare correlation, causation, and potential lurking variables in bivariate data.

Placement

This problem string can be used to introduce or support students' understanding of the notions of correlation, causation, and lurking variables. Use this problem string to give students experience reasoning about scenarios that may seem to imply causation but in fact have lurking variables that explain the correlation.

You can use this problem string to introduce or support the work in textbook Lesson 9.4 Bivariate Data and Correlation.

Guiding the Problem String

This problem string begins with a pair of statements that deal with smoking, one that incorrectly implies causation and another that correctly states causation. Use these two statements to introduce or support students' notions of causation, correlation, and lurking variables. You could also introduce the statistics terms explanatory and response variables. The next two problems state correlations and ask about the potential causation. The last three problems state correlations, where the students are asked to state the potentially faulty causation and refute it.

Throughout the string, press students for justification. Restate imprecise language with precise mathematical vocabulary as necessary. For each problem, quickly sketch a graph of a few points that represent positive or negative correlation.

A fun fact to throw in: The color of a chicken's ears is the same color as the eggs it lays—white or brown. Correlation? Causation?

About the Mathematics

An association between two variables is called correlation. A strong correlation may exist between two sets of data, but this does not necessarily imply a causal relationship. Drawing conclusions from data requires such things as properly designed and executed experiments using randomization to help rule out lurking variables. For example, people who smoke are more likely to have alcoholism but that does not mean that smoking causes alcoholism. There are other variables in play like depression, anxiety, susceptibility to addictive behavior, family background, etc. The statement that smoking causes an increase in the risk of developing lung cancer, on the other hand, has been rigorously studied and lurking variables accounted for.

Important Questions

Use the following as you plan how to elicit and model student strategies.

- *What does correlation mean? What does correlation not mean?*
- *What does it mean to correlate positively? Negatively?*
- *What does causation mean?*
- *What is a lurking variable?*

- *Give an example of a lurking variable?*
- *Why do we need to consider causation when variables are correlated?*

How would you summarize some of the things that came up in this string today?

- *Correlation means that there is an association between variables.*
- *Variables are correlated positively if one variable increases and the other variable increases.*
- *Variables are correlated negatively if one variable increases and the other variable decreases.*
- *If data are highly correlated, a straight line will model the data points well.*
- *A strong correlation may exist between two sets of data, but this does not necessarily imply a causal relationship.*
- *Lurking variables need to be accounted for.*

Sample Final Display

Your display could look like this at the end of the problem string:

Smoking causes alcoholism.

correlation
not causation

Smoking causes an increase in the risk of developing lung cancer.

correlation
causation

As levels of CO_2 have increased, obesity levels have increased. Do increased levels of CO_2 cause increased rates of obesity?

correlation
not causation

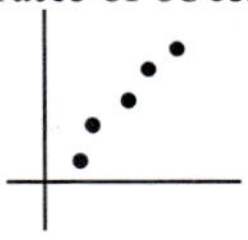

As more lemons are imported from Mexico, the number of highway crashes in the USA decrease.

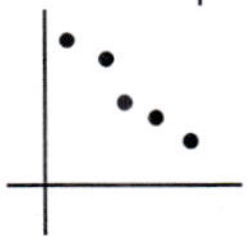

correlation
not causation

Ice cream sales have decreased as sunscreen sales have decreased. Does that mean that ice cream causes sunburns?

correlation
not causation

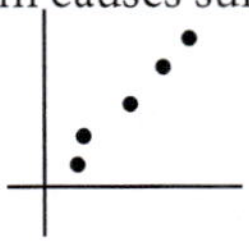

Students who are tutored tend to get worse grades than students who are not tutored.

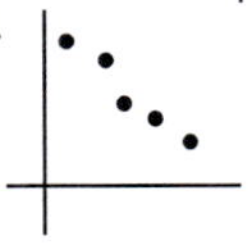

correlation
not causation

The number of times the pilot switches on the seat belt sign increases as the number of drinks spilled on an airplane increases.

correlation
not causation

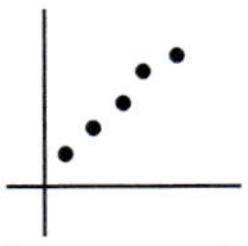

(continued)

Facilitation Notes

This version of the problem string lists short notes for important teacher moves during the string. After you've done the string yourself and studied the relationships involved, you might make similar notes for the things you want a reminder of or deem important.

Smoking causes alcoholism.	*What's the difference between correlation and causation?* *Present the scenario. Does smoking cause alcoholism?* *This statement implies both correlation and causation. True? No.* *What are some potential lurking variables?* *What is a quick sketch of correlation, positive or negative?*
Smoking causes an increase in the risk of developing lung cancer.	*What about this statement—correlation, causation, both?* *Make a quick sketch of positive correlation.*
As levels of CO_2 have increased, obesity levels have increased. Do increased levels of CO_2 cause increased rates of obesity?	*Read statement and question. Discuss.* *Both variables have increased over time, but one hasn't caused the other.* *Make a quick sketch of positive correlation.*
As more lemons are imported from Mexico, the number of highway crashes in the USA decrease.	*Repeat. Negative correlation.* *One hasn't caused the other.*
Ice cream sales have decreased as sunscreen sales have decreased. Does that mean that ice cream causes sunburns?	*Repeat. Positive correlation.* *What's the correlation during a hot season? Lurking variables?*
Students who are tutored tend to get worse grades than students who are not tutored.	*Repeat. Negative correlation.* *Who is typically getting tutored. Is this statement even true?* *What about in affluent, highly competitive areas? Depends on sample.*
The number of times the pilot switches on the seat belt sign increases as the number of drinks spilled on an airplane increases.	*What is turbulence? Why would drinks get spilled?* *Positive correlation.*

For Display

Smoking causes alcoholism.

Smoking causes an increase in the risk of developing lung cancer.

As levels of CO_2 have increased, obesity levels have increased. Do increased levels of CO_2 cause increased rates of obesity?

As more lemons are imported from Mexico, the number of highway crashes in the USA decrease.

Ice cream sales have decreased as sunscreen sales have decreased. Does that mean that ice cream causes sunburns?

Students who are tutored tend to get worse grades than students who are not tutored.

The number of times the pilot switches on the seat belt sign increases as the number of drinks spilled on an airplane increases.

Problem Strings at a Glance

Lesson 1.0

Division as Ratio

Page 1

4 × ___ = 20

4 × ___ = 21

4 × ___ = 23

×__

5, 20, ____

×__

5, 22.5, ____

Term #	Term	
1	12	)×__
2	6	

Term #	Term	
1	12	)×__
2	9	

Lesson 1.1

Difference versus Removal

Page 6

42 − 37

61 − 4

505 − 398

9 − 6 = 3

6 − 9 = −3

6 − (−4)

−4 − 6

−4.03 − (−1.82)

Lesson 1.2

Rebound

Page 12

$u_0 = 1.0$ and $u_n = 0.64 \cdot u_{n-1}$

$u_0 = 3.0$ and $u_n = 0.64 \cdot u_{n-1}$

$u_0 = 1.0$ and $u_n = 0.8 \cdot u_{n-1}$

$u_0 = 1.0$ and $u_n = 0.4 \cdot u_{n-1}$

2, 1.5, 1.125, 0.844

Lesson 1.3

Difference or Ratio

Page 17

4, 2, ___

1, 5, ___

3.2, 6.4, ___

0.1, 0.01, ___

Lesson 1.4

Missing Terms

Page 22

1, 64, ...

1, ___, 64, ...

1, ___, ___, 64, ...

5, ___, ___, 40, ...

−2, ___, ___, −250, ...

1, ___, ___, ___, 81, ...

Lesson 1.5

Finding Slope

Page 25

(0, 5) (2, 9)

(0, 5) (2, 1)

(−14, 5) (−10, −5)

(−14, 5) (−10, 15)

(−37, −2) (−40, −5)

(−37, −2) (?, ?)

Lesson 1.6

Equations of Lines 1

Page 31

rate of 2, start at 0 distance

rate of 2, (3, 7)

rate of 3, (2, 4)

Lesson 1.7

Equations of Lines 2

Page 37

rate of 3, (20, 64)

rate of −2, (1, 6)

rate of −5, (−10, 35)

Lesson 1.8

Equations of Lines 3

Page 43

x	y
0	27
1	21
2	15

x	y
−3	−2
0	4
3	10

x	y
3	25
4	20
5	15

x	y
2	5
6	17
10	29

Problem Strings at a Glance (continued)

Lesson 1.9

Equations of Lines 4

Page 46

x	y
−1	1
0	0
1	−1
2	−2

x	y
−2	3
−1	2
0	1
1	0

x	y
−2	1
−1	0
0	−1
1	−2

x	y
0	−2
1	−1
2	0
3	1

Lesson 2.0

Solving for *y*

Page 51

$2x-4y=16$

$-3x+9y=-15$

$8x-2y=10$

$\frac{2}{5}x-\frac{1}{10}y=\frac{1}{4}$

Lesson 2.1

Solving Systems of Linear Equations

Page 58

$5x-y=-2$
$3x+y=-6$

$5x-y=-2$
$2x-y=1$

$5x-y=-2$
$10x-2y=-4$

$5x-y=-2$
$x-2y=5$

$5x-y=-2$
$5x-y=0$

Lesson 2.2

Substitution or Elimination?

Page 67

$x=\frac{3}{2}y-4$
$y=-\frac{3}{2}x+7$

$2x-3y=-8$
$3x+2y=14$

$16x+2y=-60$
$-16x-5y=54$

$-3x+y=-17$
$y=2x-11$

$y=-2x-2$
$-2x+4y=12$

Lesson 2.2

Think Before You Eliminate

Page 73

$56x-10y=25$
$-56x+5y=-20$

$-3x-10y=5$
$7x+5y=-20$

$36x+45y=412$
$-18x+90y=21$

$2x-y=5$
$-6x+7y=60$

$-4x+5y=12$
$7x-8y=5$

Lesson 2.3

How Many Solutions?

Page 76

How many possible solutions are there for systems of equations that form:

- a line and a line
- a line and a parabola
- a line and an exponential function
- a line and a cubic equation
- a line and a circle
- a line and a hyperbola
- a parabola and a parabola
- a parabola and a hyperbola

Lesson 2.4

What's Your Solution?

Page 80

$x = -4$

$x \leq -4$

$x > -4$

$2x - 4 = 0$

$y = 2x - 4$

$y < 2x - 4$

$y \geq 2x - 4$

$y > 2x - 4$
$x > -4$

Lesson 3.1

Function Notation

Page 87

$(-2, 5)$

$f(2) = -3$

$f(5) = -9$

$f(1) = \_\_\_$

$f(\_\_\_) = 0$

$f(x) = \_\_\_\_\_\_\_\_$

Lesson 3.2

Combining Functions

Page 94

$f(x) = 2x + 3$
$g(x) = 4$
$f(x) + g(x)$
$h(x) = -5$
$f(x) + h(x)$
$^*w(x) = -2x$
$^*f(x) + w(x)$

*optional problems

Lesson 3.4

Translations and the Quadratic Family

Page 100

$y = x^2$
$y = (x - 200)^2$
$y = x^2 - 750$
$y = (x + 1300)^2$
$y = x^2 + 24{,}000$

Lesson 3.5

Reflections and the Square Root Family

Page 107

$y = \sqrt{x}$
$y = -\sqrt{x}$
$y = \sqrt{-x}$
$y = -\sqrt{-x}$
$y = -\sqrt{x - 15}$
$y = -\sqrt{x + 300}$
$y = \sqrt{-x} - 15$
$y = \sqrt{-x} + 300$
$^*y = \sqrt{-x + 25}$

*optional problem

Lesson 3.6

Dilations and the Absolute-Value Family

Page 110

$y = |x|$
$y = 0.01|x|$
$y = 50|x|$
$y = -|x|$
$y = |-x|$
$y = 2|x + 15| - 60$
$y = -\frac{1}{2}|x - 28| + 1$
$y = 10|x + 12| + 300$

Lesson 3.7

Transformations

Page 119

$y = -\frac{1}{3}(x - 12)^2 - 12$
$y = 1.5|x + 30| + 15$
$y = 30\sqrt{-x} - 100$

Lesson 4.0

Exponent Relationships 1

Page 121

$2^3 \cdot 2^2 = 2^6$
$2^a \cdot 2^b = 2^{a \cdot b}$
$\frac{2^6}{2^2} = 2^3$
$\frac{2^a}{2^b} = 2^{\frac{a}{b}}$
$\frac{1}{3^2} = \frac{1}{9}$
$\frac{2^2}{2^5} = 2^{-3}$
$\frac{1}{2^a} = 2^{-a}$
$(2^3)^2 = 2^6$
$(2^a)^b = 2^{a \cdot b}$

Lesson 4.1

Exponential Functions and Transformations

Page 128

$y = 2^x$

$y = 12 \cdot 2^x$

$y = -5^x - 25$

$y = 13^{-x} + 200$

$y = 0.9^x + 75$

$y = 0.5^{x+20}$

$y = 2^{-x-20}$

Lesson 4.2

Exponential Decay

Page 131

$y = 30(0.8185)^x$

$y = 34(0.8185)^x$

$y = 25(0.8185)^x$

$y = 30(0.75)^x$

$y = 30(0.95)^x$

$(0, 32)\ (1, 24)\ (2, 18)\ (3, 14)\ (4, 10)\ (5, 8)$

Lesson 4.3

Exponent Relationships 2: Rational Exponents

Page 134

$4^{\frac{3}{2}}$

$25^{\frac{3}{2}}$

$27^{\frac{4}{3}}$

$9^{\frac{3}{2}}$

$8^{\frac{2}{3}}$

$16^{\frac{3}{4}}$

Lesson 4.4

Exponent Relationships 3: More Rational Exponents

Page 136

$3^{12} = (\_)^3$ *

$3^{12} = \left(3^6\right)^{\_}$ *

$9^{\frac{3}{2}} = (\_)^3$ *

$9^{\frac{3}{2}} = (\_)^{\frac{1}{2}}$ *

$4^x = 16$

$2^{2x} = 16$

$2^{2x} = 2$

$4^x = 2$

$9^x = 27$

$125^x = 25$

*optional problems

Exponent Relationships 4

Page 142

$\frac{x^{15}y^{-2}}{x^{16}y^{-3}} = \frac{1}{xy}$

$\left(\frac{x^{100}z^{-48}}{y}\right)^{-1} = \frac{yz^{48}}{x^{100}}$

$\left(\frac{x^3y^{-14}z^2}{x}\right)^{-\frac{1}{2}} = \frac{xy^7}{z}$

Lesson 4.5

Inverse Functions

Page 149

$f(4) = \_\_$

$f^{-1}(2) = \_\_$

$f(\frac{1}{2}) = \_\_$

$f(\_\_) = 0$

$f(\_\_) = 1$

$f(\frac{1}{4}) = \_\_$

$f^{-1}(3) = \_\_$

$f^{-1}(x) = \_\_$

Lesson 4.6

Introducing Logarithms

Page 156

$L\ 2, 8 = 3$

$L\ 5, 25 = 2$

$L\ 3, 81 = 4$

$L\ 2, 16 = \_\_\_$

$L\ 9, 81 = \_\_\_$

$L\ 5, 125 = \_\_\_$

$L\ 11, 11 = \_\_\_$

$\log_8 64 = \_\_\_$

$\log_{42} 42 = \_\_\_$

$\log_{10} 1{,}000 = \_\_\_$

$\log_{16} 4 = \_\_\_$

$\log_a c = b \Leftrightarrow$

Lesson 4.8

Sequences or Series?

Page 161

$a_0 = 4, a_n = a_{n-1} + 2$

$a_0 = 4, a_n = a_{n-1} \cdot 2$

$a_0 = 16, a_n = a_{n-1} - 2$

$a_0 = 16, a_n = a_{n-1} \cdot 0.5$

Lesson 5.0

Solving Quadratic Equations 1

Page 163

$x^2 + 4x - 5 = 0$

$x^2 - 4x - 5 = 0$

$-x^2 + 4x + 5 = 0$

$-x^2 - 4x + 5 = 0$

Lesson 5.1

Graphing Quadratic Functions 1

Page 169

$y = x^2$

$y = 3x$

$y = x^2 + 3x$

$y = x(x+3)$*

$y = x^2 - 3x$

*optional problem

Graphing Quadratic Functions 2

Page 175

$y = x^2$

$y = 6x$*

$y = x^2 + 6x$

$y = x^2 - 6x$

$y = x^2 + 6x + 7$

$y = x^2 - 6x + 7$

$y = x^2 + 4x + 5$

*optional problem

Lesson 5.2

Complete the Square

Page 178

	x	-2
x		
-2		

	___	___
___	x^2	$3x$
___	$3x$	9

	___	___
___	x^2	$5x$
___	$5x$	

	___	___
___	x^2	$-5x$
___	$-5x$	

$x^2 + 20x +$ ____

$x^2 + 5x +$ ____

$x^2 - 5x +$ ____

$x^2 + bx +$ ____

Solving Quadratic Equations 2

Page 184

$x^2 = 49$

$(x+2)^2 = 49$

$(x+3)^2 + 13 = 49$

$(x-2)^2 - 36 = 0$*

$(x+1)^2 - 2 = 0$

$(x-4)^2 - 8 = 0$

*optional problem

Lesson 5.3

The Quadratic Formula

Page 190

$\frac{-4 \pm 8}{2}$

$\frac{-4}{2} \pm \frac{8}{2}$*

$\frac{-4 \pm 8}{8}$

$\frac{-4 \pm \sqrt{8}}{2}$

$\frac{-6 \pm \sqrt{27}}{3}$

$\frac{-6 \pm \sqrt{27}}{9}$

Optional string with complex numbers:

$\sqrt{4}, \sqrt{-4}$

$(\sqrt{8}), \sqrt{-8}, \sqrt{-18}$

$\frac{2 \pm \sqrt{-8}}{-2}$

$\frac{2 \pm \sqrt{-8}}{-8}$

$\frac{-3 \pm \sqrt{-18}}{6}$

$\frac{-3 \pm \sqrt{-18}}{3}$

$\frac{-3 \pm \sqrt{-18}}{2}$

*optional problem

Lesson 5.4

Solving Quadratic Equations 3

Page 195

$x^2 + 4x + 5 = 0$

$x^2 - 4x + 5 = 0$

$-x^2 - 4x - 5 = 0$

$-x^2 + 4x - 5 = 0$

Problem Strings at a Glance (continued)

Lesson 5.5

Solving Quadratic Equations 4

Page 201

$x^2+3x+2=0$

$3x^2+5x+20=0$

$x^2-4x+1=0$

$-4.9x^2+76.2x+9.67=0$

$x^2+8x=9$

Lesson 5.8

Focus-Directrix: Solving for *f*

Page 204

$f=\frac{11}{5}$

$\frac{1}{f}=\frac{5}{11}$

$4f=8$

$4f=\frac{1}{8}$

$\frac{1}{4f}=\frac{1}{8}$

$\frac{1}{4f}=8$

$\frac{1}{4f}=-2$

Using the Focus-Directrix Definition

Page 208

Focus $(0,0)$ directrix : $y=-2$

Focus $(0,0)$ directrix : $y=-5$

Focus $(1,2)$ directrix : $y=10$

Focus $(1,2)$ directrix : $y=6$

Focus $(1,2)$ directrix : $y=4$

Focus $(1,2)$ directrix : $y=3$*

Focus $(1,2)$ directrix : $y=2.5$

Focus $(1,2)$ directrix : $y=2.1$

*optional problem

Lesson 6.1

Polynomials

Page 211

$y=x^3$

$y=2x$

$y=x^3+2x$

$y=x^3-2x$

$y=x^2$

$y=x^3+x^2$

$y=x^3-x^2$

$y=x^3+x^2-2x$

$y=x^3-x^2-2x$

Lesson 6.2

Using Zeros

Page 215

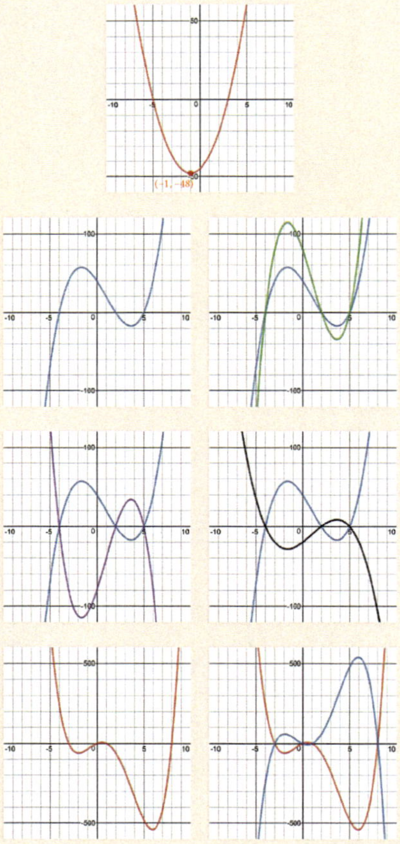

Lesson 6.3

Polynomial Viewing Windows

Page 222

$f_1(x)=0.1(x+1)(x+1)(x-2)(x+20)$

$f_2(x)=-0.1(x-4)(x+4)(x-33)$

$f_3(x)=-0.1(x+2)(x+3)(x-3)^2(x-21)$

$f_4(x)=(x-3)(x-20)(x+22)$

$f_5(x)=0.5x^3(x+1)(x-1)(x+50)$

Lesson 6.4

Polynomial Division

Page 229

	$2x^2$	__	4
__	$2x^3$		
-2		$6x$	

	$3x$	$+1$
__	$3x^3$	
__	$6x^2$	
__		-5

	__	__	__
x	$4x^3$	$-5x^2$	$3x$
$+1$	$4x^2$	$-5x$	3

$(4x^3-13x^2+5x-6)\div(x-3)$

$(6x^3+11x^2-x-6)\div(2x+3)$

Lesson 6.5

The Rational Parent Functions

Page 237

$1 \div 2$

$1 \div 4$

$1 \div 10$

$1 \div 1{,}000$

$1 \div 1$

$1 \div \frac{1}{2}$

$1 \div \frac{1}{4}$

$1 \div \frac{1}{10}$

$1 \div \frac{1}{1{,}000}$

$1 \div -1$

$1 \div -3$

$1 \div -9$

$1 \div -1{,}000$

$1 \div -\frac{1}{4}$

$1 \div -\frac{1}{8}$

$1 \div \frac{1}{1{,}000}$

$y = \frac{1}{x}$

$y = \frac{1}{x^2}$

Lesson 6.6

Rational Functions

Page 241

$\frac{x^2+x}{x+2}$

$y=x-1$

$\frac{x^3+x}{x+2}$

$\frac{x^4+x}{x+2}$

$\frac{2x}{x+2}$

$\frac{2x}{x^2-4}$

$\frac{2x}{x^3-1}$

Lesson 6.7

Rational Expressions

Page 252

True or False?

$\frac{36}{60}=\frac{30+6}{30+30}=\frac{6}{30}$

$\frac{36}{60}=\frac{2^2 \cdot 3^2}{2^2 \cdot 3 \cdot 5}=\frac{3}{5}$

$\frac{x+2}{x+5}=\frac{2}{5}, x \neq 0$

$\frac{2(x+15)}{5(x+15)}=\frac{2}{5}$

$\frac{x^2+5x+6}{x^2-4}=\frac{x+3}{x-2}$

$\frac{x^2+5x+6}{x^2-4}=\frac{5x+6}{-4}$

$\frac{x+3}{x-2}+\frac{x-2}{x+3}=1$

Lesson 6.8

Dividing Rational Expressions

Page 255

cheese (c.)	3
pizza	2

cheese (c.)	2
pizza	$\frac{1}{5}$

cheese (c.)	8
pizza	$\frac{4}{5}$

cheese (c.)	2
pizza	$\frac{a}{b}$

x^2	
$\frac{x+1}{x}$	

x^2-1	
$\frac{x+1}{x}$	

$\frac{x^2-5x+6}{x^2-5x-6} \div \frac{x^2-4}{x+1}$

Lesson 6.9

Solving Rational Equations

Page 261

$\frac{17}{4}=x$

$\frac{4}{17}=\frac{1}{x}$

$\frac{x}{2}=x+1$

$\frac{2}{x}=\frac{1}{x+1}$

$\frac{4}{2x+3}=\frac{1}{x}$

Lesson 7.1

Special Right Triangles

Page 264

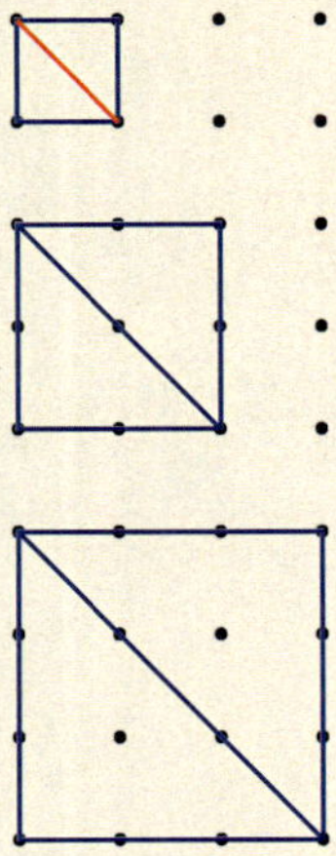

In a 45°, 45°, 90°, what if the:
side length is 6, hypotenuse?
side length is 21, hypotenuse?
side length is x, hypotenuse?

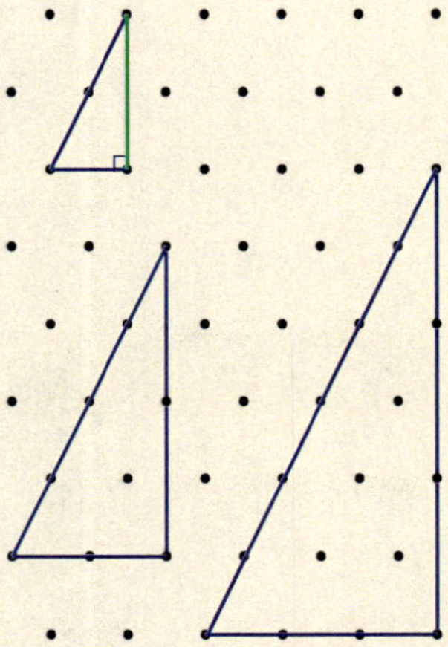

In a 30°, 60°, 90°, what if the:
short side is 6, hypotenuse? long side?
short side is 21, hypotenuse? long side?
short side is x, hypotenuse? long side?

Lesson 7.2

Using Reference Angles

Page 268

Find the sine, cosine, and tangent of the angle created by these points. What is the angle?

(3, 4)

(−3, 4)

(−3, −4)

(3, −4)

Lesson 7.3

Quadrant Angles

Page 278

Find the sine, cosine, and tangent of:

0°

90°

180°

270°

Lesson 7.4

Radian Measure

Page 282

π

$\frac{\pi}{2}$

$\frac{3\pi}{2}$

$\frac{\pi}{4}$

$\frac{2\pi}{3}$

$\frac{7\pi}{4}$

$\frac{11\pi}{6}$

$\frac{5\pi}{3}$

$\frac{5\pi}{6}$

$-\frac{3\pi}{4}$*

*optional problem

Lesson 7.5

Sinusoid Viewing Windows

Page 286

$y = \sin x$*

$y = \sin x + 9$

$y = 50 \sin 2x$

$y = 0.01 \sin 0.5x$

$y = \sin 0.1 - 4$

$y = \sin(x - 1.5\pi)$

Follow-up problem string:

$y = \cos x$*

$y = 2\cos x - 10$

$y = \cos\left(\frac{x - 5\pi}{10}\right)$

$y = 0.1\cos(0.2x)$

$y = 75\cos 4x$

$y = \cos(x + 1.5\pi)$

*optional problems

Lesson 7.6

A Bouncing Spring Follow Up

Page 293

$$h(t)=0.652+0.076\cos\left[\frac{2\pi(t-0.43)}{0.8064}\right]$$

$$h(t)=0.8+0.076\cos\left[\frac{2\pi(t-0.43)}{0.8064}\right]$$

$$h(t)=0.5+0.076\cos\left[\frac{2\pi(t-0.43)}{0.8064}\right]$$

$$h(t)=0.652+0.1\cos\left[\frac{2\pi(t-0.43)}{0.8064}\right]$$

$$h(t)=0.652+0.05\cos\left[\frac{2\pi(t-0.43)}{0.8064}\right]$$

$$h(t)=0.652+0.076\cos\left[\frac{2\pi(t-0.23)}{0.8064}\right]$$

$$h(t)=0.652+0.076\cos\left[\frac{2\pi(t-0.63)}{0.8064}\right]$$

$$h(t)=0.652+0.076\cos\left[\frac{2\pi(t-0.43)}{0.5}\right]$$

$$h(t)=0.652+0.076\cos\left[\frac{2\pi(t-0.43)}{0.9}\right]$$

Lesson 7.7

Solving Trigonometric Equations

Page 297

$\sin x=0$

$\cos x=0$

$\sin x=\frac{1}{2}$

$\cos x=\frac{1}{2}$

Lesson 8.1

Probability

Page 305

What's the probability, with two 6-sided dice or one 12-sided die, of rolling a total of:

exactly 6?

exactly 12?

exactly 2?

exactly 4?

What else has probability of $\frac{3}{36}$?

Lesson 8.2

Multiplication Rules of Probability

Page 312

Which has the greater probability?

- 90% then 10% or 10% then 90%
- ¼ then ½ or ½ then ¼
- 84% then 75% or 75% then 84%
- 24% then 5% or 5% then 24%
- 20% then 32% or 30% then 22%

(For contexts for each question see the full lesson.)

Lesson 8.3

Addition Rules of Probability

Page 320

Probability That People Like Abby's Social Media Posts:

P(Grandmother)

P(friend)

P(Mom)

P(brother)

P(Mom and Grandmother)

P(Mom or Grandmother)

P(friend and brother)

P(friend or brother)

P(brother or Mom)

Lesson 8.4

Bivariant Independence

Page 327

For the survey results below, find the interval likely to contain the true proportion:

	Side Item	
	Salad	Fries
Soda	12	8
Diet Soda	10	20

Diet soda and salad, n = 50

Diet soda and salad, n = 100

Diet soda and salad, n = 1,000

Diet soda and salad, n = 10,000

Lesson 9.1

Experimental Design

Page 330

Which popcorn is best?

The front door test

The cafeteria probe

The movie popcorn study

Design a better popcorn study*

(Contexts for each question can be found in the full lesson.)

*optional problem

Lesson 9.2

Normal Distribution

Page 333

For the data of $\mu = 497$ cm, $\sigma = 99$ cm for adult male giraffes:

- What range of heights fall in the middle 68 percent?
- How tall is an adult male giraffe in the 99.85th percentile?
- How tall is an adult male giraffe in the 2nd percentile?
- An adult male giraffe is 200 cm tall. What percentile?
- Find the probability that a random adult male giraffe is taller than 695 cm? Shorter than 695 cm?

Lesson 9.3

Confidence Intervals

Page 341

$n = 9, \overline{x} = 170, \sigma = 9$

$n = 100, \overline{x} = 170, \sigma = 9$

$n = 1{,}000, \overline{x} = 170, \sigma = 9$

$n = 100, \overline{x} = 170, \sigma = 8$

$n = 100, \overline{x} = 170, \sigma = 7$

$n = 100, \overline{x} = 170, \sigma = 1$

Lesson 9.4

Correlation versus Causation

Page 348

Correlation or Causation?

- Smoking causes alcoholism.
- Smoking causes an increase in the risk of developing lung cancer.
- As levels of CO_2 have increased, obesity levels have increased. Do increased levels of CO_2 cause increased rates of obesity?
- As more lemons are imported from Mexico, the number of highway crashes in the USA decrease.
- Ice cream sales have decreased as sunscreen sales have decreased. Does that mean that ice cream causes sunburns?
- Students who are tutored tend to get worse grades than students who are not tutored.
- The number of times the pilot switches on the seat belt sign increases as the number of drinks spilled on an airplane increases.